РОСТ КРИСТАЛЛОВ
ROST KRISTALLOV
GROWTH OF CRYSTALS

VOLUME 10

Growth of Crystals

Volume 10

Edited by

N. N. Sheftal'
Institute of Crystallography
Academy of Sciences of the USSR, Moscow

Translated by
J. E. S. Bradley
Senior Lecturer in Physics
University of London

 CONSULTANTS BUREAU · NEW YORK AND LONDON

The Library of Congress catalogued the first volume of this title as follows:

Growth of crystals. v. [1]⁻
  New York, Consultants Bureau, 1958–

  v. illus., diagrs. 28 cm.

  Vols. 1, 3–   constitute reports of 1st–   Conference on Crystal
Growth, 1956–   v. 2 contains interim reports between the 1st and
2d Conference on Crystal Growth, Institute of Crystallography,
Academy of Sciences, USSR.
  "Authorized translation from the Russian" (varies slightly)
  Editors: 1958–   A. V. Shubnikov and N. N. Sheftal'.

  1. Crystals–Growth. I. Shubnikov, Alekseĭ Vasil'evich, ed. II.
Sheftal', N. N., ed. III. Consultants Bureau Enterprises, inc., New
York. IV. Soveshchanie po rostu kristallov. V. Akademiīa nauk
SSSR. Institut kristallografii.

QD921.R633   548.5   58-1212

Library of Congress Catalog Card Number 58-1212
ISBN-13: 978-1-4613-4258-8   e-ISBN-13: 978-1-4613-4256-4
DOI: 10.1007/978-1-4613-4256-4

The original Russian text, published for the Institute of Crystallography of the
Academy of Sciences of the USSR by Nauka Press in Moscow in 1974, has been
corrected by the editors for this edition.

PREFACE

This tenth volume completes the first series of "Growth of Crystals," which began in 1957. The sources of the volumes are as follows: for Vol. 1, the 1st All-Union Conference on Crystal Growth; for Vol. 3, the 2nd; and for Vols. 5 and 6, the 3rd; Vols. 7 and 8 reported the International Symposium on Crystal Growth at the Seventh International Crystallography Congress, and Vol. 9 the 1969 symposium on crystal growth dedicated to E. S. Fedorov; Vols. 2, 4, and 10 did not originate in conferences.

The main problem that largely occupied the conferences and symposia and also the intermediate volumes was that of real crystal formation, as well as the relation of crystal growth theory to practical crystal production.

This tenth volume, which completes this first series, is to a considerable extent a survey. It contains more extensive theoretical and experimental original papers, as well as some shorter papers dealing with particular but important aspects of real crystal formation.

The volume opens with a paper by V. V. Voronkov, which deals with the structure of crystal surface in Kossel's model. The model as proposed by Kossel is extremely simple. It deals qualitatively with the basic trends in the growth of an idealized crystal in its own vapor at absolute zero, and naturally does not allow one to perform quantitative studies on complex real processes.

Voronkov's paper to some extent summarizes recent theoretical developments of Kossel's model, and deals in detail at the level of current physical ideas with the structure of the surface of a Kossel crystal over a wide range of real conditions, including melts. For certain cases the theory can provide quantitative results; for instance, it has given in very close agreement with experiment values for the partition coefficients of certain impurities for growth of silicon crystals from the melt.

The paper by V. V. Puchkov and L. D. Kislovskii deals with the production and structure of a liquid in a new original treatment, which is of interest to specialists on crystal growth; it takes into account the variability in the electronic structure of atoms and molecules when the state of aggregation changes, and this enables them to consider condensation as a process occurring in a number of successive stages, with the production of metastable formations in the initial stages of crystal growth.

Borovinskii's paper is the first to deal systematically in a theoretical fashion with the mechanism by which microscopic nonuniformities near the surface influence the growth of a homopolar crystal, particularly the nucleation rate for two-dimensional nuclei. Such studies are very important for elucidating the effects of defects on crystal growth, as will be clear from the numerous papers on real crystal formation appearing in earlier volumes.

Papers by R. N. Sheftal' and R. N. Sheftal' with L. A. Borovinskii deal with epitaxy. The first is a review, but differs from existing fundamental reviews, which contain abundant factual evidence of theoretical and experimental profiles, in that the epitaxial growth mechanisms are divided into two types, which are considered as essentially different in regard to formation of the oriented layer. The first type is epitaxy under conditions of vacuum condensation from a molecular beam, while the second type covers all other cases of epitaxy.

The structural correspondence in heteroepitaxy, extremely important under conditions of ordinary crystallization, plays no part during vacuum deposition. Particular attention should be given to the electron micrographs of thin cleaved NaCl crystals from the work of the Japanese authors Shinozaki and Sato, which are given in the review. The author of the Soviet review has observed on the photographs a submicroscopic step structure in the cleaved surface. The steps form right reentrant angles, whose symmetry in the present case is subject to a four-fold axis, and this is considered to control the orientation of the growing layer during vacuum condensation.

The second paper deals with a semiquantitative theory of the formation and deposition of complexes formed by a vacuum condensation under optimal conditions, which will give a single-crystal layer.

The long article by N. A. Pangarov, a representative of the Bulgarian scientific school founded by Kaishev and Stranski, deals with the production of a preferred orientation in crystallites in the electrodeposition of metals, and also with the formation of twins in electrocrystallization. The author assumes, and one is forced to agree with this, that crystallization on a structureless substrate results in a preferred orientation controlled by the supersaturation, which also determines the twinning probability. In electrocrystallization, the analog of the supersaturation is the overvoltage, which is the factor that can be controlled very precisely. Elementary calculations of the energy of formation for a two-dimensional nucleus have been performed for various crystallographic faces, which determine the axis of preferred orientation of the three-dimensional crystals, which enables one to determine the probability of formation of a crystal with a given orientation in relation to the overvoltage. Similarly, one can determine the twinning probability, which increases with the overvoltage. Very close control is possible for pulse overvoltage methods, and the simultaneous use of cinephotography has made possible the identification of the conditions for formation of single crystals of various orientations as well as twins. These experiments have produced that unusual situation, agreement with calculations. This agreement is of considerable interest, although it cannot be considered as entirely free from accident at the present stage, and it does show that crystallization at large overvoltages on structureless substrates begins with the formation of two-dimensional nuclei.

The paper by B. M. Bulakh considers the crystallization of cadmium sulfide from nonstoichiometric medium. This deals with the crystallization mechanisms for complexes and it is shown that the character of the complexes determines the morphology of the resulting crystals.

The next article, by B. M. Bulakh and N. N. Sheftal', deals with simulation of these processes, and four different models are considered. The first two are the most simple to prepare and analyze, since they deal with the assembly and the disruption of complexes of different types that are involved in crystallization and dissolution.

Three papers by Kleshchev and others are interesting primarily because of the clarity of the photographs that illustrate the relation between the internal structure of a crystal and the crystallization conditions on the one hand and the features of the growth pyramids on the other, which determine the entry of the external medium, which is the principal way in which the medium influences the growth processes.

The paper by L. G. Lavrent'eva summarizes her numerous publications on the perfection of a growing layer in relation to the deviation of the substrate from exact orientation corresponding to the principal faces of the crystal. These studies have been made mainly on crystals of gallium arsenide and germanium.

The paper by N. G. Sokolova and N. D. Lyubalin deals with the debatable question of the roles of reentrant and convex corners in the growth of twins with the diamond structure. It is shown that the crystallization conditions may give preference to growth in a reentrant angle (at low supersaturations) or in a convex one (at high supersaturations).

The paper by S. A. Grinberg deals with the crystallization mechanism of germanium from solution in molten gold, considering the motion of the droplets of gold in the temperature gradient field on the surface of the germanium plate, which is either in vacuum or hydrogen. The trends in the droplet motion are presented very clearly, and these clearly simulate to some extent the motion of liquid inclusions in crystals.

The paper by L. A. Zadorozhnaya deals with the domain structure of antimony sulfoiodide crystals in the scanning electron microscope, and the domains can be visualized under the beam, and the movements of them can be followed.

The paper by L. A. Krater and I. V. Frishberg deals with the mass growth of crystals from the vapor; the subject of the "Growth of Crystals" series is, strictly speaking, restricted to single crystals, but this study is of basic interest, for it combines the physical approach with the crystal morphology outlook, and also it deals with the visual observation of the motion of the medium during crystallization, and it shows that this motion is unique and arises from the interaction with the set of growing crystals, influencing their structure and morphology.

The paper by N. N. Sheftal' summarizes his studies in recent years in real crystal formation; the phenomena of preferential growth are considered, together with phenomena similar to artificial epitaxy, growth via complexes, the relation of equilibrium form to the shape of the perfect crystal, the natural selection of crystals for uniformity, and finally phenomena explained by long-range interaction.

The paper by Bevz et al. deals with some new evidence on the effects on the shape from dislocations in silicon crystals, and this supplements the article by the same workers in "Growth of Crystals," Vol. 9.

The paper by Yu. M. Smirnov and E. S. Fal'kevich deals with the effects of supercooling on germanium crystal shape, together with the advantages as regards growth from a planar crystallization isotherm coincident with a (111) face, which influences crystal nonuniformity, which is a topic that has long been discussed in the literature.

The paper by Budevskii et al. deals with a survey of the major results on electrocrystallization obtained by Bulgarian workers in the last two or three years; a particular feature of these studies is the extremely good agreement between theoretical calculations and the experimental results.

The paper by N. N. Sheftal' and A. N. Buzynin deals with some results on a very simplified simulation of crystal growth processes taking into account particle migration in the adsorption layer, which involves movement to the most favorable site for incorporation. This approach has provided theoretical confirmation of certain principles put forward in papers on the growth of perfect crystals using qualitative considerations and experimental evidence.

The memorial to A. S. Shein, an outstanding crystallographer of engineering background, takes note of his untimely death. At the start of the Second World War, he showed how piezoelectric elements could open the way to general use of synthetic single crystals in the Soviet Union, namely crystals of Rochelle salts, for which he then developed a method of commercial production.

The volume ends with a postscript which characterizes (mainly from foreign materials) the development of crystal growth problems in recent years, together with prospects in the near future.

The English edition in addition includes the following: a paper by A. A. Shternberg dealing with polymorphism and polytypism, which is closely related to papers by the same author in Vols. 5 and 9, as well as a paper by N. N. Sheftal' on the energy fluxes through a crystal, which completes the work of this author on the causes of crystal symmetry (published in Vol. 4), and also a paper by N. N. Sheftal' and E. G. Kolomyts on the evolution of the final growth forms of crystals in response to incorporation of the medium (first published in Russian in Acta Phys. Acad. Sci. Hung.).

Taken together, this volume gives a complete representation of Soviet theoretical works on real crystal formation.

In conclusion I would like to express my indebtedness to Nauka Press for the considerable and constant attention to the publication of this series since 1957.

I am also indebted to Plenum Publishing Corporation for their invariably careful attention to the translation of "Growth of Crystals," which began with the first volume of this publication.

N. N. Sheftal'

CONTENTS

STRUCTURE OF A CRYSTAL SURFACE AND KOSSEL'S MODEL

V. V. Voronkov

Crystal growth is governed firstly by heat and material transfer in the volumes of the crystal and medium, and secondly by the macroscopic properties of the phase interface; for instance, some of the most important characteristics of a surface are the free energy, the growth rate as a function of supersaturation or supercooling, and the impurity trapping coefficient. These macroscopic surface properties in turn are determined by the structure of the surface on an atomic scale. Kossel's model is a simple one for the transition layer between phases, and it is often used in analyzing processes at surfaces; however, it is fairly difficult to calculate the above macroscopic characteristics even within the framework of this simple and clear model; some basic problems have so far remained unsolved. On the other hand, Kossel's model enables one to understand many qualitative features of surface structure, and sometimes to obtain quantitative agreement with experiment.

Kossel's Model

First of all we consider the interface between a crystal and the vapor, both containing only one type of atom. Kossel's model is based essentially on the following assumptions: (1) the spatial lattice near the surface is completely undistorted relative to the volume of the crystal; (2) the interaction energy for the atoms is made up of the energies of pair interaction for reasonably close neighbors; these energies are dependent only on the distances between the atoms, being independent of the position relative to the surface; (3) the free energy of the crystal as dependent on the lattice vibrations is governed only by the total number of particles in the crystal and is independent of the surface configuration.

In the simplest case, we take into account only the interaction energy of nearest neighbors $\varepsilon < 0$ (the sign of ε corresponds to attraction between the atoms). Within the bulk of the crystal, each bond from each atom has energy $\varepsilon/2$, since ε is shared between two neighbors. Some of the bonds are broken for the surface atoms, and we denote the number of atoms in the crystal by N_1, while the number in the vapor is N_2, the number of broken bonds for all surface atoms being N_{12}. The energy related to the surface atoms is the same as if they were in the volume but with $N_{12} \cdot \varepsilon/2$ subtracted. Therefore, the free energy of the crystal−vapor system for a given surface configuration consists of the bulk free energy $F_1(N_1)$ of the crystal, the bulk free energy of the vapor $F_2(N_2)$, and the surface energy $E = -N_{12} \cdot \varepsilon/2$.

We further consider a part of the surface of given orientation, which is bounded by a planar fixed edge. The crystalline and vapor phases are in thermodynamic equilibrium. From the macroscopic viewpoint we have a planar surface with a vector $\mathbf{n}$ for the normal and a specific surface free energy $\sigma(\mathbf{n})$. From the microscopic viewpoint, the surface configuration is continually changing by attachment and detachment of particles, together with surface diffusion. Then the sum of the bulk free energies $F_1(N_1) + F_2(N_2)$ is constant at equilibrium (when

1

one has a constant total volume and constant number of particles $N_1 + N_2$), so that the equilibrium surface structure is determined by E, i.e., by the relation between N_{12} and the surface configuration, together with the energy $W = -\varepsilon/2$ per broken bond. The specific free surface energy $\sigma(n)$ is determined by the statistical sum over all possible surface configurations

$$\sigma(n) = -\frac{kT}{S} \ln \sum_{\text{config.}} \exp\left(-\frac{E}{kT}\right), \tag{1}$$

where S is the area of the relevant part of the surface.

The entire statement is readily extended to this case if one takes into account the interactions other than those between nearest neighbors; we denote the vector joining two interacting atoms by r, and the interaction energy by $\varepsilon(r)$. There are several types of such linking vectors r, which differ in magnitude or direction. If the total number of broken bonds of type r is $N_{12}(r)$, the contribution from these to the surface energy is $N_{12}(r)\varepsilon(r)/2$; the total surface energy E is the sum over all types of linking vectors:

$$E = \sum_r N_{12}(r) W(r), \tag{2}$$

where $W(r) = -\varepsilon(r)/2$ is the surface energy per broken bond of type r.

Equilibrium Surface Structure of a

Crystal − Vapor Interface

The preceding section shows that the equilibrium structure is determined by the way in which $N_{12}(r)$ is dependent on the surface configuration; for a subsequent analysis, we restrict ourselves to the case where the crystal has a simple lattice. The various surface configurations include one such that the surface coincides with a crystallographic plane $P(n)$ perpendicular to vector n. We will call this configuration the basic one for brevity. In a simple lattice, each bond from an atom in direction r is accompanied by a bond of equivalent energy going from the same atom in the opposite direction $-r$, so when the atom links to the basic configuration, the total number of broken bonds can alter only on account of lateral broken bonds parallel to $P(n)$. The same applies to removal of an atom from the basic configuration. For this reason, the surface properties are determined by which vectors r lie in the plane $P(n)$.

We consider first of all the most important case, where in plane $P(n)$ there lie at least two nonparallel bond vectors; such surfaces are usually called close-packed. Here attachment of atoms to the basic configuration increases E; if we attach N atoms, the increment in E will be minimal if the lateral bonds of these atoms are as far as possible mutually satisfied, i.e., if the atoms form a compact array. The increment in E is here related to the boundary between this array, which forms a step. As we have two nonparallel lateral bond vectors, one has lateral broken bonds running from a step of any orientation. The increment in E is proportional to the perimeter of the array as regards order of magnitude, i.e., it increases with N as $N^{1/2}$, so the close-packed surface at a sufficiently low temperature virtually retains the basic configuration, i.e., is atomically smooth and contains only a few adsorbed atoms and vacant sites (Fig. 1a).

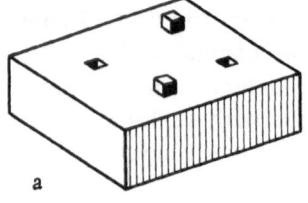 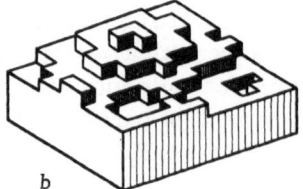

Fig. 1. Crystal surface: a) atomically smooth; b) rough.

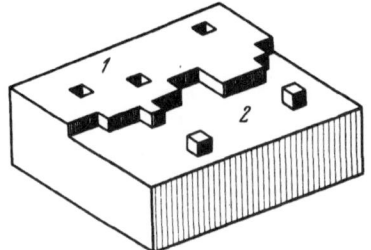

Fig. 2. An adsorbed monolayer on a smooth surface at low temperatures: 1) layer almost complete; 2) rarefied two-dimensional gas. The boundary between these phases is represented by a step.

If the temperature is high enough, the surface becomes atomically rough, as in Fig. 1b, i.e., there are many atoms attached to the basic configuration and removed from it, so the transition region between the crystal and the vapor covers several crystallographic planes $P(n)$. A rigorous analysis of the surface structure would require us to calculate the statistical sum of (1), which is very complex as a mathematical task; however, to estimate the type of structure we need only consider those possible configurations that are produced by monolayers of atoms attached to the basic configuration. Barton and Cabrera [1] use this approach. Statistically speaking, an adsorbed monolayer is equivalent to a two-dimensional model [2]. If the temperature is low, the adsorbed layer may be in one of two states [2]: either it represents low-density two-dimensional gas of single adsorbed atoms, or else it consists of an almost filled monolayer with isolated vacant sites (Fig. 2). The density of the two-dimensional gas and the equal concentration of vacant sites in the filled layer both increase with temperature, and at a certain temperature T_{cr} the densities of the two-dimensional phases become identical. Figure 3a shows the degree of filling for the adsorption sites for a two-dimensional gas with a square surface lattice [2]. The important point is that this quantity is small almost up to T_{cr} itself, but rises rapidly over a small range near T_{cr}. Therefore, we can speak qualitatively of T_{cr} as the temperature for transition from a smooth surface structure to a rough one. The values of T_{cr} for simple lattices have been derived [2] by formal analysis of the statistical sums for adsorbed layers. However, there is [3] a very simple and clear approach for deriving T_{cr}, for which purpose we first of all discuss the properties of a step for $T < T_{cr}$, when the surface is atomically smooth. A section of a step of given orientation is bounded by fixed ends, and macroscopically is represented as a straight line with the vector $\mathbf{m}$ for the normal lying in the plane $P(n)$, having a specific edge free energy $\alpha(m)$. Microscopically, the step takes all possible configurations (Fig. 2), so α is a statistical sum of the form of (1) over all possible configurations for the length of step. The only difference from (1) is that S has to be replaced by the step length, while E is replaced by the edge energy of the step, which is defined in our case (simple lattices) only by the lateral broken bonds using a formula of the form of (2). By analogy with the previous, we introduce the basic configuration of the step, the step then coinciding with the atomic series $R(m)$ perpendicular to vector $\mathbf{m}$. Attachment of an atom to the basic configuration can affect E only on account of broken bonds parallel to $R(m)$. If at least one bond vector $\mathbf{r}$ lies along $R(m)$, then E is increased by attachment of an atom to the

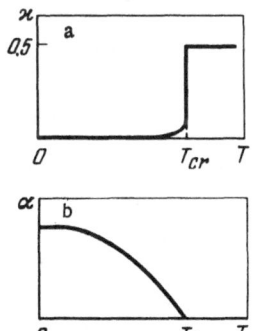

Fig. 3. Temperature dependence of: a) degree of filling $\varkappa$ for adsorption sites for the two-dimensional gas; b) specific free energy α of a step neglecting roughness.

Fig. 4. Chain of atoms
attached to the basic
configuration of a step.
The atoms in the layer
bounded by the step are
shown. These atoms
lie on a surface in the
plane of the figure.

basic configuration (such a step is called close-packed). If several atoms are attached, the increase in E will be minimal when the atoms form a compact chain (Fig. 4). The increase in the energy is here related to the boundaries of the chain, namely two kinks (one of these we call positive, and the other we call negative, Fig. 4). We denote this energy increment by 2β, with β independent of the number of atoms in the chain and having the meaning of the energy per kink. If we take into account only the interaction between nearest neighbors in the $R(\mathbf{m})$ series, then $\beta = W(\mathbf{r})$, where $\mathbf{r}$ is the vector joining these nearest neighbors.

The structure of a close-packed step is essentially different from that of a close-packed surface; in the second case, the formation of an array of N atoms increases E the more the larger N (in accordance with the $N^{1/2}$ law), so the surface is not rough if T is small. In the first case, the chain of N atoms increases the energy by 2β no matter what N is; in other words, the energy of a step is determined only by the number of kinks. If the section $R(\mathbf{m})$ contains M_0 atoms, and if the step has M^+ and M^- kinks, then the energy of the length of step is $\alpha_0 M_0 + 2M\beta$, where α_0 is the energy of the basic configuration per atom of the $R(\mathbf{m})$ series. The configurational entropy is $k \ln \frac{M_0!}{(M!)^3 (M_0 - 2M)!}$. We minimize the free energy with respect to N to get the equilibrium kink density ρ:

$$\rho = \frac{2M}{M_0} = \frac{2 \exp(-\beta/kT)}{1 + 2 \exp(-\beta/kT)} \tag{3}$$

and the specific free energy of the step α is

$$\alpha = \alpha_0 - kT \ln(1 + 2 \exp(-\beta/kT)). \tag{4}$$

As unit of distance in (3) and (4) we have taken that between nearest neighbors in the $R(\mathbf{m})$ series.

This ρ is finite for any $T > 0$, i.e., a close-packed step is rough for $T > 0$.[*]

It is simple to calculate α [1, 3] by taking account of kinks not only of unit depth, but also of double, triple, etc. depth (Fig. 2). The result does not differ essentially from (4). The value of α decreases as T increases on account of the configuration entropy of the step, and it be-

---

[*] In deriving (3) we have taken into account all possible configurations for a given M; by analogy with a surface, we might have restricted ourselves only to those configurations that are produced by chains of atoms attached directly to the basic configuration. A chain of atoms and an empty section (with no attached atoms) may be considered as two different one-dimensional phases. It is well known that phases exist separately in one-dimensional systems at $T > 0$, being always mixed [4]. Therefore, even in this approximation the step is rough.

TABLE 1. Temperature for Transition from a Smooth Structure
to a Rough One for the Closest-Packed Surface

Type of surface and lattice	kT_e/W	kT_{cr}/W
{001} simple cubic lattice	1.13	1.13
{111} face-centered cubic lattice	1.89	1.82
{111} diamond lattice	0.71	—

comes zero at a certain point T_e (Fig. 3b), which means that formation of an array of atoms on a smooth surface at $T > T_e$ results not in an increase in the free energy of the system but in a decrease, i.e., an atomically smooth surface is unstable above T_e relative to the formation of steps on it. Then T_e has the clear meaning of the temperature for transition from a smooth surface to a rough one. It is found [3, 5] that the values of T_e for steps of different orientation are roughly the same; Table 1 gives values of T_e and T_{cr} as calculated in the nearest-neighbor interaction approximation. For simple lattices, these two different estimates of the transition temperature are almost the same (in particular, T_e and T_{cr} are exactly equal for a square surface lattice). The value of T_e differs from T_{cr} in that it is readily calculated for more complex systems; for instance, values have been calculated for a diamond lattice [3] and for a two-component medium [6].

The temperature for transition to a rough structure is proportional to W (the energy per broken bond); W is directly related to the heat of sublimation per atom H. The energy of the crystal per atom is $\nu\varepsilon/2 + 3kT$, where ν is the number of nearest neighbors in the lattice, while the enthalpy per atom in the vapor is $5kT/2$ (if the vapor consists of single atoms). Then the heat of sublimation is $H = \nu W - kT/2$, or $W \approx H/\nu$. A notable point is that $kT_e/\nu W$ (i.e., kT_e/H) is practically the same for all types of lattice, and is shown by the table to be about 0.17. The ratio of the melting point T_0 to H/R lies in the range 0.03-0.05 for metals and semiconductors, so the crystal—vapor interface close-packed in the nearest-neighbor interaction approximation is certainly atomically smooth up to close to the melting point.

We now consider a surface such that the plane $P(n)$ contains only one bond vector between nearest neighbors (first-order neighbors) together with a bond vector not parallel to this between neighbors of order i (the surface energy per such bond is W_i). In that case, the approximation of kinks of unit depth represented by (4) is well obeyed for the most close-packed step; we substitute in (4) that $\beta \approx W$ and $\alpha_0 \approx W_i$, and use the fact that $W_i \ll W$ to get T_e in the form

$$\frac{W_i}{W} = \frac{2kT_e}{W} e^{-\frac{W}{kT_e}} \tag{5}$$

From (5) we see that the transition temperature is only very slightly dependent on W_i; for instance, T_e becomes less by an order of magnitude than the value for the most close-packed surface only if $W_i/W \approx 10^{-4}$, so a surface close-packed via bonds for first and second order, or first and third order, remains atomically smooth up to very close to the melting point. If however, the surface is close-packed only via second-order bonds, then the transition temperature is less by a factor W/W_2 (i.e., by an order of magnitude) than that for the most close-packed surface, and it lies below the melting point. Such a surface is rough in the range from T_e to T_0.

So far we have considered close-packed surfaces; now we consider other cases. If the $P(n)$ plane contains no bond vector, attachment of atoms to the basic configuration, or removal of them, does not affect E; such faces are always rough.

Finally, the $P(n)$ plane may contain only one bond vector (or several vectors parallel to one another). Here the attachment of compact chains of atoms extending along the bond vector

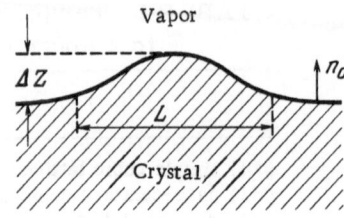

Fig. 5. Local displacement of a planar surface due to passage of a particle from the vapor into the solid.

increases the energy only via the edges of the chain, so such a surface resembles a close-packed step in being rough for T > 0, and in fact such a surface consists of an echelon of close-packed steps on some close-packed surface.

Singular and Nonsingular Surfaces

In this section we show that for a surface of given orientation n_0 there exists a unique relationship between the growth rate as a function of supersaturation and the form of σ (n) in the neighborhood of n_0; if one adds ΔN new particles to the crystal, one increases the surface free energy F_s, and under supersaturation conditions, the change in F_s is accompanied by change in the bulk free energy $\Delta F_v = -\delta\mu \Delta N$, where $\delta\mu$ is the deviation of the chemical potential of the vapor from the equilibrium value ($\delta\mu = kT \ln(1 + s)$, where s is the relative supersaturation). The sign of the total free energy change $\Delta F_s + \Delta F_v$ governs the type of growth, namely whether this is continuous or involves overcoming free-energy barriers. From the macroscopic viewpoint, attachment of new particles to the crystal indicates displacement of a small part of the surface (Fig. 5). We denote by a the distance between adjacent P(n_0) planes. To elucidate the growth mechanism we need to know ΔF_s for essentially microscopic displacements, when $\Delta z \lesssim a$ (Fig. 5); however, the changes ΔF_s for microscopic and macroscopic displacements are closely related, so we first calculate ΔF_s for macroscopic ones. We introduce a coordinate system with its z axis along n_0, the normal to the unperturbed surface, and with the axes x_1 and x_2 lying in the plane P(n_0). The value of F_s changes on account of the production on a previously planar surface n_0 of parts with a different orientation n and a different area:

$$\Delta F_s = \int \left(\frac{\sigma(n)}{\cos\theta} - \sigma(n_0) \right) dx_1 \, dx_2, \tag{6}$$

where θ is the inclination of the displaced surface element with relation to the initial planar surface. To calculate ΔF_s we need to consider the behavior of σ (n) in the region of the orientation n_0; we can set the orientation n in advance by means of the projections of n on the x_1 and x_2 axes, which we denote by n_1 and n_2.

We consider first of all a close-packed smooth surface n_0; then the surface n inclined to n_0 at the small angle $\theta \approx (n_1^2 + n_2^2)^{1/2}$ is stepped (Fig. 6). Up to terms of second order in θ, the value of σ (n) is

$$\sigma(n) = \sigma(n_0) + a^{-1}\alpha(m) \sqrt{n_1^2 + n_2^2}, \tag{7}$$

where a is step height (i.e., interplanar distance), while m is the vector for the normals to the steps. We see from (7) that n_0 represents a singular point in σ (n), where that function has a node (Fig. 7). When $n_2 = 0$, the value of n_1 takes the meaning of the angle of rotation of the

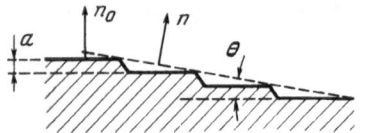

Fig. 6. Stepped structure of the surface n inclined at a small angle ϑ to the close-packed surface n_0.

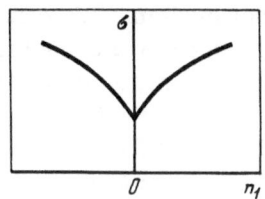

Fig. 7. Specific free surface energy σ as a function of the angle n_1 relative to a close-packed surface.

surface round the x_2 axis. The angular derivative $\partial \sigma / \partial n_1$ has a discontinuity at $n_1 = 0$ (Fig. 7), which is proportional to α. A surface of this type has been termed singular [7-9]. We now estimate the ΔF_s due to local displacement of the surface. If the linear dimension of the displaced part is of the order of L, while the displacement itself is of order Δz, then $n_1 \sim n_2 \sim \Delta z / L$, and so (6) and (7) give for a singular surface that

$$\Delta F_s \sim \alpha \mid \Delta z \mid L/a. \tag{8}$$

The concept of a singular surface arises solely from the behavior of $\sigma(\mathbf{n})$ near $\mathbf{n}_0$, and is in no way restricted to the framework of this model. In addition to singular orientations, in principle one can have regions of variation in $\mathbf{n}$ in which $\sigma(\mathbf{n})$ is an ordinary smooth function. The corresponding surfaces are generally called nonsingular. Near a given nonsingular orientation $\mathbf{n}_0$, one can expand $\sigma(\mathbf{n})$ as a series in n_1 and n_2 up to the second order inclusive:

$$\sigma(n_1, n_2) = \sigma(0, 0) + \sum_{i=1}^{2} \frac{\partial \sigma}{\partial n_i}(0, 0) n_i + \frac{1}{2} \sum_{i, k=1}^{2} \frac{\partial^2 \sigma}{\partial n_i \partial n_k}(0, 0) n_i n_k. \tag{9}$$

We calculate now ΔF_s for a nonsingular surface, first of all noting that the terms $\partial \sigma / \partial n_i(0, 0) n_i$ in (9) can be discarded as physically unimportant, since the addition to $\sigma(\mathbf{n})$ of terms of the form $(\mathbf{b} \mathbf{n})$ (where $\mathbf{b}$ is a constant vector) does not alter the total surface energy [10]. We substitute (9) into (6) and use the fact that $\cos \theta = (1 - n_1^2 - n_2^2)^{1/2}$ to get

$$\Delta F_s = \frac{1}{2} \int \left(\sum_{i, k=1}^{2} \sigma_{ik} n_i n_k \right) dx_1 \, dx_2, \tag{10}$$

where

$$\sigma_{ik} = \sigma(0, 0) \delta_{ik} + \frac{\partial^2 \sigma}{\partial n_i \partial n_k}(0, 0). \tag{11}$$

This σ_{ik} is a two-dimensional tensor and characterizes the stability of the planar surface $\mathbf{n}_0$ in relation to small local displacements; in particular, a two-dimensional displacement of the surface involving rotation of the individual parts around the x_2 axis produces a ΔF_s determined by $\sigma_{11} \cdot n_1^2$. In the isotropic case (i.e., when $\sigma(\mathbf{n}) = \text{const}$), we simply have σ replacing σ_{11}, so the latter is the effective surface free energy for a given two-dimensional displacement. We will subsequently denote this by σ_{eff}. The meaning of σ_{22} is analogous. At equilibrium (when $\Delta F_v = 0$), the surface should be stable, i.e., $\Delta F_s > 0$, and so σ_{11} and σ_{22} must be positive.

We now estimate the order of ΔF_s for displacement of part of a nonsingular surface; we substitute $n_i \sim \Delta z / L$ into (10) to get

$$\Delta F_s \sim \sigma_{\text{eff}} (\Delta z)^2. \tag{12}$$

We now consider the microscopic displacements. This feature is directly related to the surface displacement mechanism. We do not restrict ourselves to any particular model and merely assume that the atoms in the system can be definitely assigned either to the crystal or to the external medium. We consider a part of the surface with integer indices that has an area S and is bound by a planar fixed contour. The crystal and the medium are in equilibrium, so the surface is macroscopically planar (apart from fluctuations). We wish to determine how the surface free energy F_s is dependent on N, the number of particles in the crystal. Let ΔN be the deviation of N from the mean value $\bar{N}$; we introduce the more convenient variable $\xi = \Delta N/qS$, where q is the density of the atoms in the crystal.

As regards macroscopic deviations, this ξ is the surface displacement in the direction of the exterior normal n_0 as averaged over S; it is therefore a generalization of the concept of surface displacement to the case of arbitrarily small displacements. Further, we will consider the displacement ξ not of the entire surface as a whole but for each macroscopic element, i.e., we introduce ξ as a function of x_1 and x_2. We denote the free energy of unit area by $f(\xi)$. When there is a displacement by one interplanar distance a, the surface attains a position equivalent to the initial one, so $f(\xi)$ is a periodic function of ξ with period a. The complete increment ΔF^- in the free energy of the system in response to a displacement $\xi(x_1, x_2)$ is determined firstly by the local change in $f(\xi)$ and secondly by the gradient in ξ along the surface when there is nonuniform displacement. The latter term can be replaced for small gradients by a series and put in the form

$$\sum_{i=1}^{2} B_i(\xi) \frac{\partial \xi}{\partial x_i} + \sum_{i,k=1}^{2} B_{ik}(\xi) \frac{\partial \xi}{\partial x_i} \cdot \frac{\partial \xi}{\partial x_k},$$

where B_i is a two-dimensional vector and B_{ik} is some two-dimensional tensor. The term linear in $\partial \xi/\partial x_i$ can be represented as the divergence of the vector with components $\int_0^{\xi} B_i(\xi) d\xi$, so this term gives zero on integrating over the entire area ($\xi = 0$ at the boundaries). Then

$$\Delta F_s = \int_S (f(\xi) - f(0)) \, dx_1 \, dx_2 + \int_S \sum_{i,k=1}^{2} B_{ik}(\xi) \frac{\partial \xi}{\partial x_i} \frac{\partial \xi}{\partial x_k} \, dx_1 \, dx_2. \tag{13}$$

The surface is stable, so we must have $\Delta F_5 > 0$ for any deviation from the mean position $\xi = 0$; for this we must have that $f(\xi) - f(0)$ for all ξ is nonnegative, and also that the principal values of tensor B_{ik} are nonnegative.

We consider first the case where $f(\xi)$ has a minimum for $\xi = 0$; we consider a surface element displaced from the mean position $\xi = 0$ by one interplanar distance a with respect to the rest of the surface. As $f(\xi)$ is periodic, the work of formation for the displaced element is determined entirely by the boundary where $f(\xi)$ differs from $f(0)$ and where there is a gradient in ξ. The boundary of the part may be considered as a step on the surface, and this step has an essentially finite width in the direction of its normal n, and so if the width is large, ΔF_s is large on account of deviation of $f(\xi)$ from $f(0)$. Formula (13) may be inapplicable to calculate $\alpha(m)$ for a step, since $\partial \xi/dx_i$ may be large near the step; nevertheless, $\alpha(m)$ should be essentially positive because $\Delta F_s > 0$ for any displacement. Surfaces of orientation n close to n_0 are stepped; $\sigma(n)$ is given by (7), and so in this case, where $f(\xi)$ has a minimum for $\xi = 0$, corresponds to a singular surface. Cahn [11] has considered the growth mechanism of such a surface in detail. If the surface moves as a whole in the direction of n_0, i.e., if ξ is the same for all elements, a free-energy barrier ($f_{max} - f(0)$) S has to be overcome, where f_{max} is the maximum value of $f(\xi)$. Therefore, normal motion is impossible if the supersaturation is sufficiently small, and the surface is metastable, with $0 < \xi < a$ in that state; growth occurs by lateral motion of the step. Steps can arise either from dislocations [1] or by two-dimensional nucleation. On the other hand, if the supersaturation is sufficiently large, the surface can move normal to itself without increase in the total free energy; the simplest example of a surface of this type, i.e., a singular surface, is a close-packed smooth surface in Kossel's model. Here the metastable state arising at the surface when the supersaturation arises is characterized by an adsorbed atom density in excess of that at equilibrium, with a reduced density for the vacant sites.

Cahn [11] clearly assumed that $f(\xi)$ must always have a minimum for $\xi = 0$: this created the illusion that all surfaces with integer indices must be singular from very general considerations, no matter whether they are close-packed or not, and whether they are rough or atomically smooth. In fact, in principle one can have the case where $f(\xi)$ does not change with ξ, i.e., $f(\xi) \equiv f(0)$; the requirement $\Delta F_s > 0$ is here met via the second (gradient) term in (13), i.e., ΔF_s is

$$\Delta F_s = \int\limits_S \sum_{i,\,k=1}^{2} B_{ik}\,(\xi)\,\frac{\partial \xi}{\partial x_i}\,\frac{\partial \xi}{\partial x_k}\,dx_1 dx_2. \qquad (14)$$

If a macroscopic displacement is involved, one can average $B_{ik}(\xi)$ with respect to ξ because $B_{ik}(\xi)$ is a periodic function of ξ of period a; then (14) becomes exactly the same as (10) for ΔF_s for displacement of a nonsingular surface. In fact, n_i (the projection of the vector for the normal of the displaced element on the x_i axis) is $\partial \xi / \partial x_i$ up to terms of higher order in n_i, so in this case, where $f(\xi) \equiv f(0)$, we have a nonsingular surface, and the value of B_{ik} averaged with respect to ξ is $\sigma_{ik}/2$. If a nonsingular surface moves as a whole, in which case one does not get gradients in ξ, the free surface energy does not alter, so supersaturation means that the total free energy will be reduced in such a motion, i.e., the surface will move as a whole in the direction of its normal n_0. This growth mechanism is called normal. Analogy with standard phenomenological linear laws for diffusion, thermal conduction, and so on indicates that the growth rate is proportional to the supersaturation or supercooling.

We thus see that the variation in F_s in microscopic displacements is uniquely related to $\sigma(n)$ near n_0; if the n_0 surface is singular, i.e., $\partial \sigma / \partial n_i$ is discontinuous at $n = n_0$, then there is a free-energy barrier for normal displacement of the surface; such a surface is displaced by a lateral motion of steps. If surface n_0 is nonsingular, in which case $\sigma(n)$ is a smooth function near n_0, there is no barrier to normal displacement, and a surface of this type moves as a whole in the direction of n_0.

The speed of a singular surface growing by a dislocation mechanism is proportional to s^2, where s is supersaturation, provided that the supersaturation is small, while it is virtually zero up to a certain critical supersaturation if a nucleation mechanism is involved [1]. Then a given small s means that singular surfaces move substantially more slowly than nonsingular ones, for which the speed is proportional to s.

The above analysis for surfaces can be transferred to steps; a close-packed step is singular in Kossel's model for T = 0, since steps of nearly the same orientation contain kinks and are analogous to stepped surfaces. The discontinuity in the angular derivative of α is $2\beta/h$, where β is the energy of a kink and h is the depth; however, a close-packed step is rough for any T > 0, and contains a finite kink density, so one would expect that the kink density and α would vary smoothly with the orientation. In fact, a detailed consideration of the angular dependence of α shows [1] that α is a smooth function of the step orientation, i.e., for T > 0 all steps are nonsingular. Analogy with (10) and (11) indicates directly that ΔF_e, the free energy increment for a step, is as follows for macroscopic displacement of a part:

$$\Delta F_e = \frac{1}{2}\int \alpha_{eff}\,\varphi^2\,dx, \qquad (15)$$

where

$$\sigma_{eff} = \alpha(0) + \frac{d^2\alpha}{d\varphi^2}(0).$$

$$(16)$$

The integration in (15) is taken over the length of the step; φ is the inclination of the displaced step element in relation to the initial rectilinear step. If a step element contains a large number of kinks, one can proceed as for a surface, with concepts for the displacement ξ and free energy density $f(\xi)$. Further, if a part of a step has fixed ends, the change ΔF_e for a microscopic displacement $\xi(x)$ is given by a formula analogous to (13):

$$\Delta F_e = \int (f(\xi) - f(0)) \, dx + \int B(\xi) \left(\frac{d\xi}{dx}\right)^2 dx.$$

$$(17)$$

However, (17) should become (15) for macroscopic displacements in view of the nonsingularity of steps, so we always have $f(\xi) \equiv f(0)$ for a step, which means F_e does not change on normal displacement, i.e., the speed is proportional to the supersaturation.

In conclusion, in this section we consider the fluctuations in surfaces and steps; the mode of fluctuation again clearly illustrates the physical difference between singular and nonsingular surfaces [10]. At equilibrium ($s = 0$), the probability of a displacement ξ for part of a surface is proportional to $\exp(-\Delta F_s/kT)$.

Formula (12) gives the order of ΔF_s for a nonsingular surface; we put $\Delta z \approx \xi$ in (12) to get that the probability of ξ is proportional to $\exp(-\sigma_{eff} \xi^2/kT)$, whence we get the rms displacement as $(\overline{\xi^2})^{1/2} \approx (kT/\sigma_{eff})^{1/2}$; for the purposes of estimation we assume that $T = 1000°K$ and $\sigma_{eff} = 100$ ergs/cm^2 (the basis for this will be given in the next section), which gives $(\overline{\xi^2})^{1/2} = 4 \cdot 10^{-8}$ cm. Then growth and dissolution, i.e., displacement of the surface parts by distances of the order of a, occur at equilibrium by a fluctuation mechanism.

If we apply (8) to a singular surface, we conclude that $(\overline{\xi^2})^{1/2}$ decreases as S increases, so a correct analysis of the fluctuations requires us to use (13) for the microscopic displacements; (13) shows that $\Delta F_s = \frac{S}{2} \frac{d^2f}{d\xi^2}(0) \xi^2$, for small displacements, so that $(\overline{\xi^2})^{1/2}$ decreases as $S^{-1/2}$, which means that a singular surface at equilibrium is strictly fixed in the mean position $\xi = 0$.

As regards the fluctuations, an analogy can be drawn between a nonsingular surface and a three-dimensional gas at the critical point, since the fluctuations in gas density at that point are also determined by the gradient term [4]. A singular surface in that respect is analogous to a gas in an ordinary (noncritical) state.

The fluctuations of a step are defined by (15) for ΔF_e. If a section of length L in a step is displaced by ξ, then $\varphi \sim \xi/L$ as regards order of magnitude. Thus, $\Delta F_e \sim \alpha_{eff} \xi^2/L$, which means that $(\overline{\xi^2})^{1/2} \approx (kTL/\alpha_{eff})^{1/2}$; although $(\overline{\xi^2})^{1/2}$ increases with the length of section l, the relative displacement $(\overline{\xi^2})^{1/2}/L$ decreases as L increases.

Classification of Surfaces in Kossel's Model

from the Viewpoint of Singularity

The previous section shows how important it is to assign a surface to one of the two possible types: singular or nonsingular; it is therefore extremely interesting to use Kossel's model to establish the surface type for various orientations. The results from the third section indicate that any close-packed surface is singular at least below the transition temperature T_e.* If the surface is not close-packed, and does not contain any bond vector in the $P(\mathbf{n})$ plane, one can attach atoms to the basic configuration without affecting E; in other words, such a surface at T = 0 is indifferent with respect to perturbations in shape and is a degenerate case of a nonsingular surface, for which $\sigma_{eff} = 0$ for any two-dimensional displacement. Such a surface is stable for T > 0 on account of the configurational entropy; we consider this effect by a reference to stepped surfaces (Fig. 6). The number of possible configurations for a step on a stepped surface is restricted by the presence of adjacent steps, so the configuration entropy of such a step is less than that for a single step, while the specific free energy is larger by a certain quantity $\tilde{\alpha}$, which is the greater the less the distance between steps. Therefore, we have a form of configuration repulsion between steps, which provides stability for the surface. The value of σ for a stepped surface with an arbitrary inclination θ (Fig. 6) is

$$\sigma = \sigma_0 \cos\theta + a^{-1}(\alpha(\varphi) + \tilde{\alpha}(\theta, \varphi))\sin\theta, \tag{18}$$

where σ_0 corresponds to a close-packed surface and φ is the angle defining the orientation of the step. We assume that the stepped surface is nonsingular, i.e., that $\tilde{\alpha}$ is a smooth function of angle; we use $\mathbf{e}$ to denote the vector directed along the steps. Consider a two-dimensional perturbation in which parts of the surface rotate around the axis $\mathbf{e}$, which corresponds to a change in distance between steps. It is clear that $\sigma_{eff} = \sigma + \partial^2\sigma/\partial\theta^2$ for such a perturbation, and that this arises entirely from $\tilde{\alpha}$; in fact, (18) gives $\sigma_{eff} = a^{-1}\left(2\frac{\partial\tilde{\alpha}}{\partial\theta}\cos\theta + \frac{\partial^2\tilde{\alpha}}{\partial\theta^2}\sin\theta\right)$. As $\tilde{\alpha}$ increases with θ, i.e., as the distance between steps decreases, we have $\sigma_{eff} > 0$, as should be the case for a stable surface. Approximate values have been calculated for $\tilde{\alpha}$ and σ_{eff} [10] for a simple cubic lattice and steps of orientations $\langle 01 \rangle$ and $\langle 11 \rangle$ on a $\{001\}$ face. It was found that σ_{eff} is proportional to θ, the value for both orientations of the step being given by the single formula

$$\sigma_{eff} \approx 10kTa_0^{-1}\rho\theta, \tag{19}$$

where a_0 is the lattice constant and ρ is the kink density (per unit length of step). We put T = 1000°K, $a_0 = 3 \cdot 10^{-8}$ cm, $\theta = 0.1$ to get that a surface composed of $\langle 11 \rangle$ steps (when $\rho \sim 1/a_0$) gives $\sigma_{eff} \approx 100$ ergs/cm², which is a fairly large value and is comparable with the specific surface energy.

Consider now a two-dimensional perturbation in which parts of the surface rotate around an axis perpendicular to $\mathbf{e}$, which corresponds to local orientation change in the steps; it is clear that σ_{eff} in such a perturbation arises solely from α_{eff}, and all steps are nonsingular for T > 0, so there is no essential difference between surfaces composed of steps without close packing, when $P(\mathbf{n})$ contains no bond vector $\mathbf{r}$, and surfaces composed of close-packed steps, when $P(\mathbf{n})$ contains bond vectors $\mathbf{r}$ running only in one direction.

We have assumed above that surfaces are nonsingular for T > 0 if they are not close packed; analogy with steps indicates that rough surfaces generally are nonsingular. To decide the question rigorously, we would have to calculate $\sigma(\mathbf{n})$ exactly, and another method of defining the type of such a surface is the calculation of the free energy for a part of the surface as a function of ξ. However, such calculations involved great mathematical difficulties, and so we will discuss the surface type problem only qualitatively.

---

*The number of close-packed surfaces is finite when one takes into account only the interactions between reasonably close neighbors; if, on the other hand, we incorporate arbitrarily remote interactions, then any surface with integer indices will be close-packed, i.e., all such surfaces will be singular for T = 0. This result was derived in [12] for the particular case of van der Waals interaction.

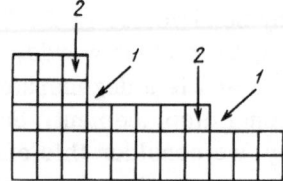

Fig. 8. Open-packed step at low temperatures: 1) deposition sites; 2) atoms in removal position.

Fig. 9. Interaction of corners: a) loss of deposition site when two kinks come together (lower kink of double depth); b) loss of removal position when two kinks come together (upper kink of double depth).

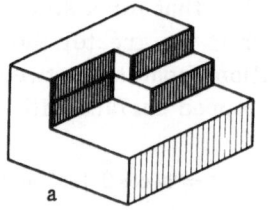 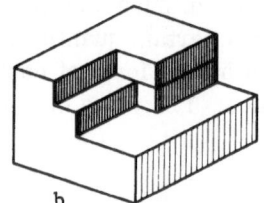

A rough surface has many deposition sites, i.e., sites at which an atom can be attached without changing E; similarly, such a surface has a large number of atoms in the removal position, i.e., atoms whose removal does not alter E. It is often assumed that such a surface can move without difficulty at a rate proportional to the supersaturation, i.e., in other words, that the surface is nonsingular. We illustrate this assertion by reference to a step not having close packing for sufficiently low T. In that case we need to take into account only those kinks that are due to the inclination relative to a close-packed step (Fig. 8). We denote the probability of attachment in unit time for an atom for a given kink of unit depth by w^+, while the probability of removal of an atom from a kink is denoted by w^-. At equilibrium, $\overset{\circ}{w}{}^+ = \overset{\circ}{w}{}^-$, where the superscript zero denotes the equilibrium value. The step has deposition sites apart from kinks of unit depth, which are produced by reentrant angles at kinks of double, triple, etc. depth (Fig. 8); the probability of attachment to such sites is also w^+. Apart from atoms at kinks of unit depth, a step will have other atoms in the removal position, such as ones at kinks of double etc. depth (Fig. 8). Each kink has one deposition site and one removal position, so the total number of deposition sites N^+ exactly equals the number of removal positions N^-. When a supersaturation is produced at a constant temperature, there is no change in w^-, while w^+ increases in proportion to the concentration of atoms in the particle reservoir (for growth from a vapor, this reservoir is the layer adsorbed on the surface adjoining the step). Then $w^+ = \overset{\circ}{w}{}^+(1 + s)$, where s is the relative supersaturation in the reservoir. The resulting flux of atoms to a step is $w^+N^+ - w^-N^- = (w^+ - w^-)N^+$. The difference $w^+ - w^- = sw^-$ is proportional to the supersaturation, so the speed of the step is proportional to s, at least for small s, where $\overset{\circ}{N}{}^+ \approx N^{+*}$; unfortunately, the problem is not so simple for surfaces. At equilibrium, a rough surface has its number of deposition sites $\overset{\circ}{N}{}^+$ equal to the number of removal positions $\overset{\circ}{N}{}^-$, which follows from the fact that $\overset{\circ}{w}{}^+ = w^-$; but a supersaturation causes N^+ and N^- to alter, and it may be that these two may change in such a way that the resultant flux of atoms $w^+N^+ - w^-N^-$ becomes zero. The surface goes over to a metastable state, i.e., although there is roughness, the surface is singular. To illustrate the variations in N^+ and N^-, Figure 9a shows two kinks coming together on different steps (the upper kink is of unit depth, while the lower one is of double depth). We have $N^+ = N^- = 2$ before these come together, whereas afterwards $N^+ = 1$, $N^- = 2$; Fig. 9b shows the reduction in N^- as the two kinks approach (beforehand $N^+ = N^- = 2$, afterwards $N^+ = 2$, $N^- = 1$). Therefore, the presence of a large number of deposition sites on the surface does not mean that the latter is not singular.

_____

* It has been shown [13] that the structure of this step without close packing does not change at all when a supersaturation appears, i.e., N^+ is exactly equal to $\overset{\circ}{N}{}^+$

We now consider the problem from another viewpoint. We consider a close-packed surface for $T < T_e$, and Fig. 3b shows that α falls as T increases, and becomes substantially less than the low-temperature value while the surface is still actually atomically smooth, i.e., when the exact theory for the adsorbed layer gives the density of the adsorbed atoms as still small, as shown in Fig. 3a. It has been shown that α is proportional to the discontinuity in the angular derivative of σ, i.e., α is a measure of the singularity. We shall see that the discontinuity in this derivative falls substantially as T increases; it is unclear what will be the value of this quantity when the roughness becomes appreciable, of course, but it is reasonable to suppose that it will fall further, and that at a certain temperature close to T_e it will become zero, which implies a qualitative transition in the state of the surface, which from singular becomes nonsingular. In principle, the problem could be solved by computer simulation. The expected transition from singularity is a second-order phase transition and so should involve a discontinuity in the specific heat, which means that the surface energy or surface roughness as a function of kT/W should have a singular point, at which the slope is discontinuous.

Extension of a Kossel's Model to a
Condensed Medium

In practice, it is more common for crystals to grow from condensed media such as melts or solutions than from vapors, so it is very important to set up models for the interfaces between crystals and condensed media. First of all we assume that all the atoms in the crystal medium system can be classified strictly as belonging to the crystal or to the medium. In the case of a dilute solution, this is an entirely reasonable assumption, while for a melt it means that atoms in the crystal persist for a long time near fixed nodes, while all other atoms drift randomly and belong to the melt. The scope for such a distinction is, of course, open to some doubt, and it must be considered as a postulate for experimental verification. We transfer from the Kossel model the view that the space lattice of the crystal is not distorted near the surface, which can take up various configurations differing in mutual disposition of the surface atoms. Analogy with Kossel's model leads us to suppose that the surface energy E is proportional to the number of broken crystalline bonds for a condensed medium also. The energy W per broken bond is here dependent not only on the energy of the interaction between atoms in the crystal but also on the interaction of the atoms in the medium one with another and with the crystal. The resulting model for the surface is entirely equivalent to Kossel's model for a crystal—vapor boundary (which takes into account interactions only between nearest neighbors), and differs from the latter only in the origin of W.* The distinctive feature of a crystal—melt interface is merely that one can scarcely meaningfully speak of single liquid atoms in the surface layer of the crystal; meaning attaches only to reasonably large groups of such atoms, which are melting nuclei. This restriction in no way affects the critical temperature T_e as found from the free energy of a step becoming zero (Table 1). To estimate W we assume that the atoms of the solvent or melt lie at the nodes of the same space lattice as the atoms in the crystal. We denote by ε_{11}, ε_{22}, and ε_{12} the interaction energies of the adjacent atoms, where subscript 1 refers to an atom in the crystal and 2 to an atom in the medium (here we take into account only interactions between nearest neighbors). The number of broken bonds, i.e., interphase bonds, is denoted by N_{12}. The interaction energy for two atoms in the crystal is the

---

* A model for a physically rough boundary [11, 14] has been discussed in the literature as well as Kossel's model for a crystal—melt surface, i.e., a model with classification of all atoms into two types; in the former, an atom or atomic layer is assigned a certain parameter, which varies continuously between two limiting values, which correspond to the two contacting phases. This parameter has no very clear meaning, and it considerably complicates the discussion of the already difficult problems arising from configurational roughness of the surface.

same as if these atoms were within the body of the crystal less the amount $N_{12} \cdot \varepsilon_{11}/2$; similarly, the interaction energy for atoms in the medium is the same as if they were in the bulk of the medium less the quantity $N_{12} \cdot \varepsilon_{22}/2$. Then the surface energy per broken bond is

$$W = \varepsilon_{12} - \frac{\varepsilon_{11} + \varepsilon_{22}}{2}. \tag{20}$$

In particular, we can assume for a melt that the interaction energy for an atom in the melt with an adjacent atom is not dependent on which phase contains this latter atom, i.e., $\varepsilon_{12} = \varepsilon_{22}$; we substitute this into (20) to get that $W = (\varepsilon_{22} - \varepsilon_{11})/2$. The resulting heat of melting H_0 is $H_0 = \nu\varepsilon_{22}/2 - \nu\varepsilon_{11}/2$, where ν is the number of nearest neighbors in the lattice, so $W = H_0/\nu$, and this estimate for a melt is analogous to the case of a vapor, where W was expressed via the heat of sublimation H.

The values of W for Ge and Si have been calculated from the known σ for Ge and α for Si [15] on the assumption that a (111) face is atomically smooth. The value of kT_0/W for these substances is 0.6, which agrees with the conception of an atomically smooth face (Table 1). It is interesting that W exceeds $H_0/4$ by a factor 1.8, i.e., any estimate of W via the latent heat of fusion is extremely rough and substantially too low.

In Kossel's model there is a restriction on the permissible values of kT/W arising from the physical requirement that $\sigma > 0$; as σ decreases with increase in kT/W for the same reason as does α, we find that it becomes zero for $kT/\nu W \approx 0.5$ [3]. If we estimate W as H_0/ν, then kT/H_0 should be less than 0.5, whereas $kT_0/H_0 \approx 1$ for most metals, which falls outside the permissible region, which means that the estimate $W = H_0/\nu$ is here even rougher than that for Ge and Si, and from this estimate one cannot draw any a priori conclusion about the surface structure. Jackson [16] has discussed the roughness of a crystal–melt boundary, and his model is completely equivalent to Kossel's for $W = H_0/\nu$; a distinctive feature of [16] is is that Jackson used as the critical roughness temperature not the exact solution for a layer (Table 1) but the high approximation $kT_{cr}/W = \nu_s/2$, where ν_s is the number of neighbors on the surface, which was obtained by the Bragg–Williams method [2]. Therefore, his conclusion [16] about roughness of the metal–melt interface was incorrect.

Growth of a single crystal from a melt is a special one from the viewpoint of non-singular surfaces; these surfaces grow more rapidly than singular ones from a vapor or solution, and so they are displaced from the phase interface, and the crystal is bounded virtually only by singular surfaces. The situation is different in a crystal–melt system; here there is usually a large temperature gradient provided by external heat sources, so the crystallization front takes the shape of the T_0 isotherm, except for parts of the singular surfaces on which a considerable supercooling is set up (Fig. 10). The front has parts of the nonsingular

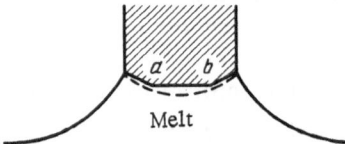

Fig. 10. Shape of crystallization front during pulling of a single crystal from the melt. The broken line shows the isotherm corresponding to the melting point T_0; ab is part of the singular surface, and the other surface is inclined to ab at several degrees, being a stepped nonsingular surface.

surfaces, in particular stepped ones, so it is very important to consider the stability of such surfaces and the values of σ_{eff} for them. Formula (19) for σ_{eff} is transferred unaltered to a crystal−melt surface; the value of ρ for a close-packed step should be high in this case, so σ_{eff} should not be too greatly dependent on the orientation of the stepped surface, i.e., for a given value of θ in relation to a singular surface. The value of σ_{eff} plays an important part, for example, in the formation of a cellular structure at the crystallization front. The cause of this structure is [17] the impurity gradient in the melt ahead of the front, which arises from impurity rejection. If there is a small periodic perturbation in the shape of the front, projections at the front enter a zone with a slightly higher T, but with a lower impurity concentration, i.e., with a higher equilibrium temperature. If the impurity gradient is sufficient, the projections are supercooled, so the perturbations in the front tend to increase, and this effect is especially pronounced when the perturbation periods are small. However, an obstacle to such perturbations is the increase in the surface free energy, which is defined by σ_{eff}; there is therefore an optimal period for the perturbations [18], which is dependent on σ_{eff} and which has the meaning of the period of the cellular structure in the first stage of development. The period actually observed for this structure in silicon [19] on surfaces with an inclination of a few degrees is in good agreement with the theoretical value derived from [19] for σ_{eff}, which shows that Kossel's model is reasonable for a crystal−melt interface. Another strong argument for this model is that it gives in correct form the impurity trapping coefficient for a stepped surface as a function of growth rate [20]. This aspect is discussed in the next section.

Impurity Partition Coefficient for a Surface

Layer of a Crystal

If a surface is atomically smooth, there are two different states for a substitutional impurity in the crystal: (1) in the bulk of the crystal and (2) in the surface layer adjoining the melt. Let K_b be the partition coefficient for the bulk of the crystal and the melt, while K_s is the same for the surface layer and the melt; if $K_b < 1$, then $K_s > K_b$, since the surface impurity atoms are in a state intermediate between the phases. The traveling steps cover the surface layer with impurity (Fig. 11), and transfer the latter ultimately to the bulk state, and it has been shown [12] that complete burial of the impurity occurs only if the steps move reasonably rapidly, namely only in the face region at the crystallization front (Fig. 10), though this in no way occurs in the region of a curved inclined front. This behavior is the cause of the impurity channels left behind by moving faces, which has been called the face effect [22, 23]. In the region of the face, the impurity concentration in a freshly covered layer is $K_s C$, where C is the atomic proportion of the impurity in the melt, and this differs from the equilibrium concentration $K_b C$, which leads to redistribution by diffusion [24]. The diffusion rate D/a is governed by the diffusion coefficient d in the crystal, and if this exceeds the normal growth rate v one can attain the equilibrium concentration $K_b C$, and the resultant trapping coefficient $K = K_b$. If, on the other hand, $v \gg D/a$, the trapped concentration remains $K_s C$ and $K = K_s$. Therefore, K is dependent on v, tending to a limit of K_s at high speeds (Fig. 12). A relationship of this form has been observed in silicon for Al and P dopes [20] and also for B [25], which enables one to determine K_s. On the other hand, K_s and K_b can be expressed in terms of the interaction energies via Kossel's model, and then K_s can be expressed in terms of K_b. As above, we will

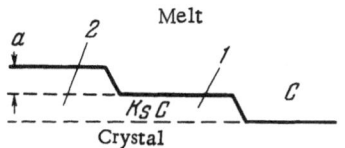

Fig. 11. Impurity trapping by a stepped surface: 1) crystal layer just laid down by step; 2) layer covered by next step.

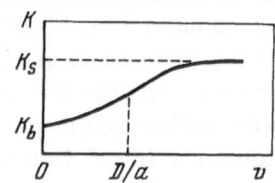

Fig. 12. Impurity trapping factor K as a function of crystal growth rate v.

assume that both phases are located at the nodes of a single lattice. We identify some one node in the volume of the crystal and one node in the volume of the melt. There is a probability K_bC that the first node is taken up by impurity while the second is not (the atomic proportion of the impurity $C \ll 1$). The probability of the converse situation is C. The ratio of these two probabilities is K_b, and the Gibbs distribution [4] indicates that this is expressed via the difference in the free energies ΔF_b for these two states:

$$K_b = \exp(-\Delta F_b/kT). \tag{21}$$

The value of ΔF_b equals the change in free energy on exchanging places between a given impurity atom in the melt and a given basic atom in the crystal; this change is made up from the change in the interaction energy ΔE_b and the entropy term $T\Delta S_b$, which arises from change in the vibrational entropy. The value of K_s is expressed by analogy with (21) via the corresponding free-energy change $\Delta F_s = \Delta E_s - T\Delta S_s$:

$$K_s = \exp(-\Delta F_s/kT). \tag{22}$$

The changes ΔE_b and ΔE_s arise from interaction of the two transferred atoms with their nearest neighbors; we denote the pair interaction energy by ε_{ik}, assigning subscripts 1 and 2 to a basic atom in the crystal and in the melt, while subscripts 1' and 2' are assigned to the impurity in these two phases. To estimate the interphase interaction energies ε_{12} and $\varepsilon_{1'2}$, one might equate these to the corresponding energies in the melt ε_{22} and $\varepsilon_{2'2}$. However, such an estimate is too rough, as we have shown in the previous section, so we make the more general assumption that $\varepsilon_{12}-\varepsilon_{22}$ and $\varepsilon_{1'2}-\varepsilon_{2'2}$ differ from zero but are identical. A node in the surface layer has λ neighbors in the melt and $\nu - \lambda$ neighbors in the crystal. Each of the λ interphase bonds produces an energy change $\varepsilon_{1'2}-\varepsilon_{2'2}$ for the transferred impurity and an energy change $\varepsilon_{22}-\varepsilon_{12}$ for the transferred basic atom; the overall change is zero, in accordance with the above assumption. Then there is a contribution to ΔE_s only from the $\nu - \lambda$ bonds directed into the crystal, and $\Delta E_s = (\nu - \lambda)(\Delta E_b/\nu)$; the entropy changes ΔS_s and ΔS_b are the same in our model. We express ΔF_s via ΔF_b, and get from (21) and (22) a formula for K_s:

$$K_s = K_b^{1-\gamma} \exp \gamma \Delta S_b/k, \tag{23}$$

where $\gamma = \lambda/\nu$ is the proportion of the bonds for a surface atom directed into the melt; $\gamma = 1/4$ for a (111) face of a diamond lattice, for example. The value of ΔS_b for exchange between an impurity in the melt and a basic atom in the crystal equals as regards order of magnitude the

TABLE 2. Observed and Theoretical Impurity Trapping Factors $\overline{K}_s$ at High Growth Rates

Impurity	K_b exp	$\overline{K}_s$	
		exp	theor
Al	0.002	0.012	0.011
P	0.35	0.75	0.70
B	0.80	1.3	1.4

difference in the entropies of melting for the host and impurity. This entropy for Ge or Si is substantially higher than the values for the usual dopes, and so we can estimate ΔS_b simply from the value H_0/T_0 for the host.

Although certain assumptions have been made, (23) describes closely the observed results for Si. A step on a (111) face in a diamond lattice is a double atomic layer, in which half of the atoms are surface ones and the other half are bulk ones, so the limiting trapping factor at high speeds (Fig. 12) is averaged over the two states and is $\overline{K}_s = 0.5(K_s + K_b)$; (23) gives

$$\overline{K}_s = 0.5 K_b (1 + K_b^{-1/4} \exp H_0/4kT_0). \tag{24}$$

This formula gives $\overline{K}_s$ close to the observed values (Table 2).

I am indebted to A. A. Chernov and D. E. Temkin for fruitful discussions and to N. N. Sheftal' for constant interest in the work.

Literature Cited

1. W. Barton, N. Cabrera, and F. Frank. In: Elementary Crystal Growth Processes [collection of Russian translations], IL, Moscow (1959), p. 11.
2. T. L. Hill. Statistical Mechanics (1956).
3. V. V. Voronkov. Kristallografiya, 11(2):284 (1956).
4. L. D. Landau and E. M. Lifshits. Statistical Physics [in Russian], Fizmatgiz, Moscow (1964).
5. V. V. Voronkov. Thesis: Theory of Surface Microstructure and Elementary Crystallization Processes [in Russian], Moscow (1967).
6. V. V. Voronkov and A. A. Chernov. Kristallografiya, 11(4):662 (1966).
7. C. Herring. The Physics of Powder Metallurgy, McGraw-Hill, New York (1951), p. 143.
8. N. B. Cabrera. Disc. Faraday Soc., 28:16 (1959).
9. N. Cabrera and R. V. Coleman. In: The Art and Science of Growing Crystals (J. J. Gilman, ed.), Wiley, New York (1963), p. 3.
10. V. V. Voronkov, Kristallografiya, 12(5):831 (1967).
11. J. W. Cahn. Acta metallurgica, 8(8):554 (1960).
12. L. D. Landau. Volume dedicated to Academician A. F. Ioffe's 70th Birthday [in Russian], Izd. AN SSSR, Moscow (1950), p. 44.
13. V. V. Voronkov. Kristallografiya, 13(1):19 (1968).
14. A. L. Roitburd. Dokl. AN SSSR, 148(4):821 (1963).
15. V. V. Voronkov. Kristallografiya, 17(5):909 (1972).
16. C. A. Jackson. In: Solidification of Liquid Metals [Russian translation], Metallurgizdat, Moscow (1962), p. 200.
17. C. Elbaum. Progr. Metal Phys., 8:203 (1959).
18. V. V. Voronkov. Fiz. Tverd. Tela, 6(10):2984 (1964).
19. M. G. Mil'vidskii and V. V. Voronkov. Fiz. Tverd. Tela, 6(12):3736 (1964).
20. V. V. Voronkov, V. P. Grishin, and Yu. M. Shashkov. Izd. AN SSSR, Ser. Neorg. Mat., 3(12):2139 (1967).
21. V. V. Voronkov. Paper at the 4th All-Union Conference on Crystal Growth, Tsakhkador: Crystal Growth and Structure [in Russian], part 2, p. 110, Izd. AN ArmSSR (1972).
22. J. A. M. Dikhoff. Solid State Electr., 1:202 (1960).
23. M. G. Mil'vidskii and A. V. Berkova. Fiz. Tverd. Tela, 5(2):513 (1963).
24. A. A. Chernov. In: Growth of Crystals, Vol. 3, Consultants Bureau, New York (1962), p. 35.
25. V. P. Grishin, G. I. Kononov, and Yu. M. Shashkov. Abstracts for the Second Conference on the Physicochemical Principles of Semiconductor Doping [in Russian], Izd. Inst. Met. im. Baikova, Moscow (1972), p. 115.

CONDENSATION MECHANISMS AND LIQUID STRUCTURE

V. V. Puchkov and L. D. Kislovskii

A great deal of information has been accumulated about the various aspects of phase nucleation on account of improved techniques such as electron microscopy, ion projectors, and various types of resonance spectroscopy, together with advanced techniques for obtaining pure substances and use of high vacuum and so on. A proper analysis of this information requires a careful study of the detailed and complex interatomic and intermolecular relationships during the formation of the structure.

In studies on changes in the state of aggregation, it is usual to consider a free atom or molecule as a structural unit that enters in unchanged form into the lattice, with no account of any change in electronic structure in the atom or molecule; the formation rate for the new phase is derived from information on the steady-state structure, although recent studies have shown that this approach is not always appropriate, since the condensation is of stepwise character, and various metastable formations arise in the initial stages of growth.

Theoretical chemistry provides a good indication of the changes in electronic structure of atoms when compounds are formed or decompose; there are also studies [1, 2] on the growth of crystals using the methods of chemical kinetics. It is now possible to represent the growth of a crystal as a chemical reaction on that basis. Here we show that it is very useful to take into account the variability of the electronic structure for atoms and molecules in order to understand a variety of processes occurring when the state of aggregation changes.

Initial Stages of Crystallization

We first consider some features of the changes in the electronic structure of atoms that occur during crystal growth; for definiteness, we assume that the growth occurs by condensation from the vapor, while the bond type in the condensate is covalent. We also assume that the electronic orbitals involved in this bond are sufficiently hybridized, this involving redistribution of the density in certain atomic orbitals. The symmetry of the electron cloud then bears a relationship to that of the array of atoms. A hybrid orbital is formed of several orbitals of the isolated atom, and differs from the latter in having a markedly directional electron density; an example is the bond between carbon atoms in diamond [3]. The atom passes from the $1s^2 2s^2 2p^2$ ground state to the $1s^2 2s 2p^3$ state, and the electron is excited from the 2s shell to a 2p level (Fig. 1). Figure 2 shows the energies for passage of a carbon atom from one state to another.

The concept of hybridization is not a purely formal idea suitable for analysis of types of compound [4-6] but has also considerable physical content; it reflects the conformity of the atomic electron structure with the structure of the surrounding system of atoms, whose properties are substantially dependent on the atomic interactions, and the system may be said to be cooperative. We list below the conditions that favor hybridization [4], and these may prove of value in evaluating the changes in atomic electron structure during atomic aggregation.

18

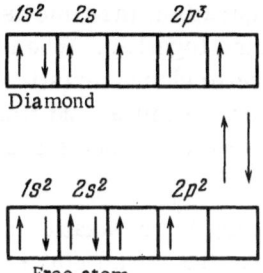

Fig. 1. Change in electronic configuration of the carbon atom in diamond.

1. Hybridization occurs for electron clouds with similar azimuthal directions.

2. The clouds to be mixed must have reasonably large densities in regions similar in radius reckoned from the nucleus as center.

3. The electrons of the contributing orbitals must have similar energies.

An atom with hybridized orbitals is in an excited state relative to the free atom; this state is stable if the atom forms part of a system such as a molecule or crystal. The energy required to produce hybridization is provided by the exothermic bond formation. Such bonds are stronger than those formed by unhybridized orbitals. As an example, Table 1 gives the relative strengths of hybrid bonds [7]. The method of calculation has some deficiencies [8], but the table is satisfactory for qualitative comparison.

The transition to the hybrid state often requires fairly considerable energy, which is termed the excitation energy of the valence state [9, 10]. This energy is involved in reconstructing the electronic structure from the isolated atom, since this structure in most cases is not optimal for setting up the bond system. The time needed to modify an electronic structure is

Fig. 2. Excitation of the valence state of the carbon atom in the graphite−diamond transition: $E_1 = 171$ kcal/mole is the atomization energy of graphite; $E_2 = 96$ kcal/mole is the hybridization energy to give the sp³ configuration; $E_3 = 94$ kcal/mole is the excitation energy of the valence state with retention of the sp³ configuration but with a different disposition of the spin and orbital moments.

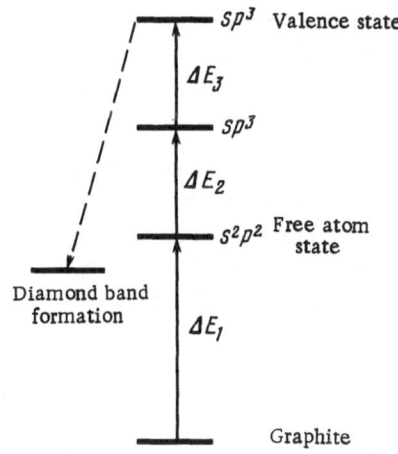

TABLE 1

Orbitals involved in hybridization	Relative strength of hybrid bond (strength 1 for s-electron bond)	Spatial configuration of hybrid bonds
sp	1.93	On a straight line
sp²	1.99	To vertices of an equilateral triangle
sp³	2.00	To vertices of a tetrahedron
dsp²	2.69	To vertices of a square
d³s	2.95	To vertices of a tetrahedron
d²sp³	2.92	To vertices of an octahedron

much greater than that for electron excitation within a given configuration; the time is determined by the probabilities for electrons in the source clouds to come together. When the electrons come together, conditions are favorable to sharing of the kinetic characteristics, and hybrid orbitals may result [11]. The greater the difference between the initial and final electronic structures of the atom, the larger the energy needed to excite the valence state and the longer the relaxation process will take.

As a rule, several collisions with other atoms will occur during the relaxation time for the electronic structure; preparatory collisions between atoms form the initial stage in setting up the chemical bond. This idea on the course of a chemical reaction differs from concepts developed in the theory of activated complexes [12, 13] in that in the latter case the reaction is considered as occurring in a single act, in which a set of interacting atoms is transformed continuously to the reaction products. The energy required to transfer the atom to the valency state is greater than the activation energy of the reaction, since the latter is a quantity averaged over the entire volume involved in the reaction.

A series of interatomic collisions disrupts the initial electronic structure of the isolated atom and combines the excited atoms into unstable complexes, in which the bond is provided by electrons in some intermediate metastable level. The scope for forming such complexes is determined by the relaxation time for the excited electronic system (the relaxation time is greater than the time between collisions) and also by the fact that the structure changes via a set of discrete steps. Therefore, some collisions may excite the atoms to fairly long-lived states, in which the atoms form complexes of varying degrees of stability. A confirmation of this is available from observations of atomic aggregation in the gas state [14, 15].

It has repeatedly been observed that crystallization is a stepwise process, where metastable phases initially arise that then give way to stable ones. This observation has been summarized as Ostwald's law of sequential reactions, which indicates that if the initial and final states are separated by a series of intermediate metastable states, the latter will follow one after the other in the order of stepwise reduction in the free energy [16]. Analogous effects occur in the growth of thin films [17-19], where first of all one finds high-temperature phases, which at a certain thickness are replaced by a low-temperature modification.

It has also been found that the highest-temperature modifications are cubic, while low-temperature ones have lower symmetry [16, 20]. The reason is probably that the large thermal vibrations of the atoms hinder the complex reconstruction of the atomic electron structure, this being frequently necessary for hybridization; therefore, one finds only the simplest structures in the high-temperature region, in which the atoms have minimally hybridized or distorted orbitals. If the bonds are provided by orbitals with pure s, p, or d states, then one gets cubic structures.

Mosaic Structure of Single Crystals

The combination of interacting atomic aggregates into complexes with a new structure may be considered as a chemical reaction between macromolecules. As it is necessary to excite all the atoms in the molecule to complete the chemical reaction [13], more energy is required for a large system and more time is needed to modify the electronic structures of the constituent atoms. There are some examples that illustrate this [12]. A theoretical calculation for atoms with s electrons shows that colliding quasimolecules containing three or four hydrogen atoms each will require activation energies of the following order in order to produce the activated complex:

$$H-H-H \qquad\qquad \begin{matrix} H-H \\ \diagdown\diagup \\ H \end{matrix} \qquad\qquad \begin{matrix} H-H \\ |\ \ | \\ H-H \end{matrix}$$

$$E_a = 8 \text{ kcal/mole}, \quad E_a = 400 \text{ kcal/mole}, \quad E_a = 100 \text{ kcal/mole}$$

The following are the probabilities p that the reaction will occur through the activation barrier:

Colliding particles	p
Atom and molecule	$10^{-2} - 10^{-4}$
Two molecules	$10^{-5} - 10^{-10}$

The scale chosen is one in which the probability p = 1 for two colliding atoms.

On this basis we conclude that under these condensation conditions there should exist some limiting size for the complex beyond which the reactivity becomes negligibly small.

This question of complex size restriction during growth has been considered elsewhere [21]; it was supposed that at a certain size the electronic levels in the resulting energy bands will lie so far above the levels of the isolated atom that it will be more probable that a new complex will form. The name given to such complexes was macromolecules, and the size of such a complex is comparable with the mean free path of the electrons.

The last stage in crystal growth is the combination of complexes of limiting size; this process is not accompanied by change in structure, and it occurs via the surface energy of the complexes. If the temperature is low and there are no impurities, the result would be a perfect single crystal, but under real conditions the condensate always contains impurities, and many of the impurity particles are rejected to the boundaries of the complexes, which interferes with matching of the mutual orientations, and the crystal acquires a mosaic structure. Provided the temperature is reasonably far from absolute zero, the vibrations facilitate the mutual displacement of the complexes, which also hinders combination and facilitates entry of impurities.

Structure and Properties of the Surface

The mosaic structure of a single crystal implies internal boundaries between the blocks; the redistribution of the electron density for the excited atom during condensation is determined, amongst other reasons, by the symmetry in the mutual disposition of the atoms; at the boundary of a block or of the crystal as a whole, the symmetry in the atomic environment is different from that from the symmetry within the volume. If the crystal has a structure stable under the existing conditions, one is forced to suppose that most of the atoms forming the blocks are also in one of stable states. Therefore, the symmetry of the structure is one of the factors governing the stability of the atomic state, and deviation from this at the boundary can result in instability in the surface atoms. One may assume that the atoms in the surface of a crystal are in one of the metastable states intermediate between the state of the isolated atom and the state defined by the form of the crystal structure.

It is at present impossible to define the detailed form of the hybrid orbitals for the surface atoms on account of lack of experimental evidence; we can only say that experiment does not confirm the assumption that the surface atoms have free valencies. ESR spectra give direct information on free radicals at surfaces, and they show that the radical concentration at an atomically clean surface in a crystal is negligibly small [22]. The assumption that surface atoms have free valencies is a consequence of the view of an atom as a structural unit whose characteristics are not dependent on interaction with other atoms.

This assumed instability for the surface atoms may be considered as a perturbation for the rest of the crystal volume, this extending to a finite depth within the crystal and being probably a cause of the specific structural and dynamic features of surface layers. Slow-electron diffraction has shown that surface layers differ in structure from the bulk of the crystal, the detailed structure depending on the number of the layer and on the temperature [23]. The mean-square displacements of the surface atoms are greater than those of the bulk atoms by

15% at low temperatures and by 30% near the Debye temperature [24]. This anomaly in properties is lost in the fifth layer. It has been found [25] that the Debye temperature is reduced for dispersed particles, while there is an increase in the anharmonicity of the atomic vibrations when one approaches the boundary of the crystal, which confirms the assumption [26] that the elastic constants are altered in the surface layer. The instability in the surface atoms is characterized by the increased specific heat when the material is dispersed [27], and also by the high chemical activities of surfaces in some substances. Small volumes of material have an anomalous reduction in the melting point [28], while atomic clumps show high mobility on clean surfaces [29], and this tendency is confirmed by the trends in the growth of individual particles during condensation [30]; from this it has been concluded that a liquid-type layer exists at the surface of a crystalline body. Nuclear magnetic resonance [31] shows that there is a mobile phase at the surface of ice.

These experimental results relate to the outer surface of the crystal, but it seems likely that they can be applied to a certain extent to the state of the atoms at the internal surfaces in a mosaic crystal, where the blocks differing in orientation meet. A real crystal consists of blocks separated by boundary layers of atoms in unstable states. The structure of these layers should differ from that of the main volume. A similar view has been taken [15, 32] of the structure of a real crystal, where it was supposed that the subcrystallites are surrounded by structureless or amorphous layers of atoms.

The extent of the change in atomic electron structure during condensation determines how far the characteristics of the surface layer differ from those of the crystal as a whole. The greater the difference in electronic configuration in the lattice and in the free atom, the thicker the surface layer and the more anomalous the characteristics. For instance, one might suppose that the cubic crystals characteristic of metals have the electronic structure of the atom only slightly different from that of the free atom. Therefore, the thickness of the surface layer of the blocks is small, and the structural difference in the atoms is slight, which may be one of the reasons for the good electrical conductivity of metals. In crystals of lower symmetry, the surface layers on the blocks should be thicker, and this should subtantially influence the electrical conductivity [33].

The Structure of a Liquid

This view of mosaic single crystals, with blocks separated by surface layers, leads us to describe melting as a process involving increased disorientation of the blocks and increase in the surface layer thickness [15]. The mosaic blocks are only 100-1000 Å in size, so the effects of the surface layer should be considerable. The mutual displacements of the blocks increase with temperature, i.e., the perturbation exerted by one on another increases. The perturbation propagates from the surface of the block into the volume, and the initial structure of the block is destroyed if the temperature is high enough. A certain relaxation time is involved in changing the electronic structure of the atoms to give a higher metastable level. The structure change may be accompanied by decomposition of the block into smaller clumps of atoms. Therefore, as the melting point is approached, one gets changes that are inverse in sequence to the changes occurring during the initial stages of crystallization.

It seems likely that such processes are the cause of the frequently observed premelting effects [34]. The block size and block-size distribution are dependent on the impurity concentration, which is related to the growth conditions; if we assume an equilibrium block size of 1000 Å, then a crystal can be formed with only 10^{14} particles of impurity per cubic centimeter to act as nuclei. The size distribution of the resulting blocks would have a sharp peak. If the impurity concentration is larger than the above, the block growth is less uniform, and the size distribution at the end of growth would be less sharp. A broad block size distribution would imply differences between the processes on melting, since it is possible for there to be considerably different structures in the groups of atoms. The resulting stepwise melting makes itself felt as anomalies in various characteristics of the material.

At the melting point, the blocks become completely isolated, with breakup or structure change in some blocks and increase in thickness of the surface layer; this is accompanied by a marked increase in the number of atoms having relatively unstable states, which leads to change in the specific volume and absorption of heat.

Temperature rise in the resulting liquid increases the number of unstable atoms and reduces the volumes of the blocks with crystalline structure; the breakup of blocks and the structural change results in polymorphism effects in the liquid [35], with a given structual modification dominant and stable in a certain temperature range. It has frequently been supposed that the sequence of dominant structures is the cause of anomalies in various properties of a liquid. Water has particularly many anomalies [36].

Most of the groups will have broken up when the boiling point is approached, and more and more unstable atoms will have passed into a state corresponding to the isolated atoms. Therefore, crystal—liquid and liquid—vapor phase transitions may be represented as spontaneous processes involving disruption or formation of collective bonds in groups of atoms involving change in the electronic system. A liquid differs from a crystal and from a vapor in containing more atoms in unstable electronic states.

There are many discussions [34-38] of the existence of blocks with crystal structure in liquids; the block structure of a liquid is particularly prominent in the hysteresis occurring in melting—solidification cycles for dispersed particles of metals [32, 39]. It appears clear from certain studies [40] that the liquid contains a large number of atoms in unstable states, since prolonged luminescence was observed from rapidly fronzen water. The high activity of liquids and solvents is also a consequence of the chemical activity of the groups of atoms with unstable electronic states.

If one hinders atoms from separating on evaporation, the liquid goes over to a supercritical state at sufficiently high temperatures and pressures; in that state, the liquid contains mainly unstable groups [41], which break up when the spinodal line is approched (Fig. 3). One assumes that this line corresponds to the conditions for formation of unstable associations, which exist also in the initial stages of crystal growth. Instability in the atomic electronic structure in the supercritical state results in various specific properties of that state: high chemical activity [42], high viscosity [43], high specific heat [44], etc. Prolonged reconstruction of the electronic system is clear for water molecules from the extremely prolonged relaxation of the high chemical activity of supercritical water when this is rapidly cooled [42].

The high viscosity in the supercritical state arises because the energy applied to change the macroscopic state is consumed in the main in changing the microscopic state of the unstable particles [43]. Several effects should occur at the surfaces of crystalline bodies, since

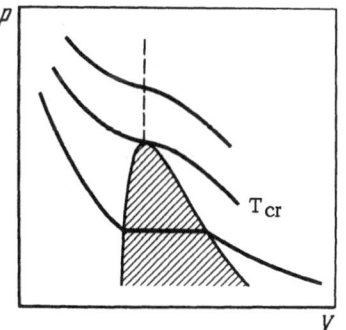

Fig. 3. The p—V diagram for the liquid—vapor system. The broken line is the one along which the liquid—vapor transition occurs; T_{cr} is the critical isotherm.

the above views imply that unstable associations occur there. In particular, surface tension in condensed media will arise because some of the energy involved in disrupting the surface is absorbed by unstable particles.

Conclusions

Some concepts from theoretical chemistry are used in describing a liquid structure and condensation.

1. The formation of compounds and certain polymorphic transitions involve change in the symmetry of the atomic electron clouds, and therefore the atom cannot be considered as an unchanging structural unit in discussing phase transitions.

2. A cooperative process is involved in atomic electron structure change accompanying the production of a new phase.

3. The structure change takes a time much greater than the period of vibration, partly because of the complexity of the relaxation processes in the electron shells of the atom and partly because the reconstruction of the atomic system is a cooperative process.

Examples have been presented to show that these concepts and their consequences can be of value in elucidating various aspects of structure formation.

Literature Cited

1. Ya. I. Frenkel'. Zh. Éksper. Teor. Fiz., 9:199 (1939).
2. Ya. B. Zel'dovich. Zh. Éksper. Teor. Fiz., 12:525 (1942).
3. W. S. Fyfe. Introduction to Solid-State Geochemistry [Russian translation], Mir, Moscow (1967).
4. S. A. Shchukarev. In: Current Problems in the Chemistry of Coordination Compounds [in Russian], Vol. 1, Izd. LGU (1966), 5.
5. S. A. Shchukarev. In: Current Problems in the Chemistry of Coordination Compounds [in Russian], Vol. 2, Izd. LGU (1968), 5.
6. V. I. Lebedev. The Significance of Ionic-Atomic Radii in Geochemistry and Chemistry [in Russian], Izd. LGU (1969).
7. E. Ya. Yanson. Coordination Compounds [in Russian], Izd. Vyssh. Shkola, Moscow (1968).
8. C. Coulson. Valence, Oxford University Press (1961).
9. S. A. Shchukarev. In: Current Problems in the Chemistry of Coordination Compounds [in Russian], Vol. 3, Izd. LGU (1970), 5.
10. S. A. Shchukarev. Inorganic Chemistry [in Russian], Vol. 1, Izd. Vysshaya Shkola, Moscow (1970).
11. L. D. Kislovskii. In: Effects of Solar Radiation on the Atmosphere and Biosphere [in Russian], Nauka, Moscow (1971), 147.
12. S. A. Shchukarev. Lectures on General Chemistry [in Russian], Izd. LGU (1964).
13. A. A. Zhukovitskii and L. A. Shvartsman. Physical Chemistry [in Russian], Izd. Metallurgiya, Moscow (1968).
14. N. N. Sheftal'. In: Growth of Crystals, Vol. 1, Consultants Bureau, New York (1959), p. 5.
15. M. Ya. Gen and Yu. P. Petrov. Usp. Khim., 38:2249 (1969).
16. A. Verma and P. Krishna. Polymorphism and Polytypism in Crystals [Russian translation], Mir, Moscow (1969).
17. A. I. Bublik and B. Ya. Pines. Dokl. AN SSSR, 87:215 (1952).
18. Yu. M. Polezhaev. Zh. Fiz. Khim., 41:2958 (1967).
19. K. L. Chopra. Phys. Status Solidi, 32:489 (1969).

20. V. K. Grigorovich. The Periodic Table and the Electronic Structure of Metals [in Russian], Nauka, Moscow (1966).
21. Yu. A. Klyachko. Zh. Prikl. Khim., 22:455 (1949).
22. V. F. Kiselev. Surface Effects in Semiconductors and Insulators [in Russian], Nauka, Moscow (1970).
23. A. Many, Y. Goldstein, and N. B. Grover. Semiconducting Surfaces, Amsterdam (1965).
24. M. Rich. Phys. Letters, 4:153 (1963).
25. A. U. MacRae and L. H. Germer. Phys. Rev. Letters, 8:489 (1962).
26. Kh. B. Khokonov. Uch. Zap. Kabardino-Balkar. Gos. Univ., No. 13, 108 (1961).
27. A. Kh. Brekher and A. A. Zhukovitskii. Zh. Fiz. Khim., 20:1459 (1946).
28. C. R. M. Wronski. Brit. J. Appl. Phys., 18:1731 (1967).
29. A. Masson, J. J. Metois, and R. Kern. Crystal Growth, Proc. Second Internat. Conf. Crystal Growth, Birmingham, 3/4:196 (1968).
30. D. W. Pashley. Phil. Mag., 10:127 (1964).
31. V. I. Kalividze, V. F. Kiselev, and L. A. Ushakova. Dokl. AN SSSR, 191:1088 (1970).
32. Yu. I. Petrov. Fiz. Met. Metalloved., 19:219 (1965).
33. L. D. Kislovskii and A. A. Abdullaev. Opt. i Spekt., 29:737 (1970).
34. A. Ubbelhode. Melting and Crystal Structure [Russian translation], Mir, Moscow (1969).
35. S. S. Urazovskii. Molecular Polymorphism [in Russian], Izd. AN Ukr. SSR, Kiev (1956).
36. W. Drost-Hansen. Industr. and Eng. Chem., 57:38 (1965).
37. A. V. Romanova and B. A. Mel'nik. Dokl. AN SSSR, 189:294 (1969).
38. Ya. I. Dutchak and N. P. Klym. Izv. VUZ, Fizika, 10:140 (1968).
39. V. A. Kotel'nikov and Yu. I. Petrov. Fiz. Tverd. Tela, 11:1391 (1969).
40. I. K. Musatov. Izv. VUZ, Fizika, 2:103 (1960).
41. V. K. Semenchenko. Selected Aspects of Theoretical Physics [in Russian], Prosveshchenie, Moscow (1966).
42. N. I. Khitarov. Geokhimiya, No. 9, p. 1130 (1967).
43. V. K. Semenchenko. Zh. Fiz. Khim., 35:2448 (1961).
44. Handbook of Thermophysical Properties of Materials [in Russian], Gosénergoizdat, Moscow (1956).

THE TWO-DIMENSIONAL NUCLEATION RATE FOR A
HOMOPOLAR CRYSTAL CONTAINING MICROSCOPIC DEFECTS

L. A. Borovinskii

Introduction

The classical kinetic treatment of layer growth of crystal faces involves the assumption that the two-dimensional nucleation is the rate-limiting step, which implies extremely low values for growth rates at low and medium supersaturations. The calculated growth rates are less by many orders of magnitude than the observed ones. The discovery of the dislocation growth mechanism would have appeared to have resolved this conflict between theory and experiment, and it would seem that the growth of the faces of the equilibrium form (G faces) would occur by the two-dimensional mechanism only at high supersaturations. However, methods have been developed for making almost dislocation-free crystals, and studies on the growth of these have shown that G faces in such crystals can grow at fairly high rates at low supersaturations. Under certain conditions, nucleation is so fast that the growth is limited not by nucleation but by bulk diffusion or by the deposition kinetics of the material during reactions in the vapor state.

There are many experimental studies that indicate that the dislocation mechanism is not a universal one; the growth rate is affected by deviations from uniformity in the crystal due to impurities and macroscopic inclusions. For instance, N. N. Sheftal' has examined various lines of experimental evidence and has shown that impurities and particles from the environment influence the growth rate and crystal shape [1, 2]. Distler et al. have shown in a series of studies using decoration of active nucleation centers [3-10] that these centers are related to impurity atoms, clumps of these, and structural defects in the surface layer. A long-range action has been demonstrated for the active centers, which can initiate nucleation out to distances of 1000-1500 Å.

Recently, Alexandrov [11-14] has shown that homoepitaxy in semiconductors is to be considered as a particular case of heteroepitaxy, since one cannot completely eliminate lack of correspondence between the lattices of the substrate and film, and also the effects of surface oxides and impurities. During the initial stages of film growth, three-dimensional nucleation predominates over two-dimensional. However, the latter process, which is examined here, cannot be ruled out entirely, and it may play a considerable part in the later stages of film development (after coalescence of the insular clumps) on the individual smooth parts of the face. Also, the orientation of the face may differ from a singular plane, in which case layerwise growth is accompanied by two-dimensional nucleation between steps, and this may here predominate over three-dimensional nucleation.

The literature carries theoretical studies of impurity effects on crystal growth and form; but the existing theoretical studies deal mainly with changes in the properties of the surface in response to adsorbed impurities (change in the specific surface free energy, reduction in the energy barriers to nucleation around particles adsorbed on the surface, and retarded growth due to adsorption of impurities at growth points [15-19]). Nucleation is affected by charge point centers, which can exist near the surface in an insulating crystal [20, 21]. The theory discussed here for the effects of impurity centers on nucleation kinetics was first briefly presented in 1969 [22]. The method proposed there has been extended and supplemented by Alexandrov and Entin [23]. Here I present a revised theory of the effects of impurities on nucleation kinetics.

Impurities, point defects, and groups of these cause microscopic deviations from uniformity even when macroscopic uniformity is maintained, which facilitates accentuated nucleation on faces; a particularly large part may be played by such impurities in the growth of semiconductor crystals, since the specifications for purity and uniformity of such crystals and films lead to the provision of growth conditions under which the above mechanisms involving adsorbed impurities have a diminished influence on the nucleation and growth, which means that impurities and point structural defects in the surface layer may come to have the predominant effect.

Here I consider the mechanism of action of dissolved impurities and structural defects as regards the two-dimensional nucleation on faces; the mechanism involves the assumption that structural defects and impurities cause fluctuations in the chemical potential of the surface. The work needed to form a nucleus will vary from point to point, and even small relative deviations of this work will give rise to individual areas in which the reduction in the energy of nucleation relative to the mean value considerably exceeds kT. Such parts of the surface act as vigorous nucleation centers.

In the first part of this study we consider the effects of fluctuations of this kind on the nucleation rate from the phenomenological viewpoint, i.e., without analysis of the causes of the fluctuations. Subsequently, we consider a homopolar semiconductor containing electrically charged impurity centers at a given concentration, and expressions are derived for the probability characteristics of the chemical potential, together with the effective energy barrier to nucleation and the ratio of the nucleation rate on the face of a crystal bearing impurities to the nucleation rate on the face of an ideal crystal.

Phenomenological Treatment of the Increase

in Nucleation Rate due to Nonuniformity

In the classical theory, the following is the number of nuclei arising on unit area of the face in unit time:

$$I = C_{\exp}\left(-\frac{A_2}{kT}\right),\tag{1}$$

where A_2 is the nucleation energy, while C is a function of temperature and the material constants (desorption energy, activation energy for surface diffusion). Chemical potential varies over the surface if there are microscopic deviations from uniformity, and therefore A_2 is dependent on the position of the nucleus on the surface. We assume that we can determine A_2 for an arbitrary position of the center of the nucleus on the face. Then we may determine A_2 for various points on the face and obtain a statistical distribution for it. If the points on the face are chosen at random, without relation to the nucleation conditions, we must consider A_2 as a random quantity with a certain distribution. To determine the flux of resulting nucleations we have to average the I of (1) over the set of possible values of A_2.

To elucidate the essence of the matter we assume that the distribution of A_2 takes a normal form with mathematical expectation $\bar{A}_2$ and dispersion σ_A^2. The probability density is

$$f(A_2) = \frac{1}{\sqrt{2\pi}\sigma_A} \exp\left\{-\frac{(A_2 - \bar{A}_2)^2}{2\sigma_A}\right\}. \tag{2}$$

With this distribution there is a finite probability of negative A_2, which is physically meaningless. We use the analogy with later nonphenomenological discussions to assume that parts of the surface for which $A_2 < 0$ have a chemical potential above that of the gas phase. In such parts of the surface, the nucleation rate for new layers must be taken as zero, i.e., (1) is inapplicable.

The averaged value of (1) is put as

$$I = C \int_0^\infty \exp\left(-\frac{A_2}{kT}\right) f(A_2)\, dA_2. \tag{3}$$

We substitute in (3) for $f(A_2)$ and integrate to get

$$I = \frac{1}{2} C \exp\left\{\frac{\sigma_A^2}{2k^2T^2} - \frac{\bar{A}_2}{kT}\right\} \cdot \left\{1 + \Phi\left(\frac{\bar{A}_2 kT - \sigma_A^2}{\sqrt{2}\sigma_A kT}\right)\right\}, \tag{4}$$

where $\Phi(z)$ is the probability integral.

If $\dfrac{\bar{A}_2}{\sqrt{2}\sigma_A} - \dfrac{\sigma_A}{\sqrt{2}kT} > 1$ and consequently $\dfrac{\sigma_A^2}{kT} < \bar{A}_2$, then

$$\Phi\left(\frac{\bar{A}_2 kT - \sigma_A^2}{\sqrt{2}\sigma_A kT}\right) \approx 1 \quad \text{and} \quad I = C \exp\frac{\sigma_A^2}{2k^2T^2} \exp\left(-\frac{\bar{A}_2}{kT}\right). \tag{5}$$

Then the nucleation rate on the face of a nonuniform crystal is greater than that on a uniform crystal under conditions giving a nucleation rate $\bar{A}_2$, and the increase in the nucleation rate is represented by the factor

$$B = \exp\frac{\sigma_A^2}{2k^2T^2}. \tag{6}$$

We will call this factor the nucleation acceleration factor. If, on the other hand, $\dfrac{\sigma_A}{\sqrt{2}kT} - \dfrac{\bar{A}_2}{\sqrt{2}\sigma_A} \gg 1$ and consequently $\dfrac{\sigma_A^2}{kT} > \bar{A}_2$, we use the expansion of $\Phi(z)$ for large z as from [24], page 211, which gives from (4) that

$$I = \frac{1}{\sqrt{2\pi}} C \frac{\sigma_A kT}{\sigma_A^2 - \bar{A}_2 kT} \exp\left(-\frac{\bar{A}_2^2}{2\sigma_A^2}\right). \tag{7}$$

The nucleation acceleration factor in that case is

$$B = \frac{1}{\sqrt{2\pi}} \frac{\sigma_A kT}{\sigma_A^2 - \bar{A}_2 kT} \exp\left(\frac{\bar{A}_2}{kT} - \frac{\bar{A}_2^2}{2\sigma_A^2}\right). \tag{8}$$

As a rule, $\bar{A}_2/kT \gg 1$, except at very high supersaturations. These relationships readily show that even small relative deviations of A_2 ($\sigma_A/A_2 \ll 1$) cause a considerable increase in the nucleation rate.

We consider as an example a (100) face of a Kossel crystal with a simple cubic lattice; we take into account only the interactions between nearest neighbors. Let φ be the work needed to separate nearest neighbors. An elementary calculation gives the following expression for the energy of formation for a two-dimensional nucleus:

$$A_2 = \frac{4\varphi^2}{kT \ln \dfrac{p}{p_\infty}} \ . \tag{9}$$

We now estimate B for values of the parameters corresponding to typical growth conditions. Let kT = 0.1 eV (T = 1160°K); $\ln p/p_\infty = {}^1/_2$ (p/p$_\infty$ = 1.65); φ = 1 eV; $\sigma_A / \overline{A}_2$ = 0.01; we use (9) to get $\overline{A}_2$ = 80 eV. In that case, (6) is applicable, and we get B = e^{32}. If on the other hand we put $\sigma_A / \overline{A}_2$ = 0.1, then (8) is applicable for the same values of the other parameters, and we get B = $e^{744.83}$.

These examples show that the microscopic nonuniformity, even with a very small dispersion in the nucleation energy, gives a very great increase in the calculated nucleation rate.

Interaction of Atoms in an Adsorbed Layer with

Impurity Centers: Energy of Formation for

Two-Dimensional Nuclei

To calculate the effects of nonuniformity on the nucleation rate under real conditions we need to know the structure of the crystal, the form of interaction between the particles, and also that of the interaction between the impurity atoms and the particles in the adsorbed layer. As an object for our subsequent calculations we selected a crystal of a homopolar quenched semiconductor containing donor and acceptor impurity centers, which macroscopically are uniformly distributed over the volume. Let 1 cm^3 contain n_d donors and n_a acceptors. We assume that the energies E_d and E_a of the donor and acceptor levels satisfy

$$E_d - E_F \gg kT, \quad E_F - E_a \gg kT,$$

where E_F is the energy of the Fermi level. Then practically all the donor and acceptor centers are in the ionized state. Each charged impurity center creates an electric field, whose potential is expressed as follows if the impurity concentration is not too high and allowance is made for screening by the free carriers:

$$\varphi(r) = \frac{q}{r} \exp\left(-\frac{r}{r_0}\right), \tag{10}$$

where q is the charge on an impurity center, which we assume to be e_0 or $-e_0$, r is the distance from an impurity center, and r_0 is the screening radius, the last being expressed by

$$r_0 = \left(\frac{\varepsilon kT}{4\pi (n^- + n^+) e_0^2}\right)^{1/2}, \tag{11}$$

where ε is the dielectric constant and n^- and n^+ are the concentrations of free electrons and holes.

We assume that the mean distance between impurity centers is much larger than the screening radius, and then we can neglect the effects of interaction between impurity centers on the volume distribution. The probability of finding any given number of impurity centers in a given volume element is therefore represented by a Poisson distribution.

Atoms of the base crystal present in the adsorption layer on a close-packed face may be weakly bound to the crystal and can act as acceptors or donors; the decision on the actual role can be taken only via quantum-mechanical calculations on the chemisorbed states on a given face. Although there is now a fairly substantial literature on the theory of chemisorbed states for foreign atoms on crystal faces (including semiconductor surfaces), there has been no solution for chemisorbed states of atoms of the same type on the faces of elemental semiconductor crystals. We therefore simply assume that these atoms in the adsorption layer are acceptors and we also assume that the probability of the ionized state for an adsorbed atom is close to unity.

The energy for interaction of an adsorbed atom with an impurity center is

$$\pm \frac{e_0^2}{r} \exp\left(-\frac{r}{r_0}\right). \tag{12}$$

Let U be the total energy for interaction of an adsorbed atom with all the impurity centers in a crystal. In practice, the only substantial interaction is with the nearest impurity centers, whose distance from the adsorbed atom does not exceed r_0 as regards order of magnitude. Around an impurity ion there may be various numbers of impurity centers at different distances, so U is a random quantity whose distribution will be deduced below.

Let μ_0 be the chemical potential of an ideal undoped crystal; then $\mu_0 + U$ is the chemical potential of the crystal with the impurities at a point on the surface where the potential energy of the adsorbed atoms is U. In other words, additional energy is needed to detach an atom from a repeating site on the surface where the potential energy of the adsorbed atoms in the electric field of the impurity centers is U, this being in addition to the work needed to break the chemical bonds, since it is necessary to transfer the atom in the adsorbed layer to a point remote from the impurity centers. This extra energy is U.

The mean chemical potential of the surface is $\mu_0 + \overline{U} = \mu_{cr}$, which defines the equilibrium conditions for the crystal in the medium. Then the effects of the impurities on the crystallization make themselves felt in particular in displacement of the phase equilibirum curve. If $\overline{U} < 0$, the impurities reduce the chemical potential, which increases the growth rate, while if $\overline{U} > 0$ there should be a reduction in the growth rate. Under otherwise equal conditions, we should find faster growth in crystals whose main impurity centers have a charge whose sign is opposite to that of the atoms in the adsorption layer. However, the effects of the impurities on the nucleation rate may be very much more important than the effects on the position of the phase equilibrium curve.

Here we consider the effects on the nucleation rate. The nucleation energy on a G face for a uniform homopolar crystal is

$$A = \frac{c\varphi^2}{\Delta\mu}, \tag{13}$$

where c is a factor of the order of unity, which is dependent on the structure of the crystal, φ is the bond energy between adjacent atoms, and $\Delta\mu = \mu_{gas} - \mu_{cr} = RT \ln p/p_\infty$ is the chemical potential of the gas medium relative to that of the crystal.

If the crystal is nonuniform, we can write the nucleation energy as

$$A = \frac{c\varphi^2}{\Delta\mu - (U - \overline{U})}, \tag{14}$$

where $\Delta\mu = \mu_{gas} - \overline{\mu}_{cr} = kT \ln p/p_\infty$ is the chemical potential of the gas phase relative to the mean chemical potential of the crystal, while $\Delta\mu - (U - \overline{U})$ is the same for the gas relative to the local chemical potential. It is clear that (14) has a meaning only for those parts of the surface for which $\Delta\mu - (U - \overline{U}) > 0$; parts on which $\Delta\mu - (U - \overline{U}) = 0$ are in equilibrium with the gas, and the energy of nucleation there becomes infinite, while the nucleation probability becomes zero. The gas is unsaturated with respect to parts of the surface on which $\Delta\mu -$ $(U - \bar{U}) < 0$. The existence of such parts may be the cause of pits on the faces of growing crystals and films. It is interesting here to note the observations of [25], which show that clumps of pits are formed in areas where the conductivity is p type, although the layer as a whole had conductivity of n type. This observation agrees with our model if the adsorbed atoms are acceptors.

On the other hand, the energy barriers to nucleation are reduced on parts of the surface where $U - \bar{U} < 0$; such points are the basic sites for two-dimensional nucleation. If the negative values of $U - \bar{U}$ are large in magnitude, the nucleation barriers may become so low that the growth is limited by bulk diffusion or other factors, not by nucleation. These parts are responsible for growth hummocks, and in decoration experiments such parts serve to attach decoration particles.

Distribution Functions and Probability Characteristics for the Potential Energies of Atoms in the Adsorption Layer in the Field of Impurity Centers

Consider an adsorbed atom in the field of impurity centers; we isolate the adjacent part of the crystal as a hemisphere with its center at the nucleus of the atom and radius R, which is so large that the numbers of donor and acceptor impurity centers in this region may be considered as equal to the mean values in the corresponding volume, while $R \gg r_0$.

We used the method of characteristic functions [26, pages 224-235] to deduce the distribution for the potential energy; first of all we consider the characteristic function for the potential energy between an adsorbed atom and one impurity center, which can lie at any point in the neighborhood. The probability that the distance of the impurity center from the adsorbed atom lies in the range from r to r + dr is $3r^2dr/R^3$; if the potential energy is given by (12), then the characteristic function takes the form

$$F_{u,R}(z) = \frac{3}{R^3} \int\limits_0^R r^2 \exp\left\{\pm \frac{ize_0^2}{r} \exp\left(-\frac{r}{r_0}\right)\right\} dr, \tag{15}$$

where the plus sign relates to an acceptor center and the minus sign to a donor.

In this region there are $N_a = \frac{2}{3}\pi n_a R^3$ acceptor enters and $N_d = \frac{2}{3}\pi n_d R^3$ donor ones; the potential energy of the adsorbed atom in the field of all the impurity centers in the region is

$$U = \sum_{i=1}^{N_a} u_a(r_i) + \sum_{x=1}^{N_d} u_d(r_k). \tag{16}$$

We have already seen that the positions of the impurity centers are completely randomly distributed and are independent, so the $u_a(r_i)$ and $u_d(r_k)$ are also statistically independent. The theorem on the characteristic function for a sum of independent random quantities gives

$$F_R(z) = [F_{u,R}^{(a)}(z)]^{Na} [F_{u,R}^{(d)}(z)]^{Nd}, \tag{17}$$

where the functions $F_{u,R}^{(a)}(z)$ and $F_{u,R}^{(d)}(z)$ are defined by (15); we substitute the expressions for

TABLE 1. Values of K(z) and L(z) as Functions of z

Function	z										
	0.0	0.1	0.2	0.3	0.4	0.5	0.6	0.7	0.8	0.9	1.0
$10^4 K(z)$	0.0	20.3	79,5	173	295	445	598	774	964	1162	1359
$10^4 L(z)$	0.0	994	1980	2955	3920	4840	5750	6640	7520	8380	9240

these functions and also for N_a and N_d, then pass to the limit $R \to \infty$, to get

$$F(z) = \exp\left\{ 2\pi n_a \int_0^\infty \left[\exp\left(\frac{ize_0^2}{r} \exp\left(-\frac{r}{r_0} \right) \right) - 1 \right] r^2\, dr + 2\pi n_d \int_0^\infty \left[\exp\left(-\frac{ize_0^2}{r} \exp\left(-\frac{r}{r_0} \right) \right) - 1 \right] r^2\, dr \right\}. \quad (18)$$

We take r_0 as our length unit, e_0^2/r_0 as our energy unit, and $(e_0^2/r_0)^{-1}$ as unit for the argument z of the characteristic function. Then the expression for the characteristic function is

$$F(z) = \exp\left\{ 3g_a \int_0^\infty \left[\exp\left(\frac{iz}{r} e^{-r} \right) - 1 \right] r^2\, dr + 3gd \int_0^\infty \left[\exp\left(-\frac{iz}{r} e^{-r} \right) - 1 \right] r^2\, dr \right\}, \quad (19)$$

where

$$g_a = {}^2/_3\, \pi n_a r_0^3 \quad \text{and} \quad g_d = {}^2/_3\, \pi n_d r_0^3 \quad (20)$$

are the mean values of the numbers of acceptor and donor impurity centers in a hemisphere of radius r_0.

The basic probability characteristics of U are determined via $\Psi = \ln F(z)$; we separate the real and imaginary parts in this function and put it in the form

$$\Psi(z) = -3(g_a + g_d) K(z) + 3i(g_a - g_d) L(z), \quad (21)$$

where

$$K(z) = \int_0^\infty \left[1 - \cos\left(\frac{z}{r} e^{-r} \right) \right] r^2\, dr, \quad (22)$$

$$L(z) = \int_0^\infty r^2 \sin\left(\frac{z}{r} e^{-r} \right) dr. \quad (23)$$

Table 1 gives values of K(z) and L(z) obtained by numerical integration.

The mathematical expectation, dispersion, and standard deviation σ_U of U are defined via Ψ:

$$M\{U\} = \frac{1}{i}\, \Psi'(0); \quad D\{U\} = \sigma_U^2 = -\Psi''(0). \quad (24)$$

These formulas give

$$M\{U\} = 3(g_a - g_d); \quad D\{U\} = \sigma_U^2 = \frac{3}{2}(g_a + g_d). \quad (25)$$

The probability density is defined from the characteristic function via

$$f(U) = \frac{1}{2\pi} \int_\infty^\infty \exp(-iUz) F(z)\, dz. \quad (26)$$

We now have to consider whether we can replace approximately the distribution for U defined by (19)-(26) by a normal distribution with the same mathematical expectation and the dispersion as in (25). The logarithm of the characteristic function for a normal distribution is

$$\Psi_n(z) = imz - \frac{\sigma^2 z^2}{2}.\qquad(27)$$

In calculating the integral of (26), it is necessary to use exact values for F(z) only for the part of the integration range in which F(z) is comparable with unity, and consequently $|\text{Re}\,\Psi(z)| = 3(g_a + g_d)K(z) \leq 1$; if in this part of the integration range we can replace $\Psi(z)$ by a function of the form of (27), then the probability density calculated from (26) will be close to that for a normal distribution. Table 1 shows that the error is not more than 3% if K(z) is represented by a parabolic function for z < 0.4, while L(z) may be considered a linear function for z < 0.5. In order to be able to neglect the deviation of K(z) from parabolic in calculating the integral of (26), and also the deviation of L(z) from linear for z > 0.4, we have to have $|F(z)| \ll 1$ for z > 0.4; since $|F(z)|$ is a monotonically decreasing function, this condition is equivalent to $|F(0.4)| \ll 1$ or

$$(g_a + g_d)\,K\,(0.4) > {}^1\!/_3,\ g_a + g_d > 10.\qquad(28)$$

We substitute (11) into (20) to get

$$g = g_a + g_d = \frac{1}{12\sqrt{\pi}}\,\frac{n_a + n_d}{(n^- + n^+)^{3/2}}\left(\frac{\varepsilon kT}{e_0^2}\right)^{3/2}.\qquad(29)$$

The values of g given by this formula for germanium and silicon show that (28) is obeyed only at relatively low growth temperatures (below 500°K for Ge and below 700°K for Si), together with high dope concentrations (10^{18} cm^{-3} or above). At higher temperatures, the smallness of the screening radius means that each point on the surface is in the field of one or two nearest impurity centers, and consequently we do not have a normal distribution for U. However, within the accuracy of this theory, the results are hardly likely to be greatly affected if we replace the complex distribution for U given by (26) by a normal distribution with the same mathematical expectation and dispersion. Therefore, in this study we will use a normal distribution for the quantity no matter whether (28) is obeyed.

The reduction in the screening radius as the temperature is raised determines not only the distribution parameters as expressed via $f(U)$ but also the correlation radius for the random function U and the dimensions of the active nucleation centers. This question has been considered in detail elsewhere [27].

Calculation of the Nucleation Acceleration Factor and the Effective Work of Formation for Two-Dimensional Nuclei

We calculate the acceleration factor by averaging (1), in which the energy of nucleation will be taken as given by (14), while the energy for interaction of an adsorbed atom with the impurity centers is assumed as having a normal distribution. The averaged expression for the nucleation rate is

$$\bar{I} = \frac{C}{\sqrt{2\pi}s_U}\int_{\bar{U} + \frac{\Delta\mu r_0}{e_0^2}}^{\infty} \exp\left\{-\frac{c\varphi^2}{kT\left[\Delta\mu - (U - \bar{U})\frac{e_0^2}{r_0}\right]} - \frac{(U - \bar{U})^2}{2s_U^2}\right\}dU.\qquad(30)$$

TABLE 2. Values of $\varkappa$ for Ge and Si as Functions of Temperature

Material	Temperature, °K										
	600	700	800	900	1000	1100	1200	1300	1400	1500	1600
Ge	1.525	0.978	0.685	0.515	0.386	0.326	0.270	—	—	—	—
Si	—	—	1.12	0.797	0.566	0.440	0.348	0.284	0.236	0.202	0.174

(The factor e_0^2/r_0 in the denominator for the first term in the exponential arises because we have transferred above to the dimensionless variable U with e_0^2/r_0 as unit of energy.)

We introduce the dimensionless parameters

$$P = \frac{c\varphi^2}{kT\Delta\mu} \tag{31}$$

$$\varkappa = \sqrt{2}\,\sigma_U\,\frac{e_0^2}{r_0\Delta\mu} = \sqrt{2\pi r_0\,(n_a + n_d)}\,\frac{e_0^2}{\Delta\mu}. \tag{32}$$

Parameter P is the nucleation energy relative to the energy of thermal motion kT; it is $10-10^2$ as regards order of magnitude. Parameter $\varkappa$ is defined by the above formula, while σ_U is expressed in (32) via (25) and (29). In calculating r_0 in (32) we may assume that the main contribution to the free-carrier concentration at typical growth temperatures arises from ionization of atoms of the base material, while the contribution from ionization of impurity centers is negligibly small. We can therefore put $n^+ = n^- = n_i$, where n_i is defined by a standard formula [28], p. 25, provided allowance is made for the temperature narrowing of the crooked band. Table 2 gives values for $\varkappa$ for the faces of Ge and Si at typical growth temperatures for supersaturation $\ln p/p_\infty = 1$ (i.e., $\Delta\mu = kT$) and for charged impurity center concentrations $n_a + n_d = 10^{17}$ cm^{-3}; as $\varkappa \approx (n_a + n_d)^{1/2}/\Delta\mu$, the data in this table readily enable one to find $\varkappa$ for other values of the supersaturation and impurity concentration.

Expression (30) may be put in the form

$$I = \frac{C}{\sqrt{\pi}}\exp(-P)\int_{-1/\varkappa}^{\infty}\exp\left[\frac{P\varkappa\xi}{1+\varkappa\xi} - \xi^2\right]d\xi. \tag{33}$$

As $I_0 = C\exp(-P)$ is the nucleation rate on a face of an ideal crystal, the nucleation acceleration factor is

$$B = \frac{1}{\sqrt{\pi}}\int_{-1/\varkappa}^{\infty}\exp\left[\frac{P\varkappa\xi}{1+\varkappa\xi} - \xi^2\right]d\xi. \tag{34}$$

The function $\Phi(\xi) = \frac{P\varkappa\xi}{1+\varkappa\xi} - \xi^2$ has a sharp peak at $\xi = \xi_m$, where ξ_m is the real root of

$$\frac{2}{\varkappa}\xi_m(1+\varkappa\xi_m)^2 = P. \tag{35}$$

In calculating the integral in (34) we can replace the lower limit of integration by $-\infty$, while $\Phi(\xi)$ is replaced by

$$\Phi(\xi) = \Phi(\xi_m) + \frac{1}{2}\Phi''(\xi_m)(\xi - \xi_m)^2, \tag{36}$$

where

$$\Phi''(\xi_m) = -\frac{2P\varkappa^2}{(1+\varkappa\xi_m)^3} - 2. \tag{37}$$

Then B is

$$B \approx \frac{1}{\sqrt{\pi}} \exp \Phi\left(\xi_m\right) \int_{-\infty}^{\infty} \exp\left\{-\left(\xi-\xi_m\right)^2 \left[\frac{P\varkappa^2}{\left(1+\varkappa\xi_m\right)^3}+1\right]\right\} d\xi = \frac{\exp\Phi\left(\xi_m\right)}{\sqrt{\dfrac{P\varkappa^2}{\left(1+\varkappa\xi_m\right)^3}+1}}. \tag{38}$$

In determining the dependence of B on P and $\varkappa$, it is convenient to consider ξ_m as an independent parameter, instead of solving (35) for it, while (35) and (38) may be considered as a parametric dependence of B on P and $\varkappa$. We take logarithms in (38) and substitute for P and $\Phi\left(\xi_m\right)$ to get

$$\ln B = \xi_m^2\left(1+2\varkappa\xi_m\right) - \frac{1}{2}\ln\left(3-\frac{2}{1+\varkappa\xi_m}\right), \tag{39}$$

$$P = \frac{2}{\varkappa}\xi_m\left(1+\varkappa\xi_m\right)^2.$$

Comparison of (33) and (34) enables us to express the mean nucleation rate as

$$I = C \exp\left[-\left(P-\ln B\right)\right]. \tag{40}$$

It is then clear that $P - \ln B = P_{eff}$ may be considered as the effective nucleation energy at the face of a nonuniform crystal as referred to kT; to show how far the nonuniformity reduces the effective nucleation energy, we give in Fig. 1 the dependence of $\ln B$ on P for $\varkappa$ of 0.1, 0.2, 0.3, 0.5, 1.0, and 2.0.

These results show that nonuniformity due to charged impurity centers greatly reduces the effective energy barriers to nucleation over a wide range in the parameters defining the growth. The nucleation is particularly accelerated at low supersaturations, i.e., when P is large. However, we must bear in mind that the acceleration has a real significance only at P_{eff} less than some critical value, since for $P_{eff} > P_{cr}$ the nucleation practically ceases, and so acceleration has only a formal meaning. We can estimate P_{cr} roughly for silicon. The factor C in (1) and (30) has the following order of magnitude:

$$C \approx 10^{24} \exp\left(-\frac{U_{diff}}{kT}\right) \text{cm}^{-2}\cdot\text{sec}^{-1}. \tag{41}$$

If we assume for silicon an activation energy of 1.57 eV [29] for surface diffusion, then at temperatures from 1200 to 1500°K

$$\exp\left(-\frac{U_{diff}}{kT}\right) \sim 10^{-6}.$$

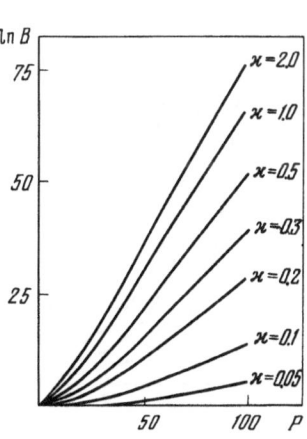

Fig. 1. Logarithm of the nucleation acceleration factor as a function of the nucleation energy expressed as a ratio to kT for various $\varkappa$.

In order for the nucleation rate to be not less than 10^4 cm^{-2} sec^{-1}, we must have

$$\exp\left(-P_{\text{eff}}\right) > 10^{-14}, \tag{42}$$

i.e.,

$$P_{\text{eff}} < P_{\text{cr}} = 32. \tag{43}$$

At larger values of P_{eff}, nucleation will be virtually unobservable, and therefore we will not observe an acceleration; the region in Fig. 1 to the right of the straight line $P = \ln B + 32$ corresponds to growth conditions under which nucleation is virtually absent.

On the other hand, the value of P does not exceed 10-15 at very high supersaturations, in which case $\ln B$ is of the order of 10, i.e., the nucleation rate increases only by around an order of magnitude. The acceleration in the nucleation is also small at growth temperatures near the melting point, on account of the marked screening action of free carriers. The ionized impurities have the largest effect on the acceleration at intermediate supersaturations, when P is somewhat larger than 10, and at temperatures not too close to the melting point. The effects of the impurities are seen as a marked reduction in the critical supersaturations for appreciable nucleation rates.

The acceleration parameter increases rapidly with $\varkappa$; at high dope concentrations $(10^{18} - 10^{19}$ cm$^{-3})$, $\varkappa$ may be rather larger than 1, and then $\ln B$ may be up to 60-80% of P at not very high supersaturations, i.e., the energy barrier is reduced by a substantial factor, and the critical value may be attained at P of 120-150, i.e., at comparatively low supersaturations.

Increase in supersaturation at a given temperature and impurity concentration reduces P and $\varkappa$, and hence produces a marked reduction in B; the character of the dependence is not very clear from the curves of Fig. 1, since the dependence of $\ln B$ on the supersaturation is expressed via the intermediate parameters P and $\varkappa$, so we give graphs for $\ln B$ as a function of supersaturation for (111) faces of silicon at certain fixed temperatures and impurity concentrations (Fig. 2). We calculated the nucleation energy for an ideal (111) face via (13), with $C = 1/4$ and $\varphi = 2.28$ eV; this formula has been derived from the classical theory, but for moderate supersaturations it does not conflict with the formula for the nucleation energy derived from the Walton—Rhodin theory. In deriving (13) it has been assumed that the number of atoms in an edge row of a nucleus is much larger than 1, and therefore one can assume that the number of atoms in the nucleus varies nearly continuously. This may not be so for nuclei around active centers so we may find a deviation from the inverse proportionality between two supersaturations for A_2 and P. However, for rough calculations this can hardly be important. To construct the curves of Fig. 2 we selected values of $\Delta\mu$ corresponding for given T and

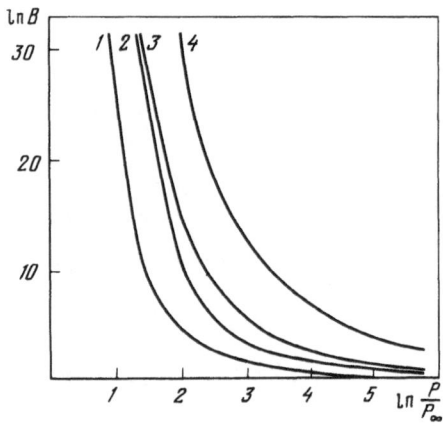

Fig. 2. Logarithm of the acceleration factor for a (111) face of silicon for the following temperatures and impurity concentrations: 1) T = 1400°K; $n_a + n_d = 10^{17}$ cm^{-3}; 2) T = 1200°K; $n_a + n_d = 10^{17}$ cm^{-3}; 3) T = 1400°K; $n_a + n_d = 10^{18}$ cm^{-3}; 4) T = 1200°K; $n_a + n_d = 10^{18}$ cm^{-3}.

$n_a + n_d$ to $\varkappa$ for which the formulas of (39) had given a tabulated dependence of $\ln B$ on P. For these values of $\Delta\mu$, we used (13) to determine P, and then from P and $\varkappa$ we found the $\ln B$ corresponding to $\Delta\mu$.

Conclusions

Interaction of ionized atoms in the adsorption layer with impurity centers near the surface of a semiconductor is the cause of fluctuations in the chemical potential at the surface; areas with low chemical potentials are active nucleation centers, while areas with elevated chemical potentials are points of preferential occurrence of etch pits. Fluctuations in the chemical potential reduce the critical supersaturation for the start of nucleation on the face of a homopolar semiconductor, while they increase the nucleation rate by several orders of magnitude. The nucleation acceleration factor has been found as a function of temperature, charged impurity concentration, and supersaturation. The acceleration factor tends to fall as the temperature and supersaturation increase, and it does not greatly exceed unity at high supersaturations and at temperatures near the melting point. On the other hand, the acceleration factor increases rapidly with the impurity concentration.

Literature Cited

1. N. N. Sheftal'. In: Growth of Crystals, Vol. 1, Consultants Bureau, New York (1959), p. 5.
2. N. N. Sheftal'. In: Growth of Crystals, Vol. 3, Consultants Bureau, New York (1962), p. 3.
3. G. I. Distler. In: Growth of Crystals, Consultants Bureau, New York (1968), p. 51.
4. G. I. Distler and V. V. Zvyagin. Dokl. AN SSSR, 174(5):1082 (1967).
5. G. I. Distler, S. A. Kobzareva, and V. S. Chudakov. Fiz. Tverd. Tela, 9(1):p. 269 (1967).
6. G. I. Distler, S. A. Daryusina, and Yu. M. Gerasimov. Dokl. AN SSSR, 154(6):1323 (1964).
7. G. I. Distler. In: Research in Surface Forces, Vol. 3 (B. V. Deryagin, ed.), Consultants Bureau, New York (1971), p. 71.
8. G. I. Distler and S. A. Kobzareva. In: Research in Surface Forces, Vol. 3 (B. V. Deryagin, ed.), Consultants Bueau, New York (1971), p. 82.
9. G. I. Distler, S. A. Kobzareva, N. K. Pokaeva, and V. S. Chudakov. In: Growth of Crystals, Vol. 8, Consultants Bureau, New York (1969), p. 125.
10. G. I. Distler. In: Growth of Crystals, Vol. 8, Consultants Bureau, New York (1969), p. 91.
11. L. N. Aleksandrov and Yu. G. Sidorov. In: Semiconductor Crystal and Film Growth Processes [in Russian], Novosibirsk (1970), p. 14.
12. L. N. Alexandrov and Iu. G. Sidorov. J. Vacuum Sci. and Technol., 8(4):571 (1971).
13. L. N. Alexandrov. Epitaxie, Endotaxie, Leipzig (1969), p. 96.
14. L. N. Alexandrov, I. G. Sidorov, and E. A. Krivorotov. Thin Solid Films, 3:395 (1969).
15. A. A. Chernov. In: Growth of Crystals, Vol. 3, Consultants Bureau, New York (1962), p. 35.
16. G. Bliznakov. Fortschr. Mineral., 51:7 (1958).
17. N. N. Sirota. In: Crystallization and Phase Transitions [in Russian], Izd. AN SSSR, Minsk (1962), p. 82.
18. D. S. Kamenetskaya. In: Growth of Crystals, Vol. 1, Consultants Bureau, New York (1959), p. 33.
19. R. Kern. In: Growth of Crystals, Vol. 8, Consultants Bureau, New York (1969), p. 3.
20. A. A. Chernov and L. I. Trusov. Kristallografiya, 14(2):218 (1969).
21. F. A. Kuznetsov, A. A. Andreeva, and T. P. Smirnova. In: Semiconductor Crystal and Film Growth Processes [in Russian], Novosibirsk (1970), Izd. SO AN SSSR, Novosibirsk (1970), p. 27.
22. L. A. Borovinskii. In: Semiconductor Crystal and Film Growth Processes [in Russian], Novosibirsk (1970), p. 44.

23.   L. N. Aleksandrov and I. A. Entin. Izv. VUZ, Fizika, No. 9, p. 34 (1971).

24.   L. M. Milne-Thomson and L. J. Comrey. Four-Figure Mathematical Tables [Russian translation], Fizmatgiz, Moscow (1961).

25.   E. I. Givargizov. Thesis: Crystallization of Germanium from the Vapor State [in Russian], Moscow (1964).

26.   B. V. Gnedenko. Textbook of the Theory of Probability [in Russian], Fizmatgiz, Moscow (1965).

27.   L. A. Borovinskii. In: Research on the Structure of Molecules, Crystals, and Crystal Nuclei [in Russian], Izd. LGPI im. A. I. Gertsen (1971), p. 44.

28.   V. M. Glazov and V. S. Zemshov. Physicochemical Principles of Semiconductor Doping [in Russian]. Nauka, Moscow (1967).

29.   H. Neumann. Phys. Status Solidi, 14(1):63 (1966).

EPITAXIAL GROWTH OF SINGLE-CRYSTAL FILMS (REVIEW)

R. N. Sheftal'

Epitaxy gives oriented single-crystal films that are now widely used in electronics and other branches of technology for a variety of devices. It is important to examine the conditions for epitaxy for new materials with valuable properties, since many such materials are involved in current researches in solid-state physics.

Researches in this area have developed vigorously in the last 10-15 years, and these have led to more extensive studies of epitaxy, such as nucleation mechanisms, film orientation and generation, and the causes of defects.

These researches have provided considerable understanding of epitaxy; much new evidence has become available on various aspects of epitaxial growth, especially from electron microscopy, which has indicated the initial stages of crystallization; some theoretical schemes have been proposed to explain epitaxy.

On the other hand, these observations have shown that epitaxial growth is very complicated and extremely sensitive to numerous detailed parameters of the process, with effects from the substrate and the deposited material. The great variety of factors makes it difficult to set up a general analytical model for epitaxy that could form the basis of detailed experimental studies on single-crystal epitaxy.

There are many detailed studies on the manufacture of epitaxial films, which have been performed mainly by trial and error, and the most important criterion in determining the scope for epitaxy has been Royer's one, which applies in the selection of an orientation of a substrate, although it is subject to many exceptions that have not been satisfactorily explained.

There is no unified approach to the phenomena of epitaxy, as is clear from surveys and major literature listings on the subject [1-4], although many valuable observations and generalizations have been made.

For instance, the book by L. S. Palatnik and I. I. Papirov on epitaxial films reveals the conflicting nature of much of the factual evidence on epitaxy, and also the discrepancies between the various theoretical schemes that have been proposed to explain the more or less general but by no means universal trends.

This justifies new attempts at systematization of epitaxial phenomena.

Epitaxy may be considered as the most general case of oriented crystallization; we have here either autoepitaxy, i.e., crystallization on the substrate of the same material, which usually differs somewhat from the deposited layer on account of differences in the production or doping conditions, or else heteroepitaxy, i.e., crystallization on a similar substrate with a related structure (this is the main case, which corresponds to the historical meaning of the term epitaxy), or finally oriented crystallization on a structureless, amorphous, or neutral substrate, which is usually smooth. In the latter case, one may get only partly oriented crystallization, namely textured films.

Here we survey the published evidence on heteroepitaxy, and the review is constructed in such a way as to illustrate the advances that have been made in understanding epitaxy and also to emphasize the problems, which if solved would enable one to predict possible epitaxial partners, together with mutual orientations and the means of producing oriented overgrowth.

We lack theoretical concepts on oriented overgrowth on account of the lack of general analytical models; there are two limiting approaches to epitaxy, the first being based on the substrate as a decisive factor, while the second involves the dimensions, energy, and formation rate of the critical nucleii.

The start of systematic studies on oriented growths relates to the period of rapid advance in methods of structural analysis, when the lattice structures of crystals were elucidated. Naturally, researches on regular overgrowths of crystals at first directed attention particularly to this lattice characteristic of the materials.

Royer [4] surveyed much experimental evidence and put forward the principle of structural correspondence between the epitaxial partners on the shared faces or rows.

By structural correspondence is meant similarity in the elementary parallelograms in the films.

Royer assumed that a restricted discrepancy in the parameters, of not more than 15%, was permissible; ($\Delta a = a_{\text{subst.}} - a_{\text{pricip.}} / a_{\text{subst.}} < 15\%$). At first, also, he assumed that it was necessary to have similar bond types and compositions in the two partners.

For instance, aqueous solutions of the alkali halides NH_4I, KI, KBr, and KCl deposit crystals on freshly cleaved muscovite although the parameter difference Δa in the (111) planes of the salt and the (001) planes of the pseudohexagonal mica are respectively 0.02 (0.4%), 0.19 (3.7%), 0.53 (10.2%), and 0.74 Å (14.3%).

Out of 2000 crystals of each salt, 96.5% were oriented for NH_4I, 69.2% for KI, and 39.0% for KBr [5].

Subsequently [3], numerous exceptions to Royer's rule were found; for instance, KI grows in an oriented fashion on NaCl, although the parameters differ by 25%; again, RbI grows regularly on NaCl although $\Delta a = 30\%$ [6].

Although the parameters do not correspond at all, also, one gets oriented overgrowth of urotropin on mica. These and other such exceptions led to the suggestion that similarity in the symmetries of the surfaces is the basic condition for oriented overgrowth, which will occur in spite of lateral lack of correspondence along one or two rows.

To retain Royer's principle in spite of these exceptions, comparisons were made of the planar nets in the interface, i.e., the principle was employed, but not for an elementary parallelogram; instead, one examined similarity in the m grids of the substrate and the n grids of the overgrowth. However, it was shown [2] that even if m and n vary only in limits from 1 to 3 and discrepancy up to 12% is considered good, various combinations of m and n allow epitaxy to occur even when the lattice parameters differ in the range 70 to 130%. Nevertheless, although Royer's principle does not always apply, structural correspondence plays a very important part, and sometimes the main one, in facilitating or making possible epitaxial overgrowth.

Historically, research on epitaxy and formulation of hypotheses proceeded at the same time as advances in studies on crystal growth. Therefore it was natural that most of the concepts from crystal growth were transferred to epitaxy. The most fundamental of these was the concept of a two-dimensional nucleus. Gibbs [7] pointed out the need for two-dimensional nucleation. Volmer [8] gave a detailed explanation of this. Stranski and Kaischev [9] characterized the equilibrium shapes of two-dimensional nuclei as islands of a monolyer with sides half those for the faces of an equilibrium three-dimensional nucleus; from this time onward the two-dimensional nucleus became an essential element in the analysis of crystal growth processes.

In 1933-34, Finch and Quarrell [10, 11] proposed the idea of surface pseudomorphism, which was based on evidence obtained from electron-diffraction studies of the initial growth stages of zinc oxide films. The concept of surface pseudomorphism amounts to the assumption that an epitaxial film at the start of formation takes up the lattice parameters of the substrate.

This hypothesis did not conflict with existing theoretical concepts and was based on experiments, and it provided a new confirmation of many existing views, and in consequence it stimulated theoretical advances on the deformation of two-dimensional nuclei.

Dankov's theory [12] covers the essence of epitaxy from the viewpoint of two-dimensional nucleation; it is assumed that epitaxy arises when the two-dimensional nuclei are formed, and Dankov considered two competing processes: formation of three-dimensional nuclei of a new phase and production of a two-dimensional nucleus A_2 on the substrate, with allowance for the deformation of the latter on account of the difference Δa in the parameters:

$$A_2^d = A_2^0 + E_d \leqslant A_3, \tag{1}$$

where A_2^d is the energy of formation for a deformed two-dimensional nucleus, A_2^0 is the same for an undeformed one, E_d is the deformation energy, and A_3 is the formation energy for a three-dimensional nucleus.

As $A_2 \ll A_3$, Bliznakov [13] gave the epitaxy condition in the simpler form

$$E_d \leqslant A_3. \tag{2}$$

Then E_d governs the formation of the new phase (oriented or unoriented) together with the rate of formation, which is proportional to $\exp(-E_d/kT)$.

Further, Dankov used the fact that the energy of formation for a three-dimensional nucleus is known ($2a^2\sigma$ for the nucleus in the form of a cube of side a) and expressed the maximum discrepancy Δa_{max} between the substrate and the film in terms of the elastic modulus and shear modulus for the crystal C_{11} and C_{12}:

$$\Delta a_{max} = \left(\frac{2a\sigma}{C_{11} + C_{12}}\right)^{1/2}. \tag{3}$$

Here Δa_{max} is not dependent on the crystal size, so if $\Delta a < \Delta a_{max}$, then the probability of two-dimensional nucleation will always be greater than that for three-dimensional case.

An extension of Dankov's work is found in [14], where the process was considered kinetically. It was assumed that the film will be deformed in order to reduce the discrepancy between the lattice constant in the shear plane, and any residual discrepancy will be compensated by a network of dislocations. Subsequently, the equilibrium deformation of an epitaxial film E_{max} was discussed [15] via three assumptions: 1) the discrepancy is one-dimensional in type, 2) the substrate is infinitely thick and undeformable, and 3) the discrepancies give rise to a series of identical edge dislocations whose Burgers vectors lie in the plane of the interface and are oriented along the direction of the discrepancy.

The equilibrium deformation for a film of thickness h is

$$E_{max} = \frac{(1-2\sigma)(2-f)\mu_0 a}{8\pi^2(1-\sigma)(2+f)\mu_a h}\ \beta\ln[2\beta(1+\beta^2)^{1/2}-2\beta^2],\tag{4}$$

where

$$f=(b-a)/a_0,\quad \beta=8\pi\mu_0 f/(1-\sigma)(1-\mu_a/\mu_b)(2+f)^2\mu_0,$$

with a_0 the distance between positions of minimum potential energy, a and b being the undeformed lattice constant of the film and substrate, μ_a, μ_b, μ_0 the shear moduli of the film, substrate, and boundary layer, and σ being Poisson's ratio, which was assumed to be the same for the film and the substrate.

Naturally of course, this theory applies when the binding forces between the substrate and the film are considerable, i.e., when the forces are such as to produce monolayer growth. However, even in such growth, as for instance Cu_2O on Cu [16] or Au on Pd [17] one finds considerable discrepancies between the experimental results and theory. For instance, the deformation of a Cu_2O film of thickness about 450 Å is 2% near the interface with the copper, and this reduces the discrepancy between the lattices to 12%. This deformation is 300 times greater than the theoretical value. For a film of gold of thickness about 100 Å on a thin palladium substrate of thickness about 400 Å, the deformation was about 0.6%, which reduced the discrepancy between the lattices to 13%. Theoretically, the film deformation should not have been 0.6% but 0.065%. The following factors [18] can be responsible for the discrepancy between theory and experiment: (1) there is partial fusion of the film and substrate; (2) the substrate may be deformed if it is thin; (3) it is difficult to generate boundary dislocations, which leads to reduction in the elastic deformation as the film thickness increases [19].

The following conclusions can therefore be drawn from one of the approaches to epitaxy on the basis of comparability in the lattice parameters in the intergrowth plane:

1. In all the theories it is assumed that epitaxial growth begins with two-dimensional nucleation.

2. The crystallization conditions that give epitaxial growth are not considered.

Bliznakov [20] used some of the above studies to develop a molecular kinetic theory of epitaxy with qualitative incorporation of the crystallization conditions.

Also, Bliznakov [21] revised Dankov's formula by introducing the supersaturation, and also corrected an inaccuracy arising from the erroneous assumption that the lengths of the edges of the two-dimensional and three-dimensional nuclei were equal (the edge of a two-dimensional one is half).

Bliznakov's formula contains an additional factor, which is given in brackets:

$$\Delta a_{max}=\left(\frac{2a\sigma}{C_{11}+C_{12}}\right)^{1/2}\ \left[1+\frac{\psi_n-\psi-\varepsilon}{kTS}\right]\times 2,\tag{5}$$

where ψ_n is the energy needed to detect a foreign particle from the substrate, ψ is the detachment energy for the adjacent particle, ε is the deformation energy referred to one particle in the peripheral row, S is the logarithm of the supersaturation factor, T is absolute temperature, and k is Boltzmann's constant; $\psi_n-\psi-\varepsilon<0$, since $\psi_n-\psi<0$ (usually $\psi_n<\psi$). Therefore, the quantity in the brackets increases with S and T, and hence Δa_{max} increases, i.e., epitaxial overgrowth is facilitated.

Then if oriented overgrowth does not occur for a given value of the discrepancy, it may still occur on raising the temperature and supersaturation; the nucleation energy decreases as ψ_n increases, so the scope for oriented overgrowth is the greater the larger the adsorption energy of the host phase in the substrate plane.

If ψ_n is such that $(\psi_n - \psi - \varepsilon) > 0$, then Δa_{max} decreases as T or S increases, i.e., the orientation is hindered; orientation is facilitated only by reducing the temperature and supersaturation, and the sequence of events occurs in the reverse order.

Dankov's theory and the Dankov—Bliznakov formula qualitatively describe correctly numerous phenomena in epitaxy, at least those where the correspondence between the substrate and the film plays the basic part.

A qualitatively different approach to epitaxy is found in theoretical studies on the kinetics of molecular processes at the surfaces of solids. Frenkel' [22] and others have made major advances here.

Frenkel' [22] made the first theoretical studies of the formation of a condensed phase on a solid; he used Langmuir's model, which is that an atom colliding with the surface of the solid persists there rather than evaporates. It was assumed that during the time spent on the surface, a pair of atoms could arise if the vapor density was appropriate, and such pairs acted subsequently as condensation centers. Frenkel' gave the following expression for the critical vapor density:

$$\gamma_{cr} = \frac{B}{\sigma_0 \tau_0} \exp\left(\frac{U}{kT}\right), \tag{6}$$

where σ_0 is the effective cross section of an adsorbed atom, τ_0 is the period of lattice vibration, U is the total energy of the adsorption of the atom and the dissociation energy for the pair of atoms, B is a constant, k is Boltzmann's constant, and T is absolute temperature.

In this model, the condensed pairs of atoms represent a two-dimensional gas, to which the transformed gas laws can be applied [23-25].

Of course, Frenkel's model for statistical formations of pairs is a great simplification; as regards energy, the best possibility is heterogeneous formation of nuclei from large numbers of atoms [26].

The theory of heterogeneous nucleation is based on the theory of homogeneous nucleation from a supersaturated vapor [27, 28]. Here, Pound and others [29-31] have given the most complete theory of condensation involving hemispherical nuclei on smooth substrates; they assumed that mobile adsorbed atoms and hemispherical groups of these exist on the surface of the substrate. In deriving the expression for the nucleation frequency, they first considered the equilibrium concentration n* of nuclei of critical size, and then multiplied this by the rate of attachment of single atoms ω^* to such a critical nucleus, which raises the stability; they also included the factor Z, which takes into account the deviation of the system from equilibrium as a result of nucleation:

$$I = Z\omega^* n^*. \tag{7}$$

The central topics in these theories are then as follows: (1) determination of the concentration of nuclei from the conditions for metastable equilibrium between complexes of various sizes and single atoms; (2) determination of the number of atoms in a critical nucleus.

Pound et al. [32] assumed that the smallest critical nucleus had the bulk properties, and on this basis they found that the shape of the nucleus is dependent on the contact angle θ [7], the relationship being

$$\sigma_S = \sigma_D + \sigma_i \cos\theta, \tag{8}$$

where σ_S, σ_D, and σ_i are, respectively, the free surface energies of the substrate, the deposit, and the interface. However, it remained an open question how far this theory is applicable when the nucleus is very small, i.e., contains only a few atoms. One can hardly in that case correctly use the free surface energy of the bulk material, and even the existence of a surface energy becomes problematical. For instance, if the nucleus consists of only 7 atoms, all of these will be at the surface. However, the Hirth—Pound theory is applicable for small supersaturations, when the critical nuclei are large.

Walton [33] and Walton and Rodin [34] extended the nucleation theory to systems with very small critical nuclei; they used the methods of statistical mechanics and kinetic theory to eliminate the droplet model and the concept of surface energy; Walton and Rhodin considered the problem via concepts for small canonical ensembles. The nuclei were treated as macromolecules, and the calculations were performed via potential energies and statistical sums. It was found that the following is the nucleation frequency for critical nuclei consisting of n* atoms:

$$I_{n^*} = N_0 R^2 a_0^2 (R/\nu N_0)^{n^*} \exp \frac{[(n^* + 1) Q_{ads} + E_{n^*} - Q_{diff}]}{kT}, \tag{9}$$

where R is the frequency (rate) with which atoms strike the substrate from the vapor (in $cm^{-2} \cdot sec^{-1}$), N_0 is the number of adsorption sites per square centimeter, ν is vibrational frequency, Q_{ads} is the binding energy of an adsorbed atom on the surface, Q_{diff} is the activation energy for this diffusion, E_{n^*} is the dissociation energy of a critical nucleus, and a_0 is the length of one diffusion step.

In this theory, the unknown is n*, which generally speaking, might be determined via the analytical expression for E_n from the Volmer-Weber theory, but this approach is not suitable for the present case of very small nuclei. Therefore, Walton and Rhodin suggested that one should take n* as varying with the supersaturation from 1 to 9 atoms. They tried by trial and error all reasonable configurations of the atoms in the nucleus, and thereby determined the dissociation energy for various small n*.

For instance, at very low substrate temperatures the limiting critical nucleus is a single atom, and then $E_1 = 0$ and the nucleation rate is

$$I_1 = R (R/\nu N_0) \exp [(2Q_{ads} - Q_{diff})/kT]. \tag{10}$$

This gives the rate of formation of pairs of atoms, and it applies only if the vapor does not dissociate before a third atom is attached. As the substrate temperature is raised, a stage is eventually reached where this relationship is not met, and the minimal stable nucleus becomes one in which there are two bonds per atom. In general there are infinitely many such configurations, but the preferred ones are those with the least numbers of atoms. For instance, a configuration with two bonds per atom can be represented by three atoms at the vertices of an equilateral triangle. Another example is a square with four atoms at the corners. The critical nucleus in this case consists of three atoms, and the nucleation rate is

$$I_s = N_0 R a_0^2 (R/\nu N_0)^3 \exp [(4Q_{ads} + E_s - Q_{diff})/kT]. \tag{11}$$

One may assume that a configuration of three atoms results in the case of a face-centered cubic metal in (111) planes parallel to the substrate; a group of 4 atoms at the corners of a square will orient on a (100) plane.

Further rise in substrate temperature means that one ultimately has the point where the stable formation has three bonds per atom, and so on. A discussion of the corresponding configurations enables one to make suggestions about the possible orientation relationship, and provides qualitative explanation of the changes in orientation with substrate temperature.

However, the Walton—Rhodin theory, which works with small groups and which is capable of predicting the varying alterations for n* < 9, is substantially restricted in application when n* > 9, since in that case the number of combinations of the forms increased greatly.

On the other hand, the Walton—Rhodin theory explains textures on unoriented substrates at high supersaturations rather better than the formation of epitaxial films.

This survey of the existing theories of epitaxy provides the following conclusions.

Theories of epitaxy involving two-dimensional nuclei, and which therefore involve mono-layer growth, are based mainly on the principle of structural correspondence.

If epitaxial growth is considered from the viewpoint of heterogeneous nucleation, many aspects remain unclear, although one does get the very important conclusion that epitaxy is possible with three-dimensional nucleation. The conclusions from this approach are rather qualitative, and they predict the trend in the process rather than the kinetics. General thermodynamic arguments explain the preferential orientation as corresponding to the lowest free surface energy; the nucleation energy for the preferred orientation should be the least, while growth rate should be the highest.

The Walton—Rhodin theory takes an intermediate position; it envisages heterogeneous nucleation, but monolayer growth. Purely geometrical arguments were applied, which clearly underestimate the effects of the periodic field of the substrate.

I consider that theories of epitaxial growth have the general disadvantage of ignoring the actual structure of the substrate.

The role of the substrate in oriented crystallization has not been completely elucidated; it has been suggested [35] that the growing crystal in epitaxy has the function of restoring the bonds in the substrate crystal broken at the surface; if such restoration occurs, one gets epitaxy.

This makes clear the particular importance of studies on the actual structure of substrates, and also of elucidating the role of the surface state in epitaxy. The surface state is determined by the following factors: (1) surface defects (steps, dislocations, point defects, etc.); (2) impurities present on the surface; and (3) concentration of preferred adsorption sites, etc.

Surface defects make a substantial contribution to the epitaxial growth mechanism, as is clear from decoration, in which the particles are preferentially deposited at edge elements in the relief and at defects. Various studies [36-44] have given excellent examples of preferential particle deposition at relief elements, which may be even as small as the edges of monomolecular steps. In fact, the high density of deposited particles at steps undoubtedly confirms the view that bond completion at the surface is important. The additional bonds arising at the corner in the step should increase the nucleation probability relative to the smooth surface.

Tolansky [45] made the first systematic studies on the cleavage surfaces of crystals, and found that such surface contains steps of various heights and densities; other studies [46-48] on rock salt showed that the cleavage surfaces had steps of height h from 30 to 500 Å and width up to 5000 Å, as well as monomolecular steps of height 2 Å.

Oriented overgrowth of silver on a carbon replica from a sylvine crystal shows [49] that the surface state greatly affects epitaxy; the structure of the silver film on the amorphous replica was exactly the same as that of a silver film deposited on the cleavage surface corresponding to the replica. In that case, the orientation is produced by the microgeometry of the cleavage surface, which is at least crudely reproduced in the carbon replica.

Palatnik et al. [50] have examined the microgeometry of cleavage surfaces and growing condensed layers of NaCl; they found that the growth surface resembles the cleavage surface in consisting of macrosteps of size up to 1000 Å and microsteps of molecular height. Also, there was a full range of steps of intermediate height. All these steps were rectangular in shape on {100} faces of rock salt.

Thompson and Cohrane [51] and later Thompson [52] suggested that oriented overgrowth is to be ascribed to linear edge elements of the relief, such as the edges of reentrant angles in cracks and steps. These edges usually lie along the principal crystallographic directions, and therefore they are capable of orienting growing crystals if the atoms of the deposited material are deposited along them.

It has also shown [53] that the step height is very important; the steps have a particularly marked effect on the formation of three-dimensional nuclei if

$$h^* \geqslant \frac{2\psi}{kTS}\left(1 - \frac{\psi_n}{\psi}\right), \tag{12}$$

where h* is the step height in relative units, ψ is the work needed to detach a structure element from a substrate of the same type, ψ_n is the work needed to detach a structure element from a foreign substrate, S is supersaturation coefficient, k is Boltzmann's constant, and T is absolute temperature.

This shows that very low steps are less active.

A detailed analysis has been given [54] for nucleation on steps, and an expression was derived for the free energy of formation for a critical nucleus at the edge of a vertical step:

$$G_i = \frac{16\pi\sigma f(\theta)}{3\Delta G_V^2}, \tag{13}$$

where $f(\theta)$ is a complex function of the contact angle, σ is the specific surface energy of the interface, $\Delta GV = -(kT/\Omega)S$, Ω is the molecular volume, and S is the supersaturation coefficient.

It was found that for $\theta < 45°$ there is in general no barrier to the nucleation at the reentrant angle; if $45° \leq \theta \leq 105°$, nuclei were formed preferentially at the reentrant angle, while for $\theta > 105°$ the predominant part was played by the frequency factor, with nuclei formed mainly on smooth parts of the surface. Then any macroscopic defect that increases the interface area will accelerate nucleation for contact angles less than 105°.

However, it remains an open question whether the cleavage steps orient nuclei of the deposited phase or whether they are merely responsible for the higher density of nuclei. The experimental evidence on this is fairly conflicting. For instance, Bassett et al. [55] have found that monatomic steps on (100) cleavage surfaces of rock salt hardly influence the orientation of nuclei formed along them. The same conclusion was drawn in [56] from a study of films of gold deposited at very high vacuum on cleaved rock salt. On the other hand, it has been shown [57] that the epitaxy of gold on silver with numerous steps is far more perfect than that on smooth faces of a silver substrate. Kleber [58] found for the crystallization of d-camphor on biotite that one gets a mirror-symmetry orientation of the d-camphor crystals on the corresponding cleavage faces of the biotite. Very interesting results have been obtained [59] in the decoration of cleavages on rock salt made under vacuum. If the steps are decorated with silver at room temperature, one finds perfect single-crystal orientation of the silver nuclei at the step; if, on the other hand, the steps are decorated with gold (also at room temperature), then the gold nuclei formed along the steps do not have the epitaxial orientation. From this one can conclude that the crystallization conditions for the epitaxial silver were optimal at room temperature, whereas the optimal conditions for epitaxy of gold were not obtained.

At first sight, the results of Robbins and Rhodin [60] appear unexpected, since they examined the nucleation of gold at very low pressures on cleaved MgO; they found that the maximum number of nuclei per square centimeter was not dependent on the deposition rate or on the substrate temperature in the range used. Their explanation was that the nucleation is controlled by point defects and impurities, and the numbers of such defects were much the same for the various substrates. Similar results have been reported [61-67] for gold on fresh rock salt surfaces and ones exposed to air.

These results do not differ essentially from those obtained by Sellat and Trillat; it appears reasonable to assume that it is the orientation and size of the nuclei rather than density that determines the epitaxial nucleation.

Here we must note that change in the deposition rate by less than an order of magnitude hardly affects the supersaturation at the surface of the substrate [68].

Recently, there have been many papers by Distler et al. [69-82, 85] on electron microscopy of real crystal surfaces; Distler used replicas from surfaces with decoration and considered that the decorating particles are selectively deposited not only at the edge elements of the relief but particularly at point defects and groups of these, which he called active centers.

Distler assigned an electrical nature to these active centers, and he related their existence to deposition of impurities together with point defects and groups of these. The orientation of the decorating particles in his experiments was the same for cleavage steps as for smooth parts of the surface, from which he concluded that the local centers initiating nucleation exist with equal probabilities on the two kinds of area [80].

Gerasimov and Distler [81] decorated cleaved surfaces with gold and found that the gold particles were located in restricted areas with (100) and (110) orientations, the latter being of lesser importance.

If gold decoration was applied to cleaved NaCl coated with carbon, as a rule the orientation of the gold nuclei persisted up to film thicknesses of 400 Å; the orientation of the gold on (100) and (110) was ascribed to active centers, which differ in symmetry and in range of action. Distler et al. [82] examined the cleaved surfaces of NaCl crystals grown to contain traces of $PbCl_2$ and found that the density of the decorating particles at the boundary between the NaCl and $PbCl_2$ phases was much greater than on the smooth parts free from impurity. The accentuated local deposition was ascribed to a double electrical layer at the interface having a negative potential higher than that found at other parts of the surface. Distler showed that coalescence occurred much earlier at points where there was a double electrical layer, which explained the perfect epitaxial orientation at these points.

On the other hand, it appears clear that the boundary between the impurity and the crystal should be under stress, which on cleavage should result in a higher step density around the impurity.

The method used by Distler et al. (electron microscopy of replicas with decoration particles) does not provide complete information about the real surface structure of the crystal, since it gives only the two-dimensional distribution of the decorated particles.

The valuable method developed by Distler should be supplemented via direct observations on the actual decorated surface in the electron microscope; the techniques for preparing such surfaces have been described in detail [83].

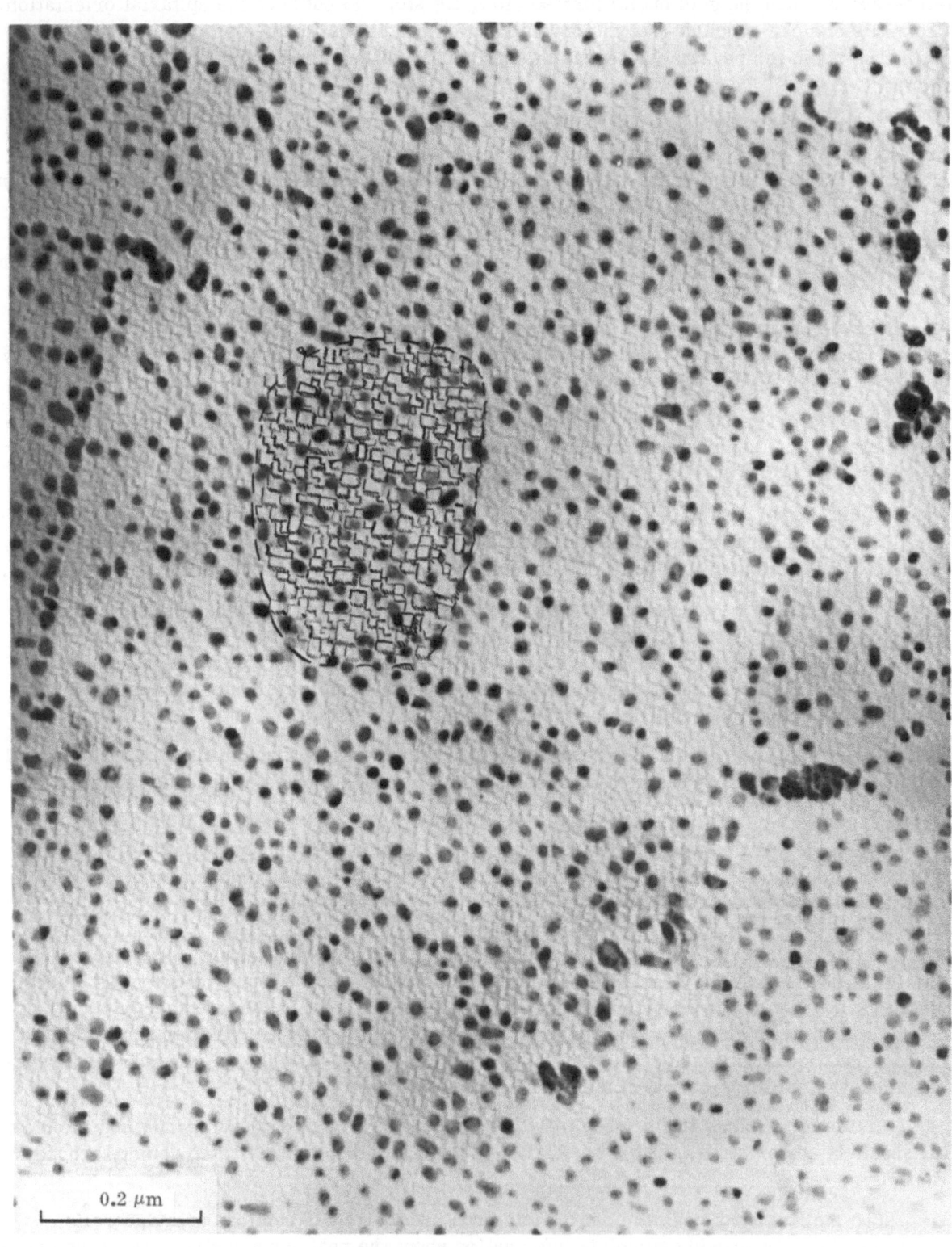

Fig. 1. Electron micrograph of a thinned cleaved fragment of rock salt bearing iron crystals
[84]. The central part of the figure has been retouched.

Very important evidence has been presented [84] on the effects of water vapor on nucleus orientation for iron on (100) surfaces of rock salt; electron micrographs were recorded from thin cleaved sections of NaCl bearing iron particles, together with electron-diffraction data on the orientation. Figure 1 shows a photograph of this kindly made available to us.

Careful examination of this photograph shows a microstructure in the cleaved surface, in particular with the steps normally seen on decorated surfaces bearing gold or silver [36-38] and regions between steps usually considered smooth [3, 45-48] and in addition a fine network of microrelief consisting of microsteps forming a pattern whose symmetry corresponds to the axial symmetry of the face on which they emerge. Figure 1 has been retouched to emphasize this.

We consider that the trihedral angles formed by these microsteps, which cover the entire cleavage surface, are responsible for the epitaxial overgrowth.

To conclude this survey, we consider the so-called orienting long-range action of crystal surfaces; Distler in 1968 [75] published results on oriented crystallization of PbSe films on mica substrates coated with amorphous films of various thicknesses. He found that oriented crystallization still persisted with films of collodian up to 1 μm thick. We [86] have published results on the orienting action for thin Bi films of material coated with intermediate layers of collodian. It was found that the action of a mica substrate extended out to 300 Å, diminishing to zero when the intermediate layer increased to 350 Å. Although our results are in general similar to those of Distler et al. the interpretations may be different; Distler explained the long-range action of the substrate in terms of electrically charged centers, which can be screened only by fairly thick layers of insulator. We assume that a thin collodian film reproduces the microrelief of the surface not only on the contact side, but also rather crudely on the outer side. Chopra [87] has attempted to reproduce Distler's results on the oriented crystallization of films of Ag, PbS, S, and SnS on rock salt and mica partly covered with amorphous films of C, SiO_x, quartz, and $BaTiO_3$; his results were negative. On the other hand, oriented overgrowth of ZnS and PbS on NaCl has been reported [88] when coatings of SiO_x or C are used. Here one must bear in mind that in none of the above experiments was attention directed to the point that orientation may arise only during the growth, not in nucleation, which is particularly so for binary compounds, for which orientation always is more pronounced in growth than in nucleation [89].

There are also differences in nature of the amorphous coatings used in these studies; collodian or similar plastic films can only serve to reproduce crudely the relief of the surface, while amorphous films of C or SiO_x condensed on the surface are themselves nonuniformly distributed, i.e., they show preferential adsorption and screening at surface features.

It seems likely that a final decision on the nature of the nucleation centers (electrical or geometrical) could come from coating the substrate with a layer less than 1 μm thick of elastic amorphous material formed independently from the substrate.

If the crystallite orientation of such an intermediate layer corresponded to epitaxial, then the only explanation would be from the viewpoint of electrically charged centers.

If epitaxial orientation did not occur, the geometrical factor would have to be considered the decisive step in the epitaxy.

Conclusions

This survey of the literature evidence gives rise to the following conclusions on the state in this area:

1. The reasons for epitaxy have not been fully elucidated.

2. Epitaxy is to be divided into two limiting types. The first is related to weak inter-action between the partners in the intergrowth plane, while the second involves strong inter-action. In the case of the first type, the structural correspondence rule is not decisive, where-as it is obligatory for the second.

3. Large deviations from structural correspondence require high supersaturations to obtain oriented overgrowth.

More detailed elucidation of epitaxy effects would appear to require decisions on the following topics: the nature of the preferential adsorption sites when the two-dimensional and three-dimensional nucleation mechanisms are involved; the mechanism of the orienting action of the substrate in each case; the role of the crystallization conditions in epitaxy; and finally how the mode of supply of the source material affects the epitaxial growth.

I am indebted to S. Shinozaki and H. Sato for the photomicrograph and permission to publish it.

Literature Cited

1. L. S. Palatnik and I. I. Papirov, Epitaxial Films [in Russian], Nauka, Moscow (1971).
2. D. W. Pashley, Adv. Phys., 5:173 (1956).
3. D. W. Pashley, Adv. Phys., 14:327 (1965).
4. L. Royer, Bull. Soc. Franc. Mineral. 51:7 (1934).
5. L. Royer, Bull. Soc. Franc. Mineral., 77:1004 (1934).
6. L. Brück, Ann. Phys., 26:233 (1936).
7. J. W. Gibbs, Thermodynamic Papers. Part I Thermodynamics [Russian translation], Gostekhizdat, Moscow (1950).
8. M. Volmer and A. Weber, Z. phys. Chem., 119:227 (1925).
9. J. Stranski and R. Kaischev, Z. phys. Chem., 26:100, 114, 312 (1934).
10. G. J. Finch and A. D. Quarrell, Proc. Roy. Soc., A141:398 (1933).
11. G. J. Finch and A. D. Quarrell, Proc. Roy. Soc., 46:148 (1934).
12. P. D. Dankov, Proceedings of the Second Conference on Metal Erosion [in Russian], Vol. 2, Metallurgiya, Moscow (1943), p. 12; Zh. Fiz. Khim., 20:843 (1946).
13. G. Bliznakov, Izv. Bolg. AN, ser. fiz., 6:301 (1956).
14. F. C. Crank and J. H. van der Merwe, Proc. Roy. Soc., A177:205, 216 (1949).
15. J. H. van der Merwe, Single-Crystal Films [Russian translation], Mir, Moscow (1966).
16. B. Borie, C. I. Sparks, and I. V. Catherart, Acta Metallurgica, 10:691 (1962).
17. J. W. Matthews, Phil. Mag., 13:1207 (1966).
18. J. W. Matthews, Physics of Thin Films [Russian translation], Mir, Moscow (1970).
19. N. Cabrera, Mem. Sci. Rev. Metallurg., 62:205 (1965).
20. G. Bliznakov [C. Blisnakow], Fortschr. Mineral., 36:149 (1958).
21. G. Bliznakov, Growth of Crystals, Vol. 5A, Consultants Bureau, New York (1968), p. 51.
22. J. Frenkel, Z. Phys., 26:117 (1924).
23. N. N. Semenov, Zh. Russ. Fiz.-Khim. Obshch, 62:33 (1930).
24. A. Dixit, Phil. Mag., 16:1049 (1933).
25. A. Dixit, Proc. Indian Acad. Sci., A48(6):330 (1958).
26. E. Zehender, Optik, 7:200 (1950).
27. R. Becker and D. W. Döring, Ann. Phys., 24:719 (1935).
28. J. H. Hollomon and D. Turnbull, Progr. Metal Phys., 4:333 (1953).
29. J. P. Hirth and G. M. Pound, Condensation and Evaporation, Oxford (1963).
30. J. P. Hirth, S. J. Hruska, and G. M. Pound, Single-Crystal Films [Russian translation], Mir, Moscow (1966).
31. G. M. Pound and J. P. Hirth, Condensation and Evaporation of Solids., Proc. Internat. Sympos., Dayton, Gordon and Breach, New York (1963).

32. G. M. Pound, M. Simnad, and L. Yang, J. Chem. Phys., 22:1215 (1954).
33. D. Walton, J. Chem. Phys., 37:1282, (1962).
34. T. N. Rhodin and D. Walton, 9th Trans. Amer. Vacuum Soc., (1962).
35. S. C. Monier, Bull. Soc. Franc. Mineral., 77:1315 (1954).
36. G. A. Basset, Phil. Mag., 3:1042 (1958).
37. C. Sella, P. Conjeaud, and J. J. Trillat, Proc. 4th Internat. Conf. on Electron Microscopy, Berlin (1958), Vol. 1, p. 508.
38. H. Bethe, Phys. Status Solidi, 2:3 (1962).
39. W. Kleber, Phys. Status Solidi, 2:823 (1962).
40. W. Kleber and R. Reinhold, Z. Kristallogr., 114:410 (1960).
41. W. Kleber and L. Ickert, Z. phys. Chem., 224:364 (1963).
42. L. S. Palatnik, V. M. Kosevich, and V. M. Moskalov, Fiz. Met. Metalloved., 16:403 (1963).
43. L. S. Palatnik, V. M. Kosevich, and V. M. Moskalov, Fiz. Met. Metalloved., 16:723 (1963).
44. V. M. Kosevich, L. S. Palatnik, and V. M. Moskalov, Fiz. Tverd. Tela, 8:8 (1966).
45. S. Tolansky, Multiple-Beam Interferometry of Surfaces and Films, Clarendon Press, Oxford (1948).
46. O. Coche and N. Wilman, Proc. Phys. Soc., 51:625 (1939).
47. L. G. Schulz, Acta Crystallogr., 5:130 (1952).
48. L. G. Schulz, Acta Crystallogr., 4:483 (1951).
49. P. Vermont and W. Dekeyser, Physica, 25:53 (1959).
50. L. S. Palatnik, V. M. Kosevich, and V. M. Moskalov, and A. A. Sokol, Growth of Crystals, Vol. 8, Consultants Bureau, New York (1969), p. 177.
51. G. P. Thompson and W. Cochrane, Theory and Practice of Electron Diffraction (1939).
52. G. P. Thompson, Proc. Phys. Soc. 61:403 (1948).
53. L. Ickert, Z. phys. Chem., 221:301, 328 (1962).
54. B. K. Chakraverty and G. M. Pound, Acta Metallurgica, 12:851 (1964).
55. C. A. Bassett, J. W. Menter, and D. W. Pashley, Proc. Roy. Soc., A246:345 (1958).
56. E. Grunbaum and J. W. Matthews, Phys. Status Solidi, 9:731 (1965).
57. J. C. Allpress and J. V. Sanders, Phil. Mag., 9:645 (1964); 10:851 (1964).
58. W. Kleber, Growth of Crystals, Vol. 5A, Consultants Bureau, New York (1968), p. 59.
59. C. Sellat and J. J. Trillat, Single-Crystal Films [Russian translation], Mir, Moscow (1966), p. 242.
60. J. L. Robbins and T. N. Rhodin, Surface Sci., 2:346 (1964).
61. S. Ino, J. Phys. Soc. Japan, 21:346 (1964).
62. J. W. Matthews, Phil. Mag., 12:1143 (1965).
63. G. G. Sumner, Phil. Mag., 12:767 (1965).
64. J. W. Matthews, J. Vacuum Sci. and Technol., 3:133 (1966).
65. S. Ino, D. Watanabe, and S. Ogawa, J. Phys. Japan, 19:884 (1964).
66. L. L. Kunin, Dokl. AN SSSR, 79:93 (1951).
67. J. W. Matthews and E. Grunbaum, Phil. Mag., 11:1233 (1965).
68. K. A. Neugebauer. In: Physics of Thin Films [Russian translation], Vol. 2, Mir, Moscow (1964).
69. G. I. Distler, S. A. Daryusina, and Yu. M. Gerasimov, Dokl. AN SSSR, 154:1328 (1964).
70. G. I. Distler, Yu. M. Gerasimov, and N. M. Borisova, Dokl. AN SSSR, 165:329 (1965).
71. G. I. Distler and S. A. Kobzareva, Fiz. Tverd. Tela, 7(1):2450 (1965).
72. G. I. Distler and L. D. Kuslovskii, Fiz. Tverd. Tela, 8:600 (1966).
73. G. I. Distler and S. A. Kobzareva, Dokl. AN SSSR, 172:77 (1967).
74. G. I. Distler, S. A. Kobzareva, N. K. Pokareva, and V. S. Chudakov, Growth of Crystals, Vol., 8, Consultants Bureau, New York (1969), p. 125.
75. G. I. Distler, Growth of Crystals, Vol. 8, Consultants Bureau, New York (1969), p. 91.
76. G. I. Distler and M. P. Fedotova, Vysokomol. Soed., 9(13):6 (1967).
77. G. I. Distler and N. M. Borisova, Kinetika i Kataliz, 8:4 (1967).

78.   G. I. Distler, Growth of Crystals, Consultants Bureau, New York (1968), p. 51.
79.   G. I. Distler, Izv. AN SSSR, ser. fiz., 32:1044 (1968).
80.   Yu. M. Gerasimov, G. I. Distler, V. M. Efemenkova, and V. E. Yurasova, Fiz. Tverd. Tela, 10:270 (1968).
81.   Y. M. Gerasimov and G. I. Distler, Naturwissenschaften, 55:132 (1968).
82.   G. I. Distler, V. N. Lebedeva, V. V. Mosivin, and E. I. Kortakova, Fiz. Tverd. Tela, 11:2390 (1969).
83.   K. Yagi and G. J. Honjo, J. Phys. Soc. Japan, 19:1892 (1966).
84.   S. Shinozaki and H. Sato, J. Phys. Chem. Solids, Suppl. 11; p. 515 (1967).
85.   G. I. Distler and B. B. Zorygin, Dokl. AN SSSR, 174:1082 (1967).
86.   R. N. Sheftal', Yu. F. Ogvin, V. N. Lutskii, and M. I. Elinson, Dokl. AN SSSR, 180:580 (1968).
87.   K. L. Chopra, J. Appl. Phys. 40:906 (1969).
88.   A. Barna, P. B. Barna, and J. F. Pocza, Thin Solid Films, 4:R32 (1969).
89.   S. A. Semiletov, Thesis: Structure and Physical Properties of Thin Films of Some Semiconductors [in Russian], Moscow (1969).

CRYSTALLIZATION OF EPITAXIAL FILMS BY VACUUM CONDENSATION: OPTIMAL CONDITIONS AND ORIENTATION MECHANISM

R. N. Sheftal' and L. A. Borovinskii

A basic condition for single-crystal overgrowth in vacuum condensation is that the substrate should be very clean and smooth, such as a cleavage plane [1].

However, the state of the surface cannot completely determine the scope for single-crystal overgrowth; another important factor is the set of crystallization conditions. It is very laborious to find the optimum conditions for crystallization of single-crystal films, and the task is further complicated by the fact that literature data on optimal conditions are conflicting even for the same substances on identical substrates. For instance, it has been stated [2-4] that single-crystal films of gold crystallize on rock salt at a variety of condensation rates and substrate temperatures, the values being given in each paper as optimal. The same may be said about the crystallization of cadmium sulfide on NaCl [5-7], since optimal conditions given by different workers vary widely and it is impossible to reproduce many of the results. There are therefore uncontrollable features in the parameters of the experiments, which arise from individual aspects of the equipment.

For instance, the crystallization temperatures are measured with thermocouples of various cross sections and various modes of attachment to the substrates; the condensation rates vary in an uncontrollable fashion; the materials in the vacuum seals may produce various gaseous products in the space; there is a marked influence on the behavior of the substrate from the inevitable adsorption of the oil, if this is used; and there are probably also many other features of such conditions.

This means that it is important to find an independent characteristic of the crystallization that would eliminate these individual features of any particular run. Such a characteristic can be found using the properties of the condensing material, especially the internal capacity to crystallize in an ordered fashion under certain conditions. To reveal this capacity clearly without interference, one should use substrates that have minimal orienting effect. Examples are smooth polished glass, polished polycrystalline materials, and single crystals whose surfaces have been made amorphous. The surface smoothness should be accompanied by chemical inertness and high melting point.

Here we describe experiments designed to determine the reproducible conditions for the most highly ordered crystallization on such substrates.

Experiments and Results

The experiments were performed in a single vacuum equipment, while in a given run the substrate temperature was measured simultaneously on orienting and nonorienting substrates

53

with a fixed condensation rate. The film structures were examined directly after formation by fast-electron diffraction in transmission and reflection.

We used gold, silver, indium, cadmium, and bismuth, while the substrates were cleaved pieces of mica and rock salt, and optically polished pyrex glass.

The methods used in the experiments were the same for all the substances, and we describe them in detail for gold.

The gold was evaporated at a pressure of $5 \cdot 10^{-7}$ mm Hg from a tungsten cone heated on the outside by a spiral; the base of the evaporator was heated more strongly than the upper part. This design provided a constant evaporation rate. Also, stabilization of the power input to the heater enabled us to work even when the voltage varied by 50%. The changes in power level were only ±1%. The substrate was heated by a platinum resistance oven placed 10 mm from the substrate holder. This had a porcelain reflector, and that provided for uniform heating of the substrates by radiation. The temperature was measured with a chromel−alumel thermocouple having a diameter of 0.2 mm, these thermocouples being attached to the working surfaces of the substrates, which were 205 mm from the evaporator, the maximum deviation from normal incidence for the atomic beam being 4°. The substrate temperature varied from run to run from 20 to 500°C by steps of 20°. The condensation rates for the various substances range from 1 to 30 Å/sec, the values being determined with a KIT-1 quartz crystal.

Polycrystalline films of gold were formed on orienting and nonorienting substrates if the crystallization temperature did not exceed 120°C; when the substrate temperatures were raised to 260°C, the single-crystal substrates gave partly oriented films, and the orientation occurred on the crystallographic planes that are epitaxial for each substrate. On the other hand, the glass substrates gave polycrystalline films. If the temperature was raised to 300°C, the degree of orientation of the crystallites on the single-crystal substrates increased, while there was a certain ordering of the polycrystalline layer even on the amorphous substrates. Crystallization of gold at 340-360°C gave clearly oriented layers of gold on the amorphous substrates, the film consisting of a complex texture, which was composed of crystallites with their {111}, {100}, and {110} planes parallel to the substrate. On cleaved surfaces of mica and rock salt, we found the most perfect films of gold with (111) orientation on mica and (100) on NaCl. If the crystallization temperature was raised to 400-450°C, the electron diffraction patterns from on nonorienting substrates gave a fairly clear orientation, which showed that such conditions were preferential for orientation. On the other hand, orienting substrates gave a deterioration in the structure of the layer, as was clear from the additional reflections from grains in twin positions.

The degree of orientation was determined from the angular width of the diffraction peaks for the polycrystalline films and from the presence of the additional reflections due to twinning,

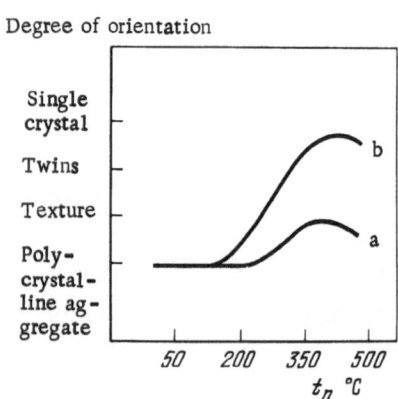

Fig. 1. Degree of perfection of gold films in relation to crystallization conditions: a) on glass; b) on orienting substrates.

packing defects, and so on for the single-crystal films; these figures for the single-crystal and amorphous substrates were plotted against the substrate temperature to demonstrate qualitative agreement between the degree of orientation on the different types of substrate at the same temperatures (Fig. 1).

Analogous results were obtained with silver, indium, cadmium, and bismuth; these substances also characteristically had agreement between the temperatures at which textures were formed on amorphous substrates and single-crystal films on orienting ones. In all the experiments the film thickness did not exceed 500 Å, so we conclude that we were examining nucleation textures, not growth ones. Similar results were obtained in the crystallization of antimony on amorphous and orienting substrates at a constant temperature but with different condensation rates [8].

Qualitative Theory

Here we give a qualitative explanation of the following experimental facts: the texture is formed over a narrow range in the deposition rate and in the substrate temperature; the crystallization conditions giving a single-crystal film on an orienting substrate are similar to those giving a textured film on an amorphous one; the single-crystal film has a definite orientation; and the crystallization conditions for an oriented film are not dependent on the material of the substrate.

We use the data of [9] to assume that the energy of the bonds between the atoms of deposited material is much greater than the energy of the bonds between the atoms of the first adsorption layer and the substrate. The bonds between the atoms of the condensing material are of ionic, metallic, or covalent type, and are very strong. On the other hand, the bonds from the film to the substrate are mainly of van der Waals character, so their energy is less than that of the others by at least an order of magnitude, and also is characterized by a comparatively slow decrease with the distance from the substrate. In fact, let the energy ε of a bond between an adsorbate and an absorbent atom at a distance r be described by the Lennard-Jones equation

$$\varepsilon = -\frac{\alpha}{r^6} + \frac{\beta}{r^{12}}, \tag{1}$$

where α and β are constants.

We take the origin of the cartesian coordinate system at the surface of the substrate, with the z axis along the normal, while the x and y axes are parallel to the substrate. We sum the energies of interaction for an atom at a distance z from the substrate, which interacts with all atoms in the adsorbent, which gives

$$\varphi(z) = -\frac{a}{z^3} + \frac{b}{z^9}, \tag{2}$$

where

$$a = \frac{\pi\alpha n_0}{12}, \qquad b = \frac{\pi\beta n_0}{45}, \tag{3}$$

with n_0 the number of particles in unit volume of the adsorbent (in the summation, the sum is replaced by an integral over the volume of the substrate crystal).

The quantities $\varepsilon(r)$ and $\varphi(z)$ can be expressed via the equilibrium distances r_m and z_m together with the minimum values for the interaction energies ε^* and φ^*:

$$\varepsilon(r) = \varepsilon^* \left\{ 2\left(\frac{r_m}{r}\right)^6 - \left(\frac{r_m}{r}\right)^{12} \right\}, \qquad (\varepsilon^* < 0), \tag{4}$$

where α and β are expressed in terms of ε^* and r_m via

$$\alpha = -2r_m^6\varepsilon^*; \quad \beta = -r_m^{12}\varepsilon^*. \tag{5}$$

Similarly, $\varphi(z)$ may be written in terms of φ^* and z_m

$$\varphi = \frac{\varphi^*}{2}\left\{3\left(\frac{z_m}{z}\right)^3 - \left(\frac{z_m}{z}\right)^9\right\}, \tag{6}$$

where the parameters a and b of (2) are related to p^* and z_m as follows:

$$a = -\frac{3}{2}z_m^3\varphi^*, \quad b = -\frac{1}{2}z_m^9\varphi^*. \tag{7}$$

$$z_m = \sqrt[6]{\frac{3b}{a}}. \tag{8}$$

We substitute for a and b in (3) and then express β/α in terms of r_m to get

$$z_m = r_m\sqrt[6]{0.8} \simeq 0.960 r_m, \tag{9}$$

i.e., the equilibrium distance of an atom from the substrate differs little from the equilibrium distance for two atoms of adsorbate and absorbent whose interaction energy is given by the Lennard-Jones formula. It is readily shown that ε^* and φ^* are related by

$$\varepsilon^* = \frac{72}{5\pi n_0 z_m^3}\varphi^*. \tag{10}$$

The quantity $1/n_0$ is the volume per atom of the substrate, which is of the order of the cube of the interatomic distance in the latter; z_m is approximately equal to the sum of the van der Waals radii for the adsorbent and adsorbate atoms, and we can assume that it is 1.5-2 times greater than the interatomic distance in the substrate. Then (10) shows that ε^* and φ^* are of the same order.

The bonds between the adsorbate atoms are much stronger than the bonds to the adsorbent, so one expects that a sufficient deposition rate will result in three-dimenional complexes of adsorbate atoms, since such complexes are more stable than two-dimensional one-layer ones.

However, although the binding energy to the substrate is small relative to the binding energy of the condensate atoms one with another, it is still fairly large by comparison with the atoms. Therefore, atoms that enter the adsorption layer are retained in it, since it is found [10] that the time spent by the atoms in the adsorption layer is defined by

$$\tau = \frac{1}{\nu}\exp\left(\frac{U_{des}}{kT}\right), \tag{11}$$

where ν is the vibrational frequency of the atoms in the adsorption layer. The basic role of the substrate in these cases is to produce the adsorption layer via its adsorption forces, and in that way to produce crystalline complexes when the parameters are suitable. On the other hand, one can assume that the energy of a crystalline complex in the field of the adsorption forces may be comparable with the surface energy of the complexes, and that surface energy (not the total bond energy) determines the form of the complexes, so the adsorption forces can substantially influence the shape of the crystalline complexes in the adsorption layer. One effect of the adsorption forces on the complexes is bound to be to flatten them.

We perform a rough calculation to explain this consideration. Let the Kossel crystal with a simple cubic lattice be in the field of the adsorption forces; we assume that the crystal is bounded only by $\{100\}$ faces, and is oriented in such a way that two of its faces are parallel to the substrate. We denote the number of particles in the edges of the crystal by n_x, n_y, and n_z; conditions of symmetry show that $n_x = n_y$ in the equilibrium form. Let ψ be the binding energy between nearest neighbors; the number of unsatisfied bonds in the surface of the crystal is $2(n_x^2 + 2n_x n_z)$, so the surface energy is

$$W_n = \psi\,(n_x^2 + 2n_x n_z). \tag{12}$$

The energy of the crystal in the field of the adsorption forces is

$$W_{\text{ads}} = -\,n_x^2 \sum_{k=1}^{n_z} \frac{a}{[z_1 + c\,(k+1)]^3} + \frac{b}{z_1^9}, \tag{13}$$

where z_1 is the distance of the first layer from the substrate and c is the lattice constant of the Kossel crystal.

We assume that we need to take into account only the energy of repulsion for the first layer, while the repulsion energy of the more remote layers is zero on account of screening.

We replace summation by integration to get

$$W_{\text{ads}} = -\,\frac{a}{2c}\,n_x \left\{ \frac{1}{z_1^2} - \frac{1}{[z_1 + c\,(n_z - 1)]^2} \right\} + \frac{b}{z_1^9}\,n_x^2. \tag{14}$$

As the calculation is rough, we neglect the term $1/[z_1 + c(n_z - 1)]^2$ relative to $1/z_1^2$ on the assumption that $c(n_z - 1) \gg z_1$. The total binding energy of the complex under these conditions is

$$W = \psi\,(n_x^2 + 2n_x n_z) - \left(\frac{a}{2C z_1^2} - \frac{b}{z_1^9} \right) n_x^2. \tag{15}$$

The condition for minimum W goes with the additional condition $n_x^2 n_z = N = \text{const}$ to give

$$n_z = n_x \left(1 - \frac{a}{2c z_1^2 \psi} + \frac{b}{z_1^9 \psi} \right). \tag{16}$$

As $-a/2c z_1^2 + b/z_1^9 < 0$ (if this were not so, the energy in the field of the adsorption forces would be positive, which would correspond to repulsion of the crystal by the substrate), so we see that $n_z < n_x$, which corresponds to a flattening of the complexes.

We consider the motion of the individual atoms and groups in the adsorption layer; if we assume that the potential energy in the adsorption layer is dependent only on z, the motion of the atoms and groups along the substrate, and the rotation around the z axis, will occur with entire freedom; in reality, of course, the energy is dependent on the x and y coordinates also. If we assume that $\varphi(z)$ is the potential energy of an adsorbed atom as averaged with respect to x and y, the total potential energy can be put in the form

$$\varphi_n\,(x, y, z) = \varphi\,(z) + g\,(x, y)\,\varphi\,(z), \tag{17}$$

where $\overline{g(x,\,y)} = 0$.

The function $g(x, y)$ is periodic for a single-crystal substrate, while it is a random function for amorphous ones. The motion of the atoms along the substrate and rotation of groups around the z axis are free only if the energy of the thermal motion is larger than the activation energy for diffusion; however, we consider that this factor is not decisive for texture formation, so in what follows we will assume that the motion along the substrate is completely free.

The crystallites in the texture are disoriented only as regards rotation around the z axis, with complete orientation around the x and y axes; therefore we consider that the decisive point as regards texture formation conditions is to analyze the conditions for rotation around the x and y axes. If we consider a crystallite outside the adsorption layer, this as a whole has six degrees of freedom; we will assume that the internal vibrational degrees of freedom are completely degenerate, which applies in this case if the vibrational frequencies are subject to the condition $h\nu \gg kT$, where h is Planck's constant.

When the crystallite approaches the substrate, the translational motion along the z axis gives way to an oscillation of the crystallite with respect to the substrate, while the rotation around the x and y axes is replaced by torsional oscillations on these axes. The degeneracy of these components of the motion leads to reduction in the residual number of degrees of freedom and to corresponding reduction in the entropy of crystallites of a given size. But simultaneously there is a fall also in the potential energy of the crystallites in the field of the adsorption forces, and correspondingly the following is the change in the free energy of a crystallite with n particles:

$$F_n = E_n - TS_n , \tag{18}$$

where E_n is the total energy and S_n is entropy; this is the result of changes in two competing factors: there is a fall in E_n as the substrate is approached, but there is an increase in the entropy term $-TS_n$, with of course the reverse trend as the substrate is left behind.

As texture formation is determined by the orientation relative to the x and y axes, we can set up the following criterion for texture formation: one gets a texture if under given deposition conditions one gets a free energy for the ensemble of crystallites that takes a minimal value with most of the crystallites in a state of zero torsional oscillation about the x and y axes. If, on the other hand, the state of minimum free energy in the ensemble has a considerable fraction of the crystallites in excited levels for the torsional oscillations around the x and y axes, then one gets a polycrystalline product.

We now consider the changes in the texture formation conditions as the deposition rate is varied for a fixed low temperature of the substrate.

At low deposition rates, the density of the atoms in the adsorption layer is small; there is equilibrium between the numbers of atoms being adsorbed and desorbed. At equilibrium, the relationship between the number of atoms adsorbed in unit time per unit surface area R is expressed in relation to the number of adsorbed atoms per unit area A by

$$A = R\tau, \tag{19}$$

where τ is defined by (11).

Two nucleation mechanisms may be involved [10, 11] when a film is produced by condensation from a molecular beam; the first involves adsorption and diffusion of the condensing atoms [11], while the second involves direct addition of atoms from the vapor or nucleation in situ [10]. The relevant mechanism in a particular case is dependent primarily on the substrate temperature and the deposition rate; if the latter is low and the substrate temperature is also low, the film grows by direct addition of atoms from the vapor. If the deposition rate is high, the heat of condensation raises the surface temperature, and this accentuates the surface diffusion of the condensed atoms. If the condensation rate is small, the incident atoms condense almost at once when they reach the surface of the substrate. The supersaturation is then below the critical value producing crystalline nuclei. The difficulty of surface diffusion hinders the growth of crystallites, and the film will consist of grains of very small size and arbitary orientation.

When the deposition rate is raised, the density of particles in the adsorption layer increases correspondingly; if the adsorbed atoms represent supersaturation only slightly exceeding the critical supersaturation, the critical dimensions of a nucleus become large, and the coefficient for the elasticity of the torsional vibrations around the x and y axes is large, while ν also is large:

$$\nu_y = \nu_x = \frac{1}{2\pi} \sqrt{\frac{\varkappa_x}{I_x}} \,,$$

(20)

where $\varkappa_x$ is the elastic coefficient for the x axis and I_x is the moment of inertia about that axis. One gets a texture if the frequency satisfies $h\nu \gg kT$.

Such complexes take up torsionally excited states in response to thermal motion only if $\varkappa_x$ decreases, which requires the complex to recede somewhat from the substrate. However, in view of the large size of the complex, the resulting increase in the potential energy is not compensated by the entropy part of the free energy, and so under these conditions most of the resulting complexes are in a state of zero torsional vibration, which corresponds to texture formation.

When the deposition rate is increased further, there is a rise in the supersaturation of the adsorbed material, and the size of the critical nuclei is correspondingly reduced; this means that the elimination of the degeneracy from the torsional oscillations is easier, i.e., a complex needs to recede less far from the substrate. Also, removal to the same distance from the substrate involves less increase in the energy of a complex in view of the smaller size. Therefore, if a complex is sufficiently small in size, the decrease in the free energy on freeing the torsional oscillations as a result of increase in the entropy is larger than the increase in the potential energy and the field of the adsorption forces as the distance from the substrate increases somewhat. The state in which the torsional oscillations are degenerate is not favored by energy. Lifting of the degeneracy leads to disorientation of the complexes, which corresponds to absence of a texture.

Similar arguments can be put forward for the change in deposition conditions with temperature at a constant deposition rate. Imagine for a certain substrate temperature and deposition rate that a texture is formed; the texture is lost on raising the temperature on account of transition to the excited state of torsional oscillation for the critical nuclei, with increased role for the entropy. On the other hand, the absence of a temperature on reducing the substrate temperature arises from cessation of nucleation on account of marked reduction in the rate of surface diffusion.

We now pass to qualitative analysis of single-crystal film formation on a single-crystal substrate, with explanation of the close correspondence of the deposition conditions needed for formation of a single-crystal film on a single crystal.

There is no appreciable effect from the relationship between the lattice parameters of the film and substrate as regards film formation when the deposition conditions are appropriate, which indicates that in these cases the periodic field of the single-crystal substrate plays only a small part. The basic part is played by the comparatively slowly decreasing van der Waals forces, which retain the atoms near the substrate. The effects of the single-crystal substrate on the orientation of the complexes are determined not by the field of the substrate but by geometrical features such as steps and reentrant angles. The free energy of a complex of critical size or larger is only slightly dependent on the orientation on account of the bulk character of the adsorption forces.

If the complexes were brought up to the substrate from outside, it is clear that the single-crystal substrate could not cause it to become oriented, and so the film would be polycrystalline, or at best textured. But in fact we have to deal with complexes that grow by the substrate itself. In the first stage of development, before a nucleus is formed, a complex consists of only a few atoms, and the role of the substrate is not restricted to merely retaining the atoms. When complexes are formed around steps, and particularly around reentrant angles, there is a reduction in the energy barrier to formation of complexes of a given size and (especially) of a given orientation. The probability of formation for a given size is proportional to $(-F_n/kT)$, where F_n is the free energy of a complex, so reduction in the energy of a complex to a quantity comparable with kT results in considerable probability of formation. The potential energy of the adsorbed atoms in the field of the adsorption forces is small by comparison with the binding energies of one adsorbed atom to another, but on the other hand it is considerably larger than the energy of thermal motion. The potential energy of an atom binding to the substrate should be larger in a reentrant angle by about a factor 2, so the probability of complex formation in such an angle is much larger than that on a smooth surface.

For purely geometrical reasons, such a growing complex takes up the shape of the reentrant angle and orients itself on it; in other words, the oriented disposition of the epitaxial partners is such as to produce the number of common or similar symmetry elements.

The decisive part is played by the geometrical structure of the substrate, not by the periodicity of the field at the substrate, which is confirmed by the fact that prior chemical polishing of the substrate, which removes large steps and reentrant angles, results not in a single-crystal film but a texture.

One reservation needs to be made when emphasizing the decisive role of the geometrical structure of the surface, and the unimportant role of the periodic field at the surface, which implies also an unimportant part for the correspondence of lattice parameters of adsorbent and adsorbate. Any discrepancy in lattice parameters makes itself felt in that the crystalline complexes on fusion will then give rise to stresses that will affect the lattice constant of the growing phase [12].

We now need to consider why similar deposition parameters are required to produce a texture or a single-crystal film correspondingly on amorphous and single-crystal substrates. The previous arguments show that the basic condition for texture formation is a large critical size for a complex, i.e., large size for the nuclei, which does not allow the complexes to pass to excited levels of torsional vibrations relative to axes parallel to the substrate. If the deposition rate is raised to the levels at which a polycrystalline deposit is formed, one gets a reduction in the critical size of the complexes. If the crystallization is performed under conditions approaching equilibrium, the critical dimensions of a nucleus should be dependent on the physicochemical nature of the substrate; in fact, the nature of the adsorbent and of the adsorbate will affect the potential of the adsorption forces, and this potential governs the lifetime of the atoms in the adsorption layer, as (11) shows, while the lifetime governs the surface density, as (19) shows, and hence the supersaturation in the adsorption layer. Therefore, at first sight it appears that the nature of the substrate should determine the critical dimensions of the nuclei and hence influence the parameters at which the texture or single-crystal film is formed. However, one must bear in mind that usually deposition is performed under conditions very far from equilibrium, which means that the lifetime of an atom in the adsorption layer is much greater than the time necessary for attachment of the atom to a growing complex. Under these conditions, the lifetime of an atom in the adsorption layer hardly influences the surface density and hence the supersaturation. This shows why the crystallization conditions for a texture are independent of the nature of the substrate.

Literature Cited

1. P. W. Pashley. Adv. Phys., 14:327 (1965).
2. D. S. Campbell and P. J. Stirland. Phil. Mag., 9:100 (1964).
3. R. F. Adamsky and R. E. Leblanc. J. Vacuum Sci. and Technol., 2:79 (1965).
4. J. W. Matthews. Appl. Phys. Letters, 7(5):131 (1965).
5. A. G. Zhdan, R. N. Sheftal', M. E. Chugunov, and M. I. Elinson. Radiotekhnika i Elektronika, 9(8):1536 (1966).
6. Z. A. Magomedov and S. A. Semiletov. Kristallografiya, 12(3):536 (1967).
7. R. Ueda and T. Inizuka. In: Growth of Crystals, Vol. 8, Consultants Bureau, New York (1969), p. 171.
8. R. N. Sheftal'. Kristall und Technik, 3(4):65 (1968).
9. G. Weaver. Physik und Technik von Sorptions und Desorptions (1963).
10. Ya. I. Frenkel'. Selected Works [in Russian], Vol. 2, Izd. AN SSSR, Moscow (1958), p. 239.
11. W. K. Burton, N. Cabrera, and F. C. Frank, Phil. Trans. Roy. Soc. London (1951), p. 243, 299.
12. R. N. Sheftal'. Kristall und Technik, 6(5):659 (1971).

CRYSTALLITE ORIENTATION IN ELECTRODEPOSITION
OF METALS

N. A. Pangarov*

Introduction

Many of the physicochemical properties of electrolytically deposited metals are determined by the crystal structure of the deposit; by the latter is usually meant the type of lattice for the metal, the crystallite size, the mutual disposition of the crystallite, and the orientation relative to the cathode substrate, as well as the type of lattice defect, the nature of any impurities, and so on.

Electrodeposition is widely used, and it requires a detailed study of the relation of properties to crystal structure; many researches have been made to elucidate the causes of particular structures. Measurements have been made on various modifications of electrodeposited metals, including the grain size, grain shape, coating luster, attachment to the substrate, and defect production.

At present, much attention is being given to the crystal structure of electrodeposited metals, especially mechanical, magnetic, and electrochemical features; particular attention is given to the effects of preferential crystallite orientation on the properties.

By preferential crystallite orientation is meant the mutual disposition of the grains relative to the substrate, with a particular crystallographic direction [hk$l$] perpendicular to the substrate for most of the grains. This direction is considered as the axis of preferential orientation.

Various terms have been used in describing crystallite orientation, such as fibrotexture, texture, epitaxy of the first kind, and one-degree orientation. In the last case, the term emphasizes the fact that the crystallites may be rotated around the preferential orientation axis at any angle, in distinction from cases where the substrate shows an epitaxial effect, where there is orientation on two axes.

Very often one finds the term degree of preferential orientation, which defines the proportion of crystallite having a given orientation axis relative to the substrate.

Theory of Crystallite Orientation in

Electrodeposition

I have shown [1] that the theory of internal stresses and the theory of geometrical selection are not able to explain all the observed trends in crystallite orientation; one needs to de-

─────────
*Bulgarian Academy of Sciences.

fine some other basic parameter that governs the preferential orientation axis. Also, a new theory is required not only to explain the existing evidence, but also to predict the conditions under which one expects crystallite orientation on electrodeposition using new axes. The theory would have to explain why one sometimes get orientation on two axes and how the substrate affects the degree of preferential orientation. Such a theory has been proposed at the Institute of Physical Chemistry of the Bulgarian Academy of Sciences on the basis of work by Stranski and Kaishev [2] and Kaishev and Bliznakov [3]. Pangarov and Rashkov [4] pointed out that the overvoltage for metal deposition is the basic factor governing the preferred orientation axis; the overvoltage is equivalent to the supersaturation in electrolytic deposition, and it determines the type of two-dimensional nucleus formed under appropriate deposition conditions. Then the basic concept of the theory is that a two-dimensional nucleus determines the preferential orientation axis, which does not alter in subsequent growth of the three-dimensional crystal. The type of coating growth (axial, lateral, or any other) is also dependent on the two-dimensional nucleation.

Volmer [5] gave the two-dimensional nucleation rate as proportional to $e^{-\frac{W_{hkl}}{kT}}$, where W_{hkl} is the energy of two-dimensional nucleation, k is Boltzmann's constant, and T is absolute temperature. Under these conditions, the nucleation rate will be highest for nuclei having the least W_{hkl}.

Calculation of the Two-Dimensional Nucleation Energy for a

Face-Centered Cubic Lattice on an Indifferent Substrate

Figure 1 shows the disposition of the lattice atoms and the energy needed to break the bonds between the first, second, third, and fourth neighbors; as an example we consider a two-dimensional nucleus of (110) type.

Figure 2 shows the disposition of the atoms in a (110) net; a is the lattice parameter, and the minimum distance between lattice atoms is $\delta = a/\sqrt{2}$. In the calculations we take into account the interaction between the first, second, third, and fourth neighbors, which lie respectively at distances of

$$\frac{a}{\sqrt{2}}, a, a\sqrt{\frac{3}{2}} , \text{ and } a\sqrt{\frac{5}{2}} .$$

The energy of two-dimensional nucleation is [6]

$$W_{hkl} = \frac{1}{2}\sum_i \varkappa_i L_i. \tag{1}$$

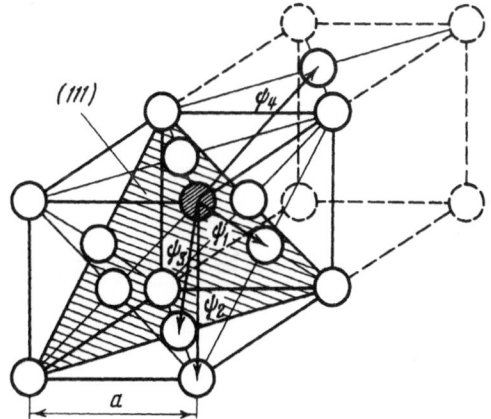

Fig. 1. Face-centered cubic lattice (parameter a).

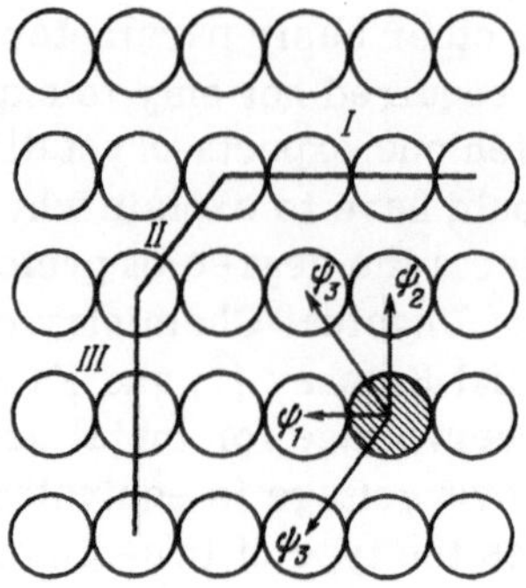

Fig. 2. Two-dimensional net on a (110) face of a
face-centered cubic lattice.

Here $\varkappa_i$ is the specific energy of edge i, whose length is L_i; equilibrium between the vapor and a two-dimensional nucleus is dependent on the mean energy required to detach an atom from a peripheral edge Φ, which may be calculated from the evaporation of peripheral atoms. At equilibrium, the energies of evaporation for all the peripheral rows in a two-dimensional nucleus should be identical.

The shape of a two-dimensional nucleus is determined as follows: the peripheral atoms must be linked in such a way that the minimum energy of detachment for each of the atoms should be at least equal to the mean work of detachment.

In the case of a face-centered cubic lattice, the energy needed to detach one atom from a position on a half crystal in (110) plane is

$$\Phi_{110}^0 = \psi_1 + \psi_2 + 2\psi_3 + \psi_0.$$

Here ψ_1 is the bond energy between two first neighbors, ψ_2 is the same for second neighbors, ψ_3 is the same for third neighbors, and ψ_0 is the energy required to detach one atom from the substrate.

The equilibrium shape of a two-dimensional nucleus will be an octagon with two edges of type I, four edges of type II, and two edges of type III (Fig. 2).

The following is the mean energy to detach one atom from a type I edge, as calculated from the evaporation of atoms from this edge:

$$\overline{\Phi}_{I} = \frac{n_I \psi_2 + 2 n_I \psi_3 + (n_I - 1)\psi_1 + n_I \psi_0}{n_I} = \Phi_{110}^\circ - \frac{\psi_1}{n_I} \ .$$

Similarly we get that

$$\overline{\Phi}_{11} = \Phi_{110}^\circ - \frac{\psi_3}{n_{11}} \ ; \quad \overline{\Phi}_{111} = \Phi_{110}^\circ - \frac{\psi_2}{n_{111}} \ .$$

At equilibrium

$$\overline{\Phi}_1 = \overline{\Phi}_{11} = \overline{\Phi}_{111}.$$

The number n_i of atoms in the peripheral rows may be calculated from the following equation in the form given by Stranski and Kaishev [2]:

$$\frac{1}{mN}(\mu - \mu_{hkl}) = \Phi_{hkl}^\circ - \overline{\Phi}, \tag{2}$$

where μ is the chemical potential of the vapor in equilibrium with the nucleus, μ_{hkl} is the chemical potential of the vapor in equilibrium with an infinitely large (hk$l$) net, m is the number of atoms in a gas molecule, and N is Avogadro's number.

We get for the numbers of atoms in peripheral rows I, II, and III that

$$n_I = \frac{\psi_1}{\frac{1}{mN}(\mu - \mu_{110})} \; ; \quad n_{II} = \frac{\psi_3}{\frac{1}{mN}(\mu - \mu_{110})} \; ; \quad n_{III} = \frac{\psi_2}{\frac{1}{mN}(\mu - \mu_{110})} \; .$$

Correspondingly, we get the lengths of the peripheral rows as

$$L_I = n_I \frac{a}{\sqrt{2}} = \frac{a \cdot \psi_1}{\frac{\sqrt{2}}{mN}(\mu - \mu_{110})} \; ; \quad L_{II} = \frac{a\sqrt{3}\,\psi_3}{\frac{\sqrt{2}}{mN}(\mu - \mu_{110})} \; ; \quad L_{III} = \frac{a\psi_2}{\frac{1}{mN}(\mu - \mu_{110})} \; .$$

The specific edge energies $\varkappa_i$ for rows I, II, and III are calculated via the mean work of detachment and are correspondingly

$$\varkappa_I = \frac{\psi_2 + 2\psi_3}{2a/\sqrt{2}} \; ; \quad \varkappa_{II} = \frac{\psi_1 + \psi_2 + 2\psi_3}{2a\sqrt{\frac{3}{2}}} \; ; \quad \varkappa_{III} = \frac{\psi_1 + 2\psi_2}{2a} \; .$$

We substitute for $\varkappa_i$ and L_i in (1) to get

$$W_{110} = \frac{B_{110}}{\frac{1}{mN}(\mu - \mu_{110})} \; , \tag{3}$$

where $B_{110} = \psi_1\psi_2 + 2\psi_1\psi_3 + 2\psi_2\psi_3 + 2\psi_3^2$; expression (3) can also be written as

$$W_{110} = \frac{B_{110}}{\frac{1}{mN}(\mu - \mu_0) - \frac{1}{mN}(\mu_{110} - \mu_0)} \; ,$$

where μ_0 is the chemical potential of the vapor in equilibrium with an infinitely large three-dimensional crystal. However,

$$\frac{1}{mN}(\mu_{110} - \mu_0) = \Phi^\circ - \Phi^\circ_{110} = C_{110} - \psi_0,$$

$$\Phi^\circ = 6\psi_1 + 3\psi_2 + 12\psi_3 + 12\psi_4,$$

where Φ^0 is the energy needed to detach one atom from a position on a half-crystal having a face-centered cubic lattice in the three-dimensional case. Then we get the following expression for the constant C_{110}:

$$C_{110} = 5\psi_1 + 2\psi_2 + 10\psi_3 + 12\psi_4.$$

Then the following form can be given to W_{hkl} for any type of two-dimensional nucleus:

$$W_{hkl} = \frac{B_{hkl}}{\frac{1}{mN}(\mu - \mu_0) + \psi_0 - C_{hkl}} \; . \tag{4}$$

For electrolytic deposition on a metal, the μ and μ_0 of (4) can be replaced by the electrochemical potentials $\mu*$:

$$\mu^* = \mu + zF\varphi,$$

TABLE 1

Energy constant	Face				
	(111)	(100)	(110)	(115)	(210)
φ°_{hkl}	$3\psi_1 + 3\psi_3 + \psi_0$	$2\psi_1 + 2\psi_2 + 4\psi_4 + \psi_0$	$\psi_1 + \psi_2 + 2\psi_3 + \psi_0$	$\psi_1 + 2\psi_3 + 2\psi_4 + \psi_0$	$\psi_1 + 2\psi_3 + \psi_0$
W_{hkl}	$3L_\perp \varkappa_\perp + 3L_\parallel \varkappa_\parallel$	$2L_\perp \varkappa_\perp + 2L_\parallel \varkappa_\parallel$	$L_\perp \varkappa_\perp + 2L_\parallel \varkappa_\parallel + L_{\parallel\parallel}\varkappa_{\parallel\parallel}$	$L_\perp \varkappa_\perp + 2L_\parallel \varkappa_\parallel + L_{\parallel}\varkappa_{\parallel}$	$L_\perp \varkappa_\perp + 2L_\parallel \varkappa_\parallel$
$\varkappa$	$\varkappa_\perp = \dfrac{2\psi_1 + 4\psi_3}{2}\dfrac{a}{\sqrt{2}}$ $\varkappa_\parallel = \dfrac{4\psi_1 + 6\psi_3}{2}\sqrt{\dfrac{3}{2}}\,a$	$\varkappa_\perp = \dfrac{\psi_1 + 2\psi_2 + 6\psi_4}{2}\dfrac{a}{\sqrt{2}}$ $\varkappa_\parallel = \dfrac{2\psi_1 + 2\psi_2 + 8\psi_4}{2a}$	$\varkappa_\perp = \dfrac{\psi_2 + 2\psi_3}{2}\dfrac{a}{\sqrt{2}}$ $\varkappa_\parallel = \dfrac{\psi_1 + \psi_2 + 2\psi_3}{2}\sqrt{\dfrac{3}{2}}\,a$ $\varkappa_{\parallel\parallel} = \dfrac{\psi_1 + 2\psi_3}{2a}$	$\varkappa_\perp = \dfrac{2\psi_3 + 2\psi_4}{2}\dfrac{a}{\sqrt{2}}$ $\varkappa_\parallel = \dfrac{\psi_1 + 3\psi_3 + 3\psi_4}{2}\sqrt{\dfrac{5}{2}}\,a$ $\varkappa_{\parallel\parallel} = \dfrac{\psi_1 + \psi_3 + 2\psi_4}{2}\sqrt{\dfrac{3}{2}}\,a$	$\varkappa_\perp = \dfrac{\psi_3}{a}$ $\varkappa_\parallel = \dfrac{\psi_2 + \psi_3}{2}\sqrt{\dfrac{3}{2}}\,a$
B_{hkl}	$3\psi_1^2 + 12\psi_1\psi_3 + 9\psi_3^2$	$\psi_1^2 + 4\psi_1\psi_2 + 6\psi_1\psi_4 + 8\psi_2\psi_4 + 2\psi_4^2$	$\psi_1\psi_2 + 2\psi_1\psi_3 + 2\psi_2\psi_3 + 2\psi_3^2$	$\frac{3}{2}\psi_1\psi_3 + 2\psi_1\psi_4 + 4\psi_3\psi_4 + \frac{1}{2}\psi_3^2 + 3\psi_4^2$	$2\psi_2\psi_3 + \psi_3^2$
$B_{hkl} = f(\psi_1)$	$3.45\,\psi_1^2$	$1.587\,\psi_1^2$	$0.211\,\psi_1^2$	$0.074\,\psi_1^2$	$0.0106\,\psi_1^2$
C_{hkl}	$3\psi_1 + 3\psi_3 + 9\psi_3 + 12\psi_4$	$4\psi_1 + \psi_2 + 12\psi_3 + 8\psi_4$	$5\psi_1 + 2\psi_2 + 10\psi_3 + 12\psi_4$	$5\psi_1 + 3\psi_2 + 10\psi_3 + 10\psi_4$	$6\psi_1 + 2\psi_2 + 10\psi_3 + 12\psi_4$
C_{hkl}	$3.804\,\psi_1$	$4.633\,\psi_1$	$5.716\,\psi_1$	$5.825\,\psi_1$	$6.716\,\psi_1$

$\psi^\circ = 6\psi_1 + 3\psi_2 + 12\psi_3 + 12\psi_4$

$\psi_1 = 1 \qquad \psi_2 = 0.125\,\psi_1 \qquad \psi_3 = 0.037\,\psi_1 \qquad \psi_4 = 0.008\,\psi_1$

where φ is electrode potential, z is ion charge, and F is Faraday number. Then (4) becomes

$$W_{hkl} = \frac{B_{hkl}}{ze_0(\varphi - \varphi_0) + \psi_0 - C_{hkl}} \,. \tag{5}$$

Here φ is the electrode potential at a certain current density, φ_0 is the equilibrium electrode potential, e_0 is the electron charge, and z is ion charge.

We see from (5) that the overvoltage $\eta = \varphi - \varphi_0$ determines the energy of two-dimensional nucleation.

In a similar way we can calculate the energies of formation for nuclei of the two-dimensional (111), (100), (113), and (210) grids of a face-centered cubic lattice; Table 1 gives the results.

Constants B_{hkl} and C_{hkl} can be calculated in relative terms using the bond energy between first neighbors as unit; in the most general case, $B_{hkl} = b_{hkl}\,\psi_1^2$ and $C_{hkl} = c_{hkl}\psi_1$, where b_{hkl} and c_{hkl} are constants whose values can be found from the change in the bond energy with distance; then one can represent W_{hkl} as a function of $(1/mN)(\mu - \mu_0) + \psi_0$, which is measured in the relative units mentioned above.

Figure 3 shows W_{hkl} for (111), (100), (110), (113), and (210) two-dimensional grids on a face-centered cubic lattice as a function of $(1/mN)(\mu - \mu_0) + \psi_0$, or, which is the same, as a function of $ze_0\eta + \psi_0$; it has been assumed that the bond energies between second, third, and fourth neighbors decrease as the sixth power of the distance. The dotted lines on the figure denote the asymptotes for the functions when the denominators in (4) and (5) tend to zero. It is clear that the (111) octahedral face gives the lowest nucleation energy at small supersaturations; as the supersaturation increases, the energy successively becomes least for (100), (110), (113), and (210).

Application of the Theory

The nucleation energy as a function of supersaturation shows how the orientation axes of the crystallites will vary in electrolytic deposition on structureless substrates as the overvoltage is increased. We get the following series:

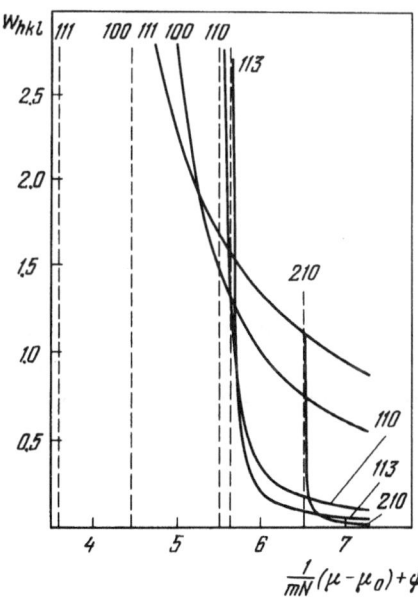

Fig. 3. Energy W_{hkl} for two-dimensional nucleation for a face-centered cubic lattice as a function of $(1/mN)(\mu - \mu_0) + \psi_0$ in units of ψ_1.

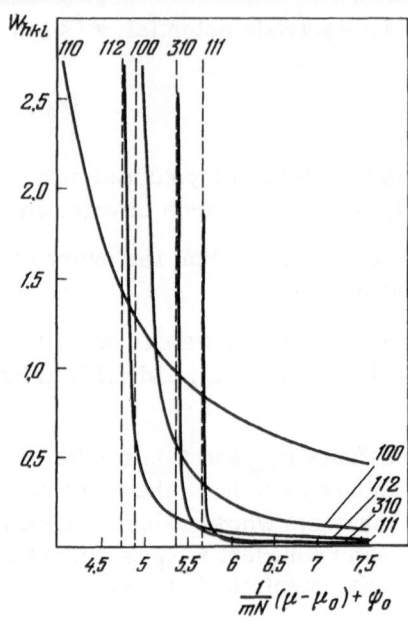

Fig. 4. Energy W_{hkl} for two-dimensional nucleation for a body-centered cubic lattice as a function of $(1/mN)(\mu - \mu_0) + \psi_0$ in units of ψ_1.

For a face-centered cubic lattice:

[111], [100], [110], [113], and [210] (Fig. 3)

For a body-centered cubic lattice:

[110], [112], [310], and [111] (Fig. 4)

For a hexagonal close-packed lattice:

[0001], [10$\bar{1}$1], [11$\bar{2}$0], [10$\bar{1}$0], and [11$\bar{2}$2] (Fig. 5)

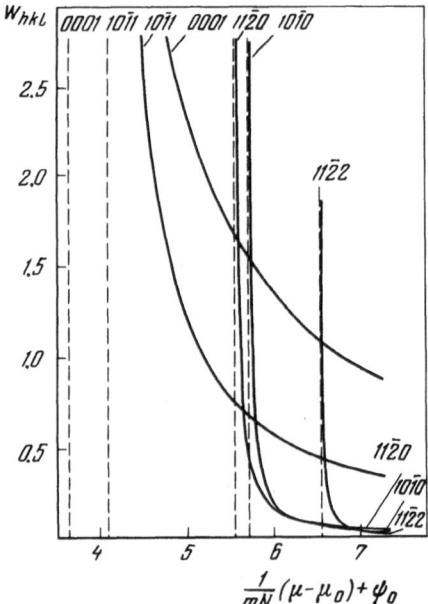

Fig. 5. Energy W_{hkl} for two-dimensional nucleation for a hexagonal close-packed lattice as a function of $(1/mN)(\mu - \mu_0) + \psi_0$ in units of ψ_1.

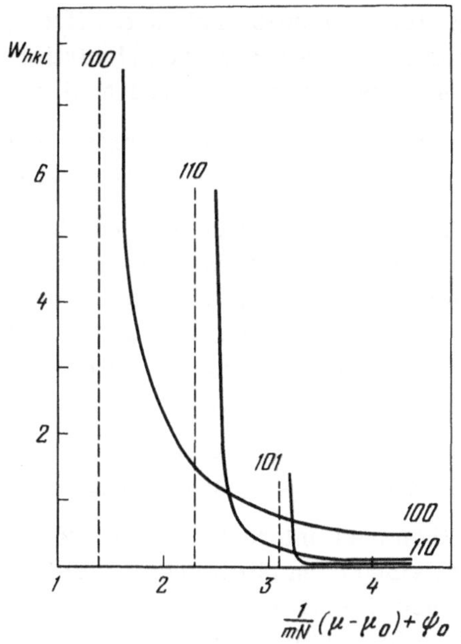

Fig. 6. Energy W_{hkl} for two-dimensional nucleation for a tetragonal body-centered lattice as a function of $(1/mN)(\mu - \mu_0) + \psi_0$ in units of ψ_1.

For the tetragonal lattice of tin:

$$[100], [110], \text{ and } [101] \text{ (Fig. 6)}$$

The theory of two-dimensional nucleation predicts some new orientation axes that have not so far been observed; for example, a [310] axis for body-centered cubic metals [7], [10$\bar{1}$1] for hexagonal close-packed metals [8], and [101] for the tetragonal lattice of tin [9].

Figure 4 shows that a [310] axis should appear over a very narrow range in supersaturation, while for tin (Fig. 6) the crystallites can take up a [101] orientation only at high overvoltages.

These theoretical curves enable one to predict under which conditions one will obtain coatings with crystallites preferentially oriented on two axes. Figures 3-6 show that this is possible only with certain strictly defined overvoltages, where the nucleation energy is the same for two faces. For instance, for the tetragonal lattice of tin (Fig. 6) one can get simultaneous orientations on [100] and [110] axes or [110] and [101] axes, whereas orientations on [100] and [101] are impossible.

Study of Orientation of Silver Nuclei on a

Platinum Substrate [10] by a Pulse

Potentiostatic Method

Figure 3 shows for a face-centered cubic lattice that the curves for W_{hkl} as a function of overvoltage lie close together at moderate overvoltages, which hinders experimental tests, especially when adsorption of various substances appreciably influences nucleation. The following two factors also seriously hinder research in experiments with face-centered metals:

1. Metals such as nickel deposit at high overvoltages, and hydrogen is then released, which prevents one from examining the effect in pure form.

2. Metals deposited at low overvoltages, such as silver, do not readily allow of electrolysis at high overvoltages, since dendrites grow or else one obtains spongy coatings inconvenient for x-ray and electron-diffraction study.

A study has been made with silver [11] of the orientation on the [111] axis, while high current densities have [12] given orientation on [100]. In these cases the silver was deposited from solutions of cyanides and 1 N silver nitrate.

A study has also been reported [13] of silver dendrite orientation; this occurs on [110] or [210] axes, which completely agrees with our theory of preferential orientation.

These studies are of isolated character and do not enable one to explain why the orientation axis of silver alters with the overvoltage; therefore, it was of interest to study the orientation of silver crystallites on structureless substrates over a wide overvoltage range. To avoid dendrite formation, one has to examine the initial stages of deposition, where one obtains only a small number of nuclei, which is feasible only if a pulse method is used.

Our cell was modified slightly from that described in [14]. Figure 7 shows the cell, in which the cathode 3 is a platinum wire of diameter 0.2 mm sealed into a glass tube; the surface of the electrode was equal to the cross section of the wire: 0.126 mm^2.

When such microelectrodes are used, it is always difficult to obtain a sufficiently smooth surface, since even microscopic defects can act as crystallization centers; also, the platinum in our case was not polished by any of the mechanical polishing methods described in the literature for metals. The most suitable way of polishing was by means of chromic anhydride, chromium trioxide, jewelers' rouge, and sodium carbonate deposited from an aqueous suspension. The substrate for the mechanical polishing was either leather or a leather substitute. The final stage of polishing was ac electropolishing. After this treatment, the surface revealed no defects on examination at ×250 under the microscope. The anode 5 was a single-crystal silver ring placed symmetrically around the cathode space. The wire was observed with an MIM-7 microscope at magnifications from 90 to 140.

The silver was deposited from a 6 N solution of silver nitrate containing 0.5 N nitric acid, the electrolyte volume being 5 ml; any impurities in the solution would interfere with examination of the pure effect, so special measures were taken to purify it. The silver nitrate was of analytical grade and was recrystallized repeatedly; the solution was made up with double-distilled water. The cell before use was treated with a hot mixture of sulfuric and nitric acids. The final purification of the solution was performed by preliminary electrolysis; as the cathode in that case we used a platinum spiral of high surface area.

The silver nuclei were produced by a pulse method; first of all, a steady potential η_1 of 5-7 mV relative to the silver anode was applied (Fig. 8); this was insufficient to produce

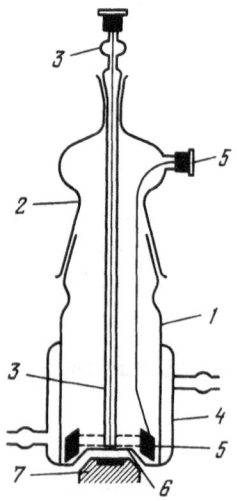

Fig. 7. The electrolysis cell: 1) body; 2) upper part with ground joint for electrode; 3) platinum wire electrode; 4) thermostatic jacket; 5) silver ring anode; 6) plane-parallel glass; 7) microscope.

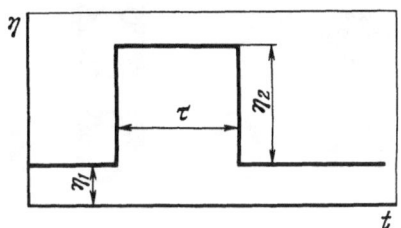

Fig. 8. Overvoltage waveform for microelec- ·
trode: η_1 initial potential; η_2 overvoltage pulse;
τ pulse length.

nuclei but adequate to cause existing nuclei to grow. Then a single rectangular pulse η_2 was applied, which produced nuclei, which subsequently grew in response to η_1.

Pulses have been used on a number of occasions to produce nuclei as in [15, 16], where a pulse generator was employed to produce silver and mercury nuclei. A pendulum has been used [17] with mechanical contacts to produce electrical pulses for generating mercury nuclei on platinum. In that case, it was found that the available electrical oscillator did not give sufficiently reproducible pulses, whereas the mechanical generator gave good pulses, but did not provide adequate voltage stability during the pulse.

We used an electronic pulse generator in conjunction with a pulse potentiostat [18], which was much better than the above generators as regards pulse stability; a single square pulse was reproduced with an accuracy of 1 mV/V, which was applied to the input of an IP-410 pulse potentiostat, whose response time was less than a microsecond. The total rise time of the pulse was less than 2.5 μsec, which was comparable with the charging time of the double layer capacitance. The pulse lengths were 5-200 msec in our tests. Figure 9 shows the block diagram of the electrical part of the equipment. All measurements were made at 25°C. The nucleation and growth were followed continuously in the microscope and by photography; the most characteristic results were recorded by cinematography.

After each run, an anode potential of 1 V was applied to the electrode, which dissolved the nuclei and cleaned the electrode.

Figure 10 shows the number of nuclei N as a function of pulse length for various overvoltages; the number increases linearly with time in the range 5-200 msec. Appropriate pulse

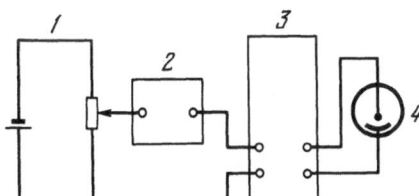

Fig. 9. Electrical system: 1) base potential source; 2) pulse generator; 3) potentiostat; 4) cell.

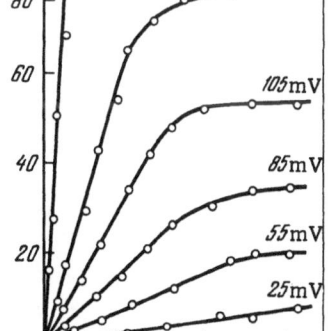

Fig. 10. Number of nuclei as a function of pulse length.

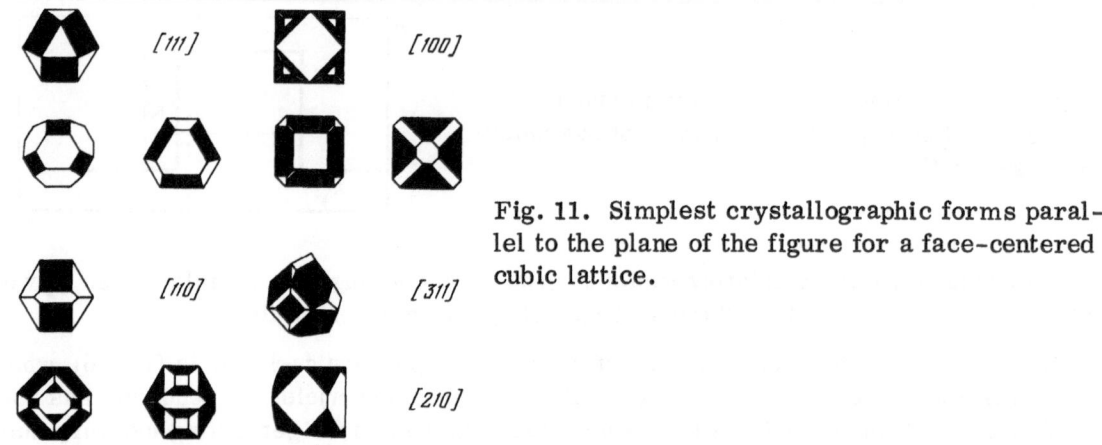

Fig. 11. Simplest crystallographic forms parallel to the plane of the figure for a face-centered
cubic lattice.

length reduction at a given overvoltage enables one to obtain isolated single nuclei, whose
orientation may be determined after growth to an appropriate size. This forms the basic
technique in our experiments.

Figure 11 shows the simplest possible forms for the crystals with various orientation
axes on the assumption that the substrate is perpendicular to the plane of the figure.

Adjustment of the overvoltage and pulse length produced isolated or grouped nuclei with
various orientations; the overvoltage range for a given orientation was that in which the fre-

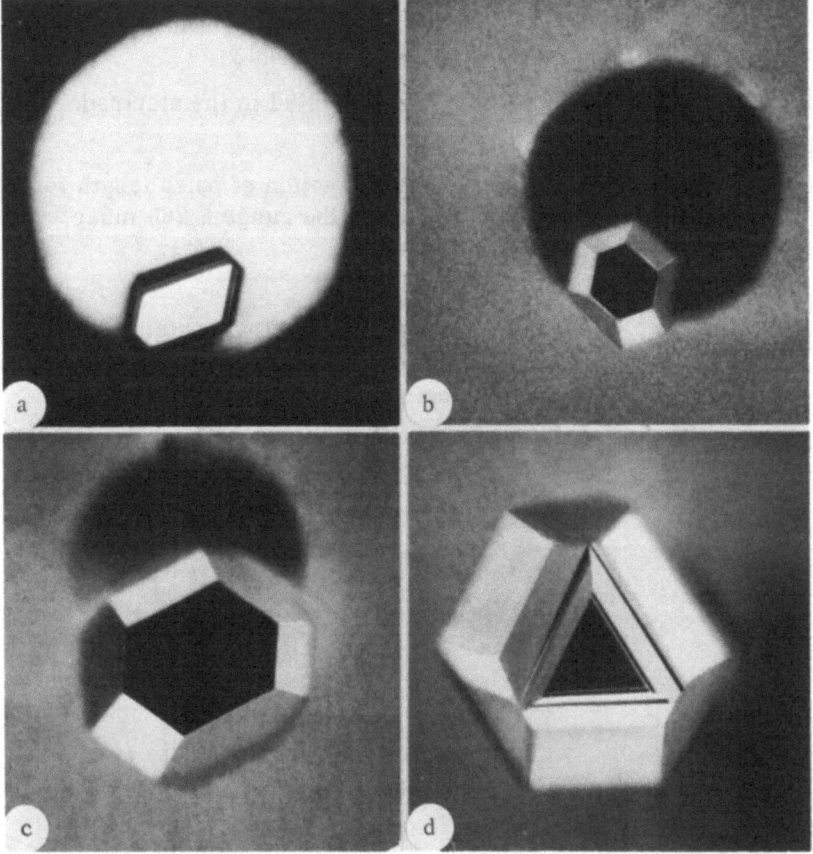

Fig. 12. Silver crystals with [111] orientation axes at various
growth stages.

Fig. 13. Silver crystals with [100] orientation axes at various
growth stages.

quency of single nuclei with a given orientation was over 90%, i.e., not less than 90% of the nuclei had the orientation on the appropriate axes.

We can give the following form to the results from numerous experiments on the orientation of silver nuclei:

(a) In the range 20-50 mV, the preferred orientation is [111] perpendicular to the substrate (Fig. 12);

(b) In the range 60-90 mV, the orientation axis is [100] (Fig. 13);

(c) [110] in the range 120-170 mV (Fig. 14);

(d) [113] in the range 180-200 mV (Fig. 15);

(e) [20] above 200 mV (Fig. 16).

These orientation axes follow very precisely in particular overvoltage ranges the sequence of relative minima in the nucleation energy for the corresponding faces shown in Fig. 3; one gets nuclei with orientation axes characteristic of two adjacent overvoltage ranges in regions where the theoretical curves intersect.

Figure 17 shows such a case for the range 50-60 mV, where there are identical frequencies for nuclei with [111] and [100] orientation axes.

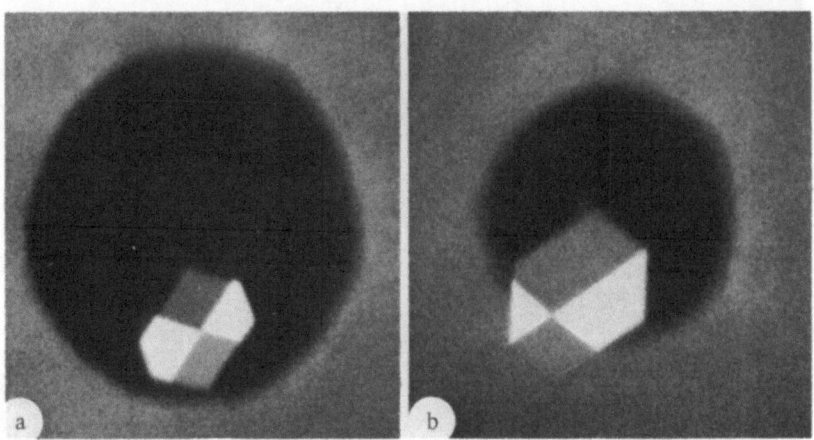

Fig. 14. Silver crystals with [110] orientation axes at various
growth stages.

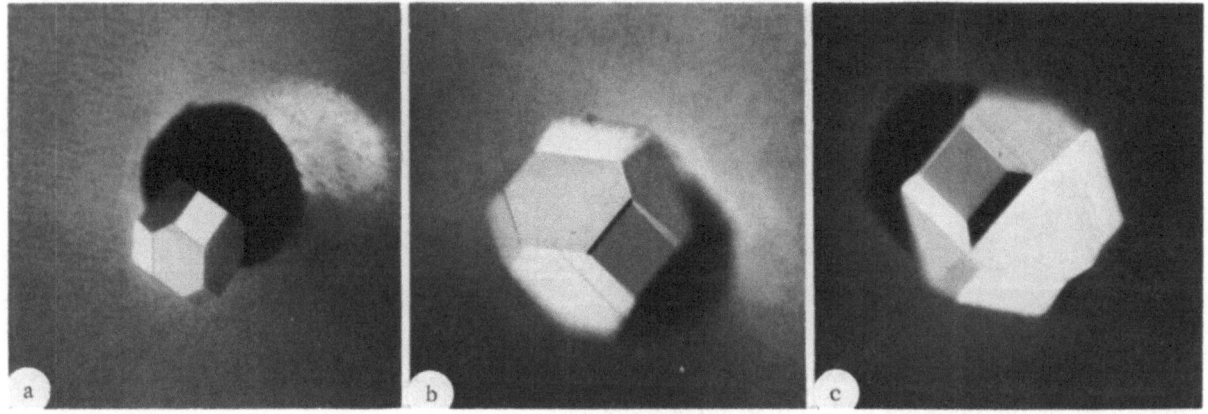

Fig. 15. Silver crystals with [113] orientation axes at various growth stages.

Fig. 16. Silver crystal with [210] orientation
axes.

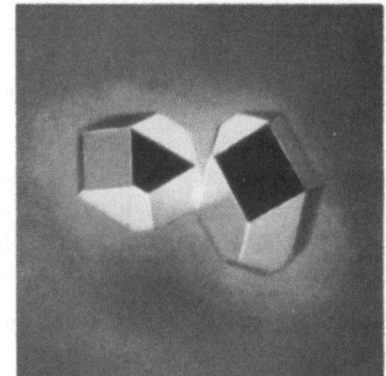

Fig. 17. Silver crystal oriented on [111] and
[100] produced at η = 55 mV.

These results show clearly that an indifferent electrode (platinum at +0.7 V relative to silver) produces an orientation for the initial nuclei determined in the main by W_{hkl}, which is a function of the overvoltage.

This study of initial nucleus orientation provides new scope for examining the effects of adsorption on nucleation, as well as on electrocrystallization generally.

Calculation of ψ_1 for Silver from the Preferred Orientation Axes [19]

The overvoltage ranges for particular texture axes enable one to calculate ψ_1 not only for crystallization from the vapor but also for electrodeposition.

Figure 18 shows W_{hkl} as a function of $(1/mN)$ $(\mu - \mu_0) + \psi_0$ for the (111), (100), and (110) faces of a face-centered cubic lattice; the method described above enables one to determine P_1 and P_2, which correspond to equally probable formation of crystals with [111] and [100] axes or with [100] and [110] axes. Then we can put

$$\frac{1}{mN}(\mu'' - \mu_0) + \psi_0 - \left[\frac{1}{mN}(\mu' - \mu_0) + \psi_0\right] = (q_2 - q_1)\,\psi_1,$$

where μ' and μ'' are the chemical potentials for the vapor in equilibrium with the nuclei at points P_1 and P_2.

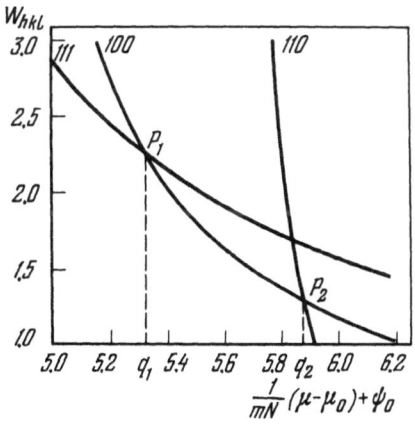

Fig. 18. Energy W_{hkl} for two-dimensional nucleation for a face-centered cubic lattice as a function of $(1/mN)(\mu - \mu_0) + \psi_0$ in units of ψ_1 for (111), (100), and (110) faces.

From this we get

$$\psi_1 = \frac{\frac{1}{mN}(\mu' - \mu'')}{q_2 - q_1} .$$ (6)

Knowing ψ_1 and assuming that ψ_2, ψ_3, and ψ_4 decrease as the reciprocal of the sixth power of the distance, one can calculate other quantities in the crystallization such as the specific surface energy σ_{hkl}, the two-dimensional nucleation energy on the same substrate W_{hkl}^k, and the number of atoms in a nucleus at a given supersaturation n_{hkl}^k.

One can determine ψ_0, the energy needed to detach a particle from a structureless substrate, from the limiting supersaturation at which the new crystalline phase begins to form on the substrate.

In this case we have

$$\psi_0 = C_{hkl} - \frac{1}{mN}(\mu - \mu_0).$$ (7)

We have shown above that μ can be replaced by μ^* in electrolytic deposition; then the formulas for ψ_1 and ψ_0 become

$$\psi_1 = \frac{ze_0\Delta\eta}{q_2 - q_1} ,$$ (8)

$$\psi_0 = C_{hkl} - ze_0\eta_0.$$ (9)

Here η_0 is the overvoltage at which the new phase begins to be deposited on a structureless substrate, while $\Delta\eta = \eta_2 - \eta_1$ is the difference in the overvoltages at P_1 and P_2.

We have found by experiment that nucleation of silver on platinum has P_1 corresponding to $\eta_1 = 0.055$ V, while P_2 corresponds to $\eta_2 = 0.125$ V; Figure 18 shows that $q_2 - q_1 = 0.53$ for a face-centered cubic lattice, and then ψ_1 for silver is

$$\psi_1 = 2.1 \pm 0.1 \cdot 10^{-13} \text{ erg}.$$ (10)

Calculation of the Specific Surface Energy
for (111), (100), and (110) Faces of Silver [19]

Table 1 gives B_{hkl} and C_{hkl} for the nuclei of types (111), (100), and (110) for a face-centered cubic lattice.

These values are quite sufficient to give σ_{hkl} for the various faces of silver, and also W_{hkl}^k for a given overvoltage, together with n_{hkl}^k.

The following is the formula for the nucleation energy:

$$W_{hkl}^k = \frac{B_{hkl}}{ze_0\eta} ,$$ (11)

since in this case $\psi_0 = C_{hkl}$

The number of atoms in a nucleus is found by dividing the nucleation energy by the detachment energy per particle:

$$n_{hkl}^k = \frac{B_{hkl}}{(ze_0\eta)^2} .$$ (12)

Table 2 summarizes the results.

TABLE 2

Face characteristic	(hkl)		
	(111)	(100)	(110)
σ_{hkl}, ergs/cm^2	570 ± 30	630 ± 32	670 ± 34
W_{hkl}^{k}, ergs for $\eta = 10$ mV	9.5×10^{-12}	4.38×10^{-12}	0.58×10^{-12}
h_{hkl}^{k} for $\eta = 10$ mV	590	270	36

The σ_{hkl} given in the table relate to vacuum, since we have neglected the effects of the electrolytes on the specific face energy; this has little effect in calculating ψ_1 and σ_{hkl}, since we have used the difference $\Delta \eta = \eta_2 - \eta_1$ corresponding to $\Delta q_2 = q_2 - q_1$; also, the final result is little affected by the choice of the power to which the distance is raised in ψ_2, ψ_3, and ψ_4. For instance, if we assume a seventh power of the distance instead of the sixth, the curves in Fig. 18 move to the left, but the difference $\Delta q = q_2 - q_1$, i.e., the relative disposition of P_1 and P_2, remains the same.

The values of W_{hkl}^{k} and n_{hkl}^{k} are given only for comparison, since we do not know how far B_{hkl}, or, which is the same, the specific edge energy $\varkappa_i$, alters in response to the electrolyte; for this reason, as one would expect, the values for n_{100}^{k} are higher than those given in [20], which were found under the same conditions for (100), where the number of atoms in a nucleus was 53.

There are reasons for assuming that this method for ψ_1 can be used in electrical deposition of metals and in research on thin films obtained by vacuum evaporation.

Twinning in Electrical Crystallization of
Face-Centered Cubic Metals

Calculation of the Twinning Probability [21]

Lattice defects are often the decisive factor governing the microstructure of electrically deposited face-centered cubic metals; twinning determines the morphology of the coating, and also many physicochemical properties, so research on twinning is an important aspect of theroetical and applied studies.

Twinning is a fairly common effect, and has repeatedly been observed, but researches on this are rather inadequate and often purely qualitative. Attention should be given to theoretical researches and to experiments on the conditions needed to obtain a single crystalline individual.

Geometrical considerations show that a twin is a regular mutual association of two crystals with a common crystallographic surface, when the orientation of one is a mirror representation of the orientation of another in the twin plane. The boundary between the two crystals is the twin boundary, and this coincides with the twin plane when the nearest neighboring atoms on the two sides of the twin boundary are symmetrically disposed. In that case one speaks of a coherent twin boundary, which is the type of boundary observed in twins obtained by electrical crystallization of metals and by deposition of metallic film by vacuum evaporation, and also in certain instances of plastic deformation.

Research on twinning in electrical crystallization began as soon as electron diffraction was applied to the structure of thin films of metals; in 1936 it was found [22] that electrolytically deposited nickel and cobalt on copper single crystals gave additional spots on electron-diffraction patterns, which were explained by twinning on (111) faces. The twin crystallites have also been found by others, as have deposition on the faces of single crystals and on poly-crystalline materials.

Kern [23] has performed a general classification of twins and the scope for formation of these in crystallization; his hypothesis about the twinning mechanism is that a defect arises on a face at a certain stage in growth, which acts as a nucleus in a twin position relative to the original crystal. General considerations have been that twin frequency is a rising function of the supersaturation, and that there is a critical supersaturation for the onset of twinning.

Calculations have been performed for the energy needed to produce a normal nucleus and a nucleus in a twin position on a (111) face in relation to supersaturation or overvoltage [21], the method of mean detachment energies [2] being employed.

Calculation of the Two-Dimensional Nucleation Energy for a (111) Face for the Normal Position. Equation (4) gives the value of W_{hkl} for any face, or alternatively (5) for the electrolytic deposition of a metal.

Nucleation on (111) is important for twinning on face-centered cubic lattices; an exact calculation shows (Table 1) that in this case B_{111} and C_{111} are respectively

$$B_{111} = 3\psi_1^2 + 12\psi_1\psi_3 + 9\psi_3^2;$$

$$C_{111} = 3\psi_1 + 3\psi_2 + 9\psi_3.$$

Here the nuclei are formed not on a structureless substrate but on their own material, i.e., a nucleus of (111) type is formed on a (111) face, and then

$$\psi_0 = 3\psi_1 + 3\psi_2 + 9\psi_3 = C_{111},$$

and (4) and (5) become

$$W_{111}^k = \frac{3\psi_1^2 + 12\psi_1\psi_3 + 9\psi_3^2}{\frac{1}{mN}(\mu - \mu_0)} , \qquad (13)$$

$$W_{111}^k = \frac{3\psi_1^2 + 12\psi_1\psi_3 + 9\psi_3^2}{ze_0\eta} . \qquad (14)$$

Calculation of the Two-Dimensional Nucleation Energy for a (111) Face in the Twin Position. When a nucleus is formed in the twin position, the energy ψ_0^t needed to detach one atom from the substrate differs from ψ_0 (that for the normal position), as Figures 19 and 20 show, by reference to an atom on a (111) face in the normal position and

Fig. 19. Normal position of an atom on a (111) face of a fcc lattice.

Fig. 20. An atom in a twin position on a
(111) face of a fcc lattice.

an atom in a twin position. In the latter case $\psi_0^t = 3\psi_1 + 3\psi_2 + 6\psi_3 + \psi'$, where ψ' is the bond energy between an atom in a twin position and and atom exactly underneath by one plane at a distance $2a/\sqrt{3}$. It is at once clear that the nucleation energy for the twin position W_{111}^t will be greater at a given saturation than that for a nucleus in the normal position:

$$W_{111}^t = \frac{3\psi_1^2 + 12\psi_1\psi_3 + 9\psi_3^2}{\frac{1}{mN}(\mu - \mu) - 3\psi_3 + \psi'},$$

(15)

Comparison of the Nucleation Energies for the Normal and Twin Positions. Equations (13) and (15) enable one to represent these nucleation energies as functions of supersaturation; the bond energy between atoms in the lattice is inversely proportional to the sixth power of the distance, so

$$\psi_2 = 0.125\,\psi_1, \quad \psi_3 = 0.037\,\psi_1, \quad \psi' = 0.053\psi_1.$$

Figure 21 shows $(1/mN)(\mu - \mu_0)$ along the abscissa, or, which is the same, $ze_0\eta$, with the unit the bond energy between first two neighbors ψ_1. Curve 1 relates to nuclei in the normal position, while curve 2 relates to the twin position. The energy for the twin position is larger at small supersaturations, but the two curves come together as the supersaturation increases.

The two-dimensional nucleation rate at a constant temperature is given by $V = \text{const}\,e^{-W_{hkl}}$, so from (13) and (15) one can calculate the nucleation probability for the twin position P as a function of supersaturation, which is

$$P = \frac{V^t}{V^t + V^n},$$

(16)

where V^t is the nucleation rate for the twin position and V^n is the same for the normal position.

Fig. 21. Energy W_{hkl} for two-dimensional
nucleation for a face-centered cubic lattice as
a function of $(1/mN)(\mu - \mu_0)$ in units of ψ_1 for
a (111) face: 1) normal nucleus; 2) nucleus
in twin position.

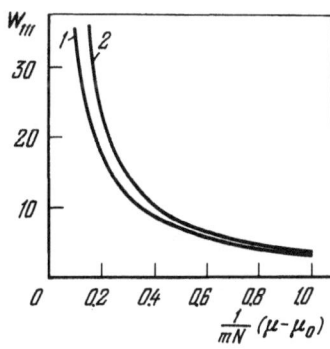

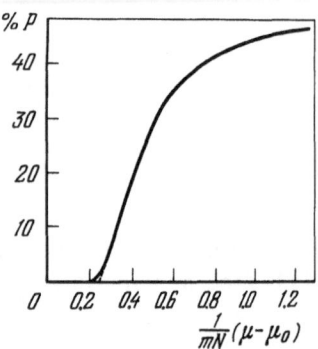

Fig. 22. Nucleation probability (%) for a two-dimensional nucleus in a twin position.

Figure 22 has the ordinate as the probability P in percent as a function of $(1/mN) (\mu - \mu_0)$.

The graph shows that nucleation in the twin position becomes possible above a certain supersaturation, and that the proportion of such nuclei then rises rapidly; the probability becomes equal to that of normal nucleation at high supersaturations.

There is only a single published study [24] of twin formation with measurement of the electrode potential, but available evidence on twinning electrical deposition agrees very well with the above theoretical results. For instance, it is known that the proportion of twin crystallites increases with the current density at first rapidly, and then more slowly [25, 26]. However, these results are only qualitative, so one needs a more detailed experimental verification. These theoretical concepts on twinning give wide scope for future experiments and theoretical extensions; checks on the curves of Figs. 21 and 22 should provide more detail on certain processes in the electrical crystallization of metals.

Preferred Orientation and Morphology of Twinned

Crystals in Electrical Crystallization of Silver

Twinning involves not only an additional energy in the twin boundary but also a disturbance in the sequence of crystalline layers, which for a face-centered packing takes the form A, B, C/A, B, C; this is ultimately reflected in the shape of the crystal, which will appear as turned through 60° in the twin plane, as is clear from the models of a and b in Fig. 23. In case 1a, the crystal has been constructed normally on an octahedral face, while in case 1b we have twinning. A characteristic property of the twins is that there is a common contact edge between two adjacent cubic faces, while there are only common vertices in the normal construction. This feature is used to identify a crystal.

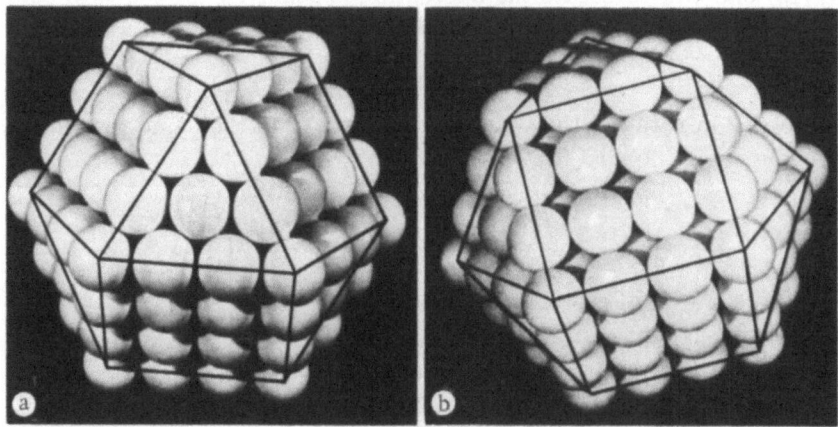

Fig. 23. Model of a crystal with a [111] orientation axis: a) normal crystal; b) twinned crystal.

We use the same initial solutions, methods, and apparatus as for the orientation of silver nuclei; a twin is morphologically more complex than a normal crystal, so each cathode pulse was selected to produce only a small number of nuclei, most often only one. This was possible only with the improved pulse potentiostatic method. The small number of crystals made it possible for each to grow to a considerable size, which greatly simplified observation under the microscope. Under these conditions, to obtain good statistics we had to perform many tests, and the processing was based on the frequency of a current of a given type of orientation.

Orientation of Twin Crystals. The overvoltage range 20-50 mV gave crystals with [111] orientation axes perpendicular to the substrate; the parts of Fig. 24 show some of the more characteristic representatives of these crystals. Comparison with the model in Fig. 23b, which represents a twin crystal of [111] orientation, clearly shows the equivalence of the morphologic features.

Overvoltages in the range 60-90 mV produced an altered orientation in the twin crystals, from octahedral to cubic; here the preferred characteristic form was a cube-octahedron with [100] orientation as shown in Fig. 25b; for comparison, Fig. 25a gives the equivalent model with the same orientation.

The curves of Fig. 3 show that at certain overvoltages one gets the same nucleation energy for the two types of nuclei, which means that one has identical probabilities for orientation of the nuclei on the two axes. In fact, experiment shows that overvoltages of 50-60 mV give with equal probabilities crystals having cubic or octahedral orientation axes, the probabilities being maximal at 55 mV. Figure 26 shows characteristic cases of two twins oriented on [100] and [111] axes.

Fig. 24. Twinned silver crystals with [111] orientation axes.

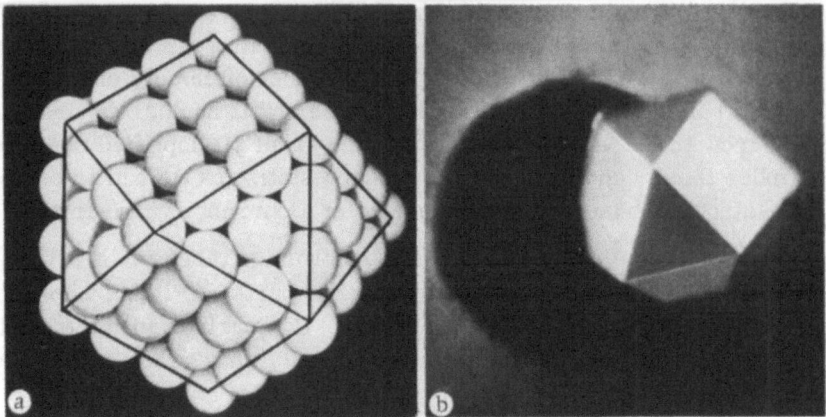

Fig. 25. Twinned silver crystal with a [100] orientation axis: a) model of twin; b) silver crystal grown on platinum microelectrode.

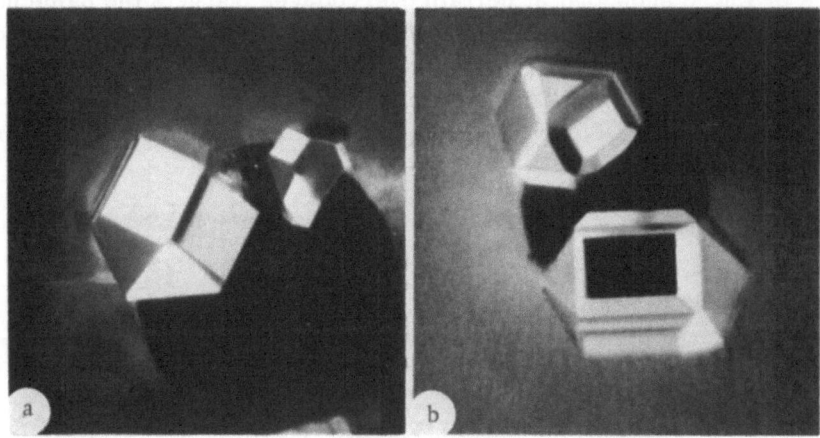

Fig. 26. Twinned silver crystals grown at η = 55 mV oriented on: a) [111]; b) [100].

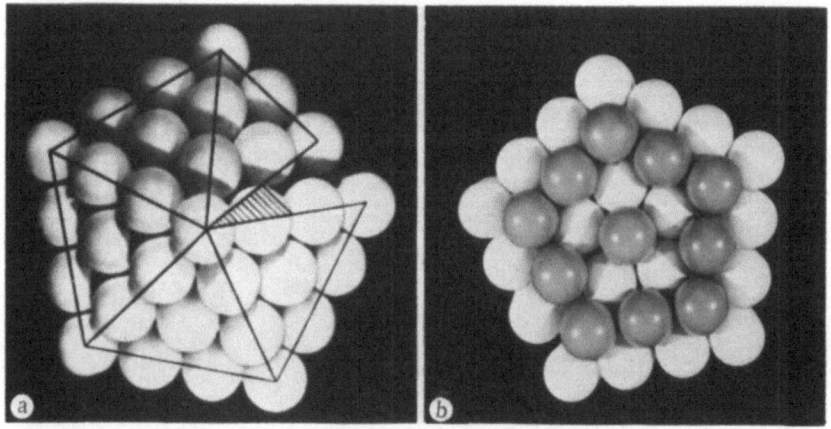

Fig. 27. Model of a crystal with fivefold symmetry and a [110] orientation axis: a) view from above; b) plane presented to the substrate.

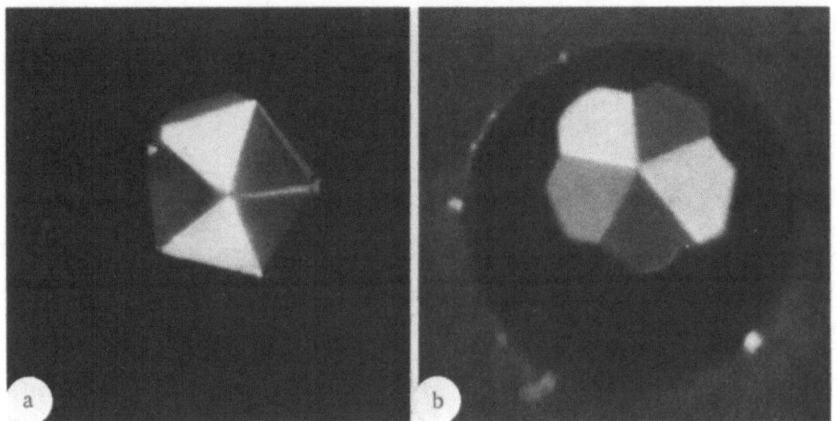

Fig. 28. Twinned silver crystals with fivefold symmetry and [110] orientation axes.

There is a close correlation between the overvoltage ranges and the orientations for twins and normal crystals; this can be explained if one remembers that twinning is a secondary process in relation to initial crystal formation, where the first layers follow the normal series of the face-centered cubic lattice, and the nucleus in its orientation is subject to the laws demonstrated in Fig. 3.

Twin Morphology at High Overvoltages. If the cathode overvoltage in a pulse is 100 mV or more, theoretical calculations lead us to expect an orientation with a (110) face parallel to the substrate; in fact, experiment shows that the crystals are oriented on the rhombododecahedron; here, however, we encounter an effect that somewhat complicates the picture, since the crystals have a much more complicated structure, with new faces and forms that are not characteristic of those observed at low supersaturations.

Figure 22 enables us to explain this, since it indicates that the twinning frequency increases rapidly with the supersaturation, the twinning probability becoming equal to the probability of normal nucleation at sufficiently high supersaturations.

Figure 27, parts a and b, shows the most characteristic form obtained at high supersaturations; the crystals have fivefold symmetry, and this form can be interpreted via the model of Fig. 27a. To construct this model we took a normal crystal with a [110] orientation axis, and on the (111) faces we produced repeated twinning. This gave a form whose morphology is shown in Fig. 27a as lying on the substrate via the face shown in Fig. 27b.

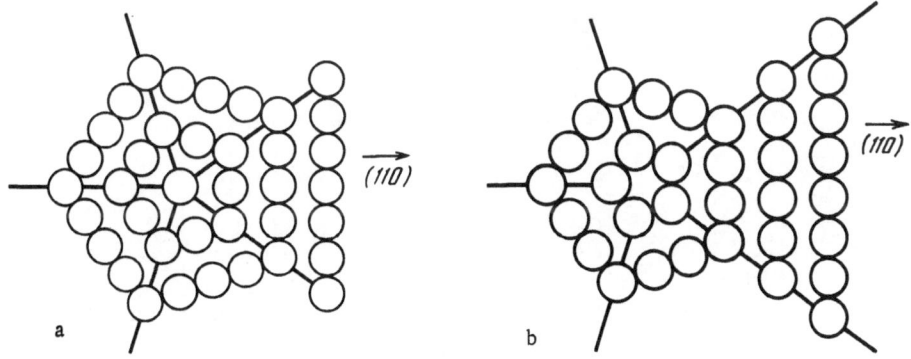

Fig. 29. Plane presented to the substrate by a crystal with fivefold symmetry (variants a and b).

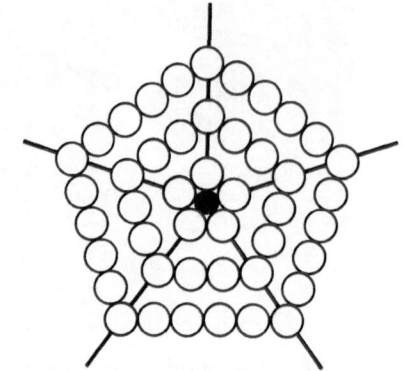

Fig. 30. Plane presented to the substrate by a crystal with fivefold symmetry (foreign atom incorporated).

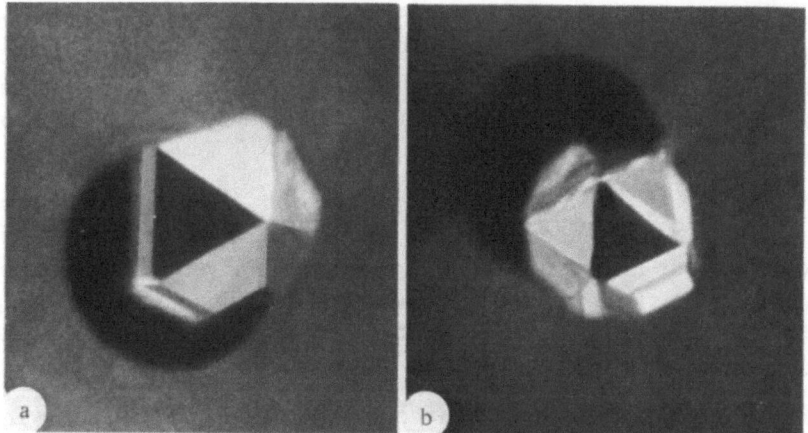

Fig. 31. Twinned silver crystals with fivefold symmetry and [111] orientation axes.

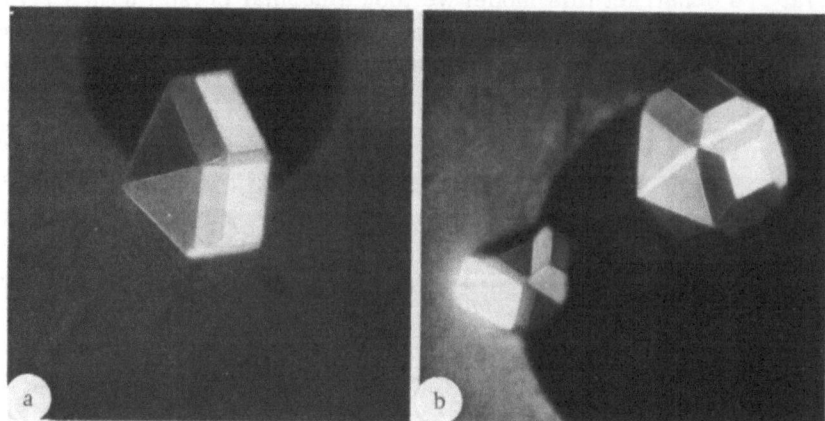

Fig. 32. Twinned silver crystals with fivefold symmetry and [100] orientation axes.

Fig. 33. Twinned silver crystals with five-fold symmetry and [111] orientation axes; single twin and twin with fivefold symmetry.

This form is built up from five separate segments, each of which has [110] orientation and is linked to its neighbor in the twin position. In the upper part there are five octahedral faces inclined to the substrate and having a common vertex, while on the sides there are five faces perpendicular to the orientation plane.

Twinning of each of the two segments produces an angle differing from the angle arising when the fiveling is geometrically symmetrical, so one ultimately gets between the first and last segments a wedge joint of angle 7.5°, which characterizes the discrepancy. This joint is clearly visible on the crystals of Fig. 28, parts a and b, and also in the model of Fig. 27a. This has been pointed out previously [27, 28], and a detailed explanation has been given for it.

Confirmation is available for this formation mechanism via secondary sequential twinning from the feature that one cannot realize a half crystal if one considers the face on which the structure exists in either of the two forms shown in Fig. 29, parts a and b, and hence it is not a primary two-dimensional nucleation structure.

Another mechanism has been proposed [29], in which fivelings are produced by impurities, which have dimensions similar to those of the atoms in the crystal, and these have time to form groups during nucleation (Fig. 30). In that model, the angle of 7.5° does not appear in any one of the segments but is distributed as 1.5° each symmetrically for each of the segments.

This explanation appears very doubtful, since our experiment shows that this fivefold form does not exist as an isolated phenomenon; although it is rarer, on account of the lower probability of repeated twinning, we have obtained such a fiveling with [111] and [100] orientations on appropriate regions at lower overvoltages, where the primary two-dimensional nuclei had cubic and octahedral faces, while twinning arises in the later stages of growth. Parts a

Fig. 34. Complex twin produced at a high voltage.

and b of Fig. 31 show fivelings with [111] orientation axes, while parts a and b of Fig. 32 show the same with [100]. Figure 33 shows a fairly uncommon but rather characteristic case of two crystals with identical orientation, one of which is a simple twin and the other is a fiveling.

Another interesting fact that confirms our explanation is that one obtains forms more complex than those so far considered when one uses very high overvoltages and longer pulse lengths, these forms having fivefold symmetry relative to each vertex. Figure 34 shows the crystal of this complex form, where it would seem that there was extensive twinning on each growing (111) face.

When large crystals of fivefold symmetry are produced, it is clear that strong growth fronts are present (Fig. 33), which begin at the junction, which indicates that this is very active. This is probably a good source of defects and dislocations, which may explain the much more rapid and coarse growth of fivelings relative to individual twins and normal crystals with fixed potentiostatic conditions.

Conclusions

The following conclusions are drawn from the above:

1. The pulse potentiostatic method enables one to use single crystal objects, which provides for more detailed study of the twinning mechanism.

2. The sequence of orientations found in twin crystals over a wide overvoltage range is in agreement with the theoretical predictions (Fig. 3).

3. There is equivalence between the overvoltage ranges in which the crystals have the corresponding orientations in the twin and normal positions, which shows that twinning is secondary relative to the primary formation of two-dimensional nuclei on a given face, it occurring at a later stage in the growth of the three-dimensional crystal.

4. The twin orientation provides a likely morphological explanation for the characteristic complicated form of crystals with fivefold symmetry, for which one invokes a mechanism of repeated sequential twinning, which occurs on a single crystal.

Twins with more complex structure are obtained at high overvoltages with longer pulses, which is due to the elevated twinning frequency, and this results in a qualitative confirmation of the theoretical relationship of Fig. 22 for the increase in twinning probability with overvoltage.

Literature Cited

1. N. Pangarov. J. Electroanalyt. Chem., 9:70 (1965).
2. I. Stranski and R. Kaischev, Z. phys. Chem., B26:100, 114, 312 (1934); Ann. Phys., 23:330. (1933); Z. Phys., 36:393 (1935).
3. R. Kaischev and G. Bliznakov, Compt. Rend. Acad. Bulg. Sci., 1:23 (1948).
4. N. Pangarov and S. Rashkov. Compt. Rend. Acad. Bulg. Sci., 13:555 (1960).
5. M. Volmer. Kinetik der Phasenbildung, Leipzig (1939).
6. H. Brandes. Z. phys. Chem., 126:200 (1927).
7. N. Pangarov and S. Vitkova. Electrochim. Acta, 11:1719 (1966).
8. N. Pangarov and S. Vitkova. Electrochim. Acta, 11:1733 (1966).
9. N. A. Pangarov and V. Mikhailova. Dokl. AN SSSR, 153:1119 (1963); Izv. Inst. Fiz. Khim. Bolg. AN. 4:111 (1964).
10. N. Pangarov and V. Velinov. Electrochim. Acta, 11:1753 (1966).
11. R. M. Bozorth. Phys. Rev., 26:390 (1925).
12. R. Glocker and E. Kaupp. Z. Phys., 24:121 (1924).

13. H. I. Matthews, S. Muttucumarana, and H. Willman. Acta Crystallogr., 14:636 (1961).

14. R. Kaishev, A. Sheludko, and G. Bliznakov. Izv. Bolg. AN, 1:137 (1950).

15. A. Sheludko and G. Bliznakov. Izv. Bolg. AN ser. fiz., 2:227 (1951).

16. A. Sheludko and M. Todorova. Izv. Bolg. AN, 3:61 (1952).

17. R. Kaishev and B. Mutefchiev. Izv. Bolg. AN, 4:105 (1954); 5:77 (1955).

18. M. Atanasov and V. Kertov. Izv. Inst. Fiz. Khim. Bolg. AN, 5:129 (1965).

19. N. Pangarov. Phys. Stat. Sol., 20:365 (1967).

20. E. Budewski, W. Bostanoff, T. Vitanoff, A. Kotzewa, and R. Kaischew. Phys. Status Solidi, 13:577 (1966).

21. N. Pangarov. Phys. Status Solidi, 20:371 (1967).

22. W. Cochrane. Proc. Phys. Soc., 48:723 (1936).

23. R. Kern. Bull. Soc. Franc. Mineral Cristallogr., 84:291 (1961).

24. Yu. M. Polukarov and Z. V. Semenova. Electrokhimiya, 2:184 (1966).

25. G. Poli and L. Peraldo. Metallurigia Ital., 51:548 (1959).

26. T. H. V. Setty and H. Sillman. Trans. Faraday Soc., 51:984 (1955).

27. H. Schlotterer. Metalloberfläche, 18:33 (1964).

28. R. W. De Blois. J. Appl. Phys., 36:1647 (1965).

29. R. Schwoebel. J. Appl. Phys., 37:2515 (1966).

CRYSTALLIZATION OF CADMIUM SULFIDE FROM VAPOR

B. M. Bulakh

The compounds of group $A^{II}B^{VI}$ are semiconductor materials widely used in making many photoelectric and electroacoustic devices; the electrophysical properties of these semiconductors have been extensively studied during the past two decades, since they are, especially cadmium sulfide, extremely interesting as regards crystallization. More methods have been described and used for making cadmium sulfide crystals than for any other substance [1-3]. Amongst these, a special place is taken by crystallization from the vapor, since methods of this type give the purest and most perfect crystals. It is important that crystallization from the vapor gives crystals of various shapes and sizes, from very thin whiskers to large single-crystal blocks. The growth forms are dependent on the conditions, and the temperature region for crystallization is extremely wide, which provides favorable conditions for studying the growth processes and mechanisms.

Here I discuss the actual growth conditions and mechanisms for cadmium sulfide single crystals produced from the vapor; particular attention is given to the effects of vapor composition on the growing crystals.

Experimental Results

A knowledge of the actual vapor composition above the growing crystal appears to be decisive in elucidating the crystallization processes [4]; this gives particular interest to growth of single crystals from the vapors of the elements in a flow of transport gas [5-9], since this method gives control of the vapor composition. This is the method I have used for growing cadmium sulfide single crystals as regards research on the growth conditions.

Crystallizer for Growing Single Crystals from the Vapor

The details of the growth method have been discussed elsewhere [10, 11]; the main disadvantage of the technique using opposing flows of transport gas is that the vapor is unevenly distributed in the reaction region. A special device (Fig. 1) was used in the present work to provide a uniform distribution of the vapors in the reaction space.

This apparatus consists of two vessels 1, which have conical recesses 2 at the end, which are used to link the vessels tightly together, this giving a closed space of constant dimensions and shape. The vessels have fins 3 on the outside, which center the crystallizer in the protective quartz tube 4. At the bottom of each vessel there are conical holes, in which fit the tubes 5 for supplying the transport gas and the vapors of the elements.

The conical screens 6 are fitted at the inlets, and these can move within the vessel, which enables one to regulate the gap between the screen and the end. The connecting end 7 of the screen has channels 8 uniformly distributed around the circumference to allow the vapors of the elements to enter the reaction space. These screens provide uniform mixing of the

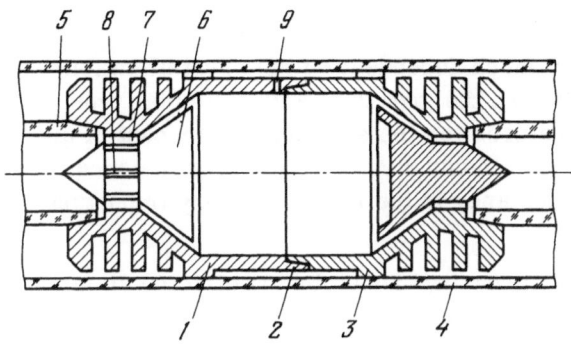

Fig. 1. Device for producing single crystals from vapor.

vapors in the reaction space. The screens also prevent the vapor of one element penetrating into the supply pipe for the other, and also eliminate blockages on cooling.

The transport gas and the unreacted vapors emerge from the vessel via holes of diameter 1.5-2 mm at points 9; the small hole diameter facilitates good mixing of the vapors in the vessel.

The vessels and screens are made of graphite; the internal diameter of the vessel is 50 mm, and the distance between the screens is 80 mm.

This design provides constant crystallization conditions and uniform distribution of the element vapors in the reaction space; the criterion for uniform vapor distribution is the uniformity in the crystal distribution together with the independence of the crystal growth region from the evaporation temperatures of the elements [11].

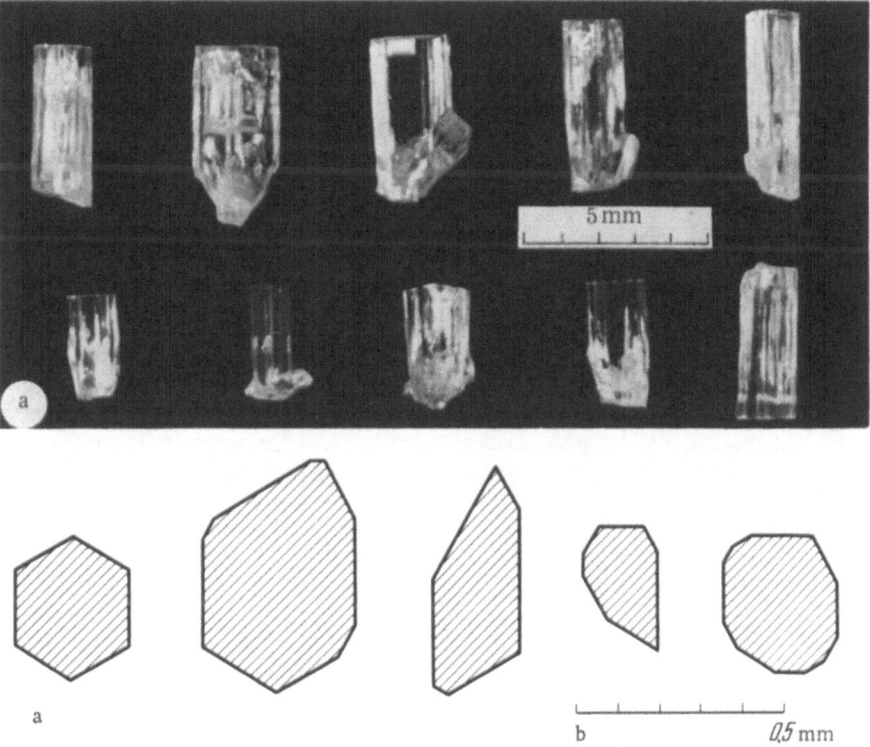

Fig. 2. Prismatic CdS crystals: a) photo; b) sections on (0001). Prisms formed by $(10\bar{1}0)$ and $(1\bar{2}10)$ faces.

Results

This closed vessel gave results that were very much dependent on the crystallization conditions; in an ordinary open quartz crystallizer [6, 7], the growth of any particular type of crystal can occur at a variety of evaporation temperatures for the elements on account of uneven distribution of the vapor along the vessel, whereas in our runs we found, for instance, platy crystals only at certain temperatures. This constancy in the results is very important in the analysis and evaluation of growth conditions [11].

We found that cadmium sulfide crystals usually grow in a nonstoichiometric medium containing an excess of cadmium, and the ratio between the elements has a substantial effect on the shape of the crystals. The ratio determined whether one obtained at a given crystallization temperature prisms, pyramids, needles, platelets, strips, or whiskers; the formation conditions for the various growth forms were as follows for crystallization temperatures of 1020-1040°C:

1. Prisms (Fig. 2). The stoichiometric ratio between the elements gave small crystals in the form of prisms with many faces, whose height was 1-3 mm and width 1-2 mm. The six-faced prisms were bounded by (10$\bar{1}$0) or (1$\bar{2}$10) faces, while the other prisms had alternating faces of these types (Fig. 2a).

2. Pyramids (Fig. 3). Pyramids grew at cadmium–sulfur ratios of 1-1.3; their heights were 3-8 mm, and width at the base was 1-3 mm. As a rule, the pyramids were hollow (Fig. 3a).

3. Needles (Fig. 4). Ratios of 1.3-1.5 gave needles of length 10-30 mm and thickness 0.1-2 mm. These resembled the prisms in having six or more faces with alternating (10$\bar{1}$0) and (1$\bar{2}$10) planes (Fig. 4a). Many of the needles were hollow .

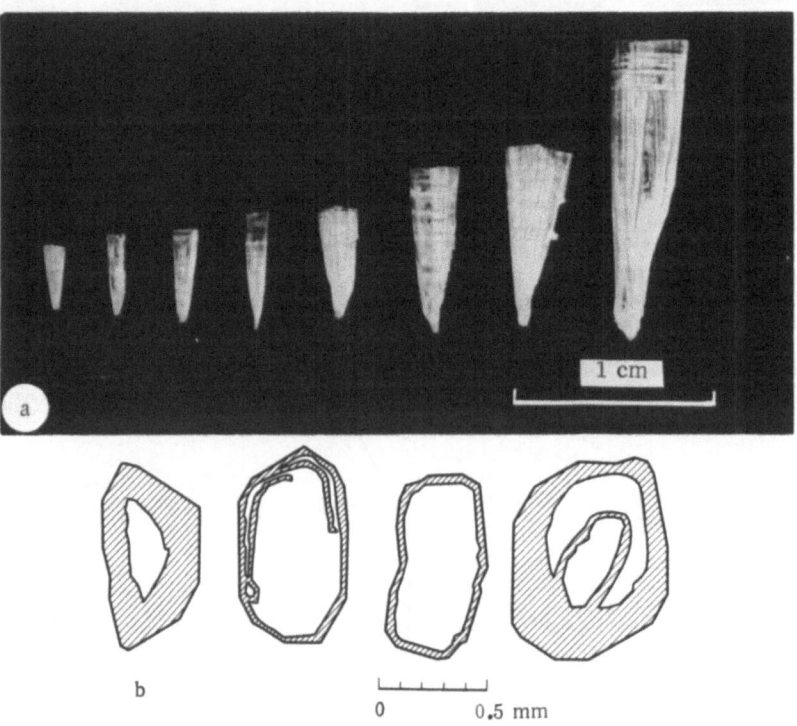

Fig. 3. CdS pyramids: a) photo; b) sections of bases.

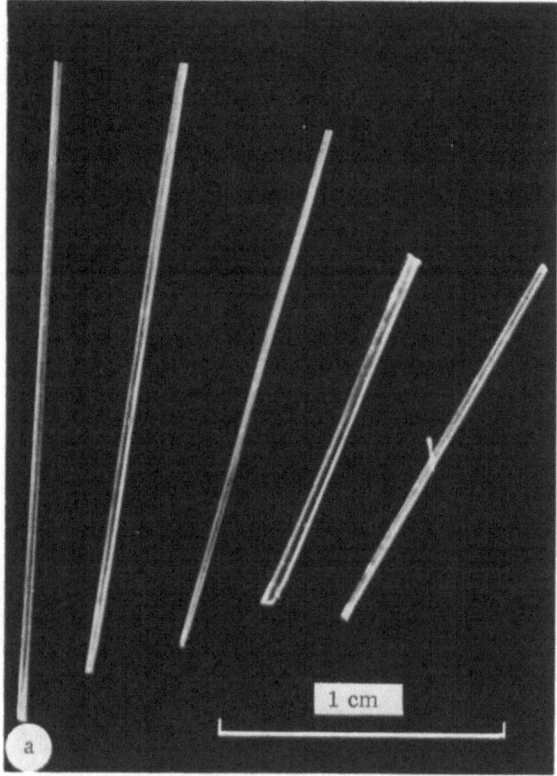

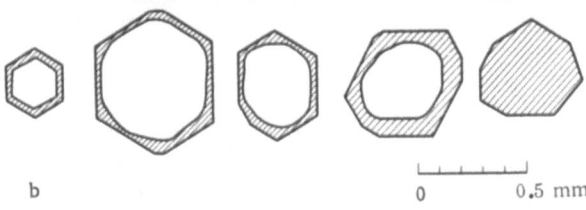

Fig. 4. CdS needles: a) photo; b) sections on (0001).

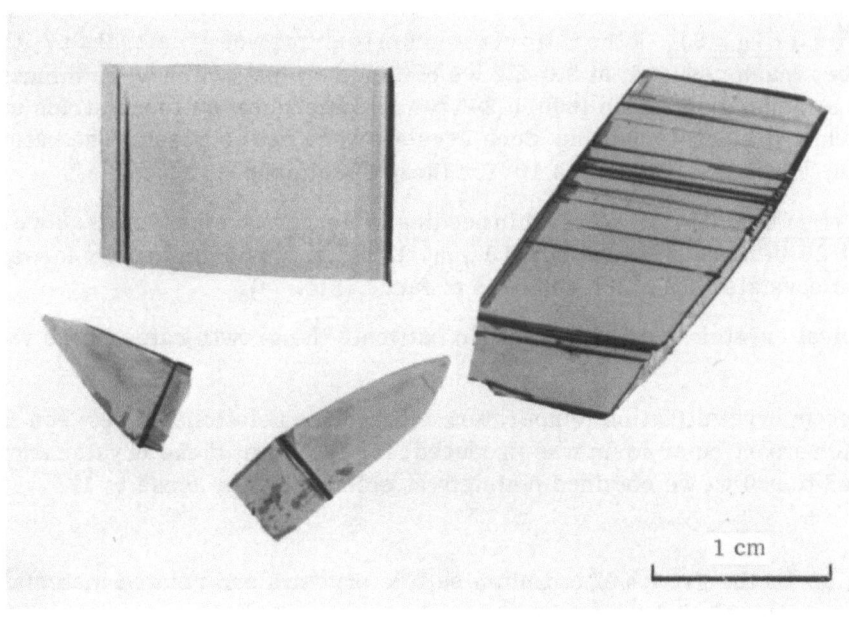

Fig. 5. CdS plates.

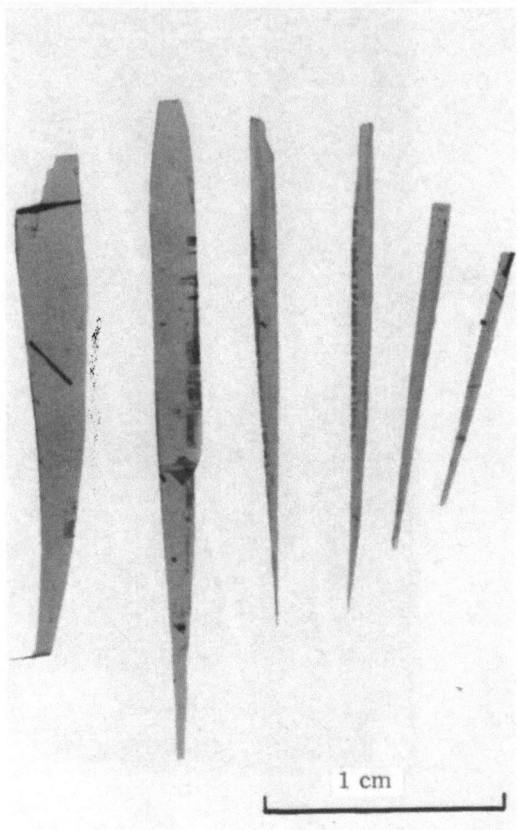

1 cm

Fig. 6. CdS strips.

4. Plates (Fig. 5). These were formed at ratios of 1.5-2.0, and they usually had broad ($1\bar{2}10$) faces. Their dimensions on [0001] were 5-10 mm, and on [1010] were 10-30 mm. The thicknesses were 10-500 μm. Usually, the platelets had ridges parallel to the c axis on the ($1\bar{2}10$) planes, but quite often there were crystals with very smooth surfaces.

5. Strips (Fig. 6). If the ratio of cadmium to sulfur was greater than 2, the yield of platelets was very much reduced; at 2.0-2.2 we obtained strips, which were thinner than the platelets (1-10 μm) and smaller on [0001] (2-5 mm). Sometimes we found strips up to 3 mm long and of thickness down to 0.2 μm. Such crystals were red or green. The ratio of linear dimensions to thickness was as high as 10^4 for these specimens.

6. Whiskers (Fig. 7). Very thin needles (whiskers) grow at ratios above 2.2; their lengths were 10-20 mm and thickness 1-10 μm. Usually, a whisker had six faces, but quite often there were crystals with other numbers of faces (Fig. 7a).

These typical crystal sizes enable one to estimate the growth rates in the various directions (Table 1).

Any change in crystallization temperature altered the relationship between the initial elements at which a particular form was produced; for instance, if the crystallization temperature was reduced to 980°C, we obtained platelets at element ratios close to 1.

Discussion

In all studies on the growth of cadmium sulfide crystals and related materials from the vapor it has been assumed that the vapor phase above the growing crystal is of stoichiometric composition [6-8]. This assumption has been based on studies showing that cadmium sulfide and analogous materials evaporate congruently [12-15].

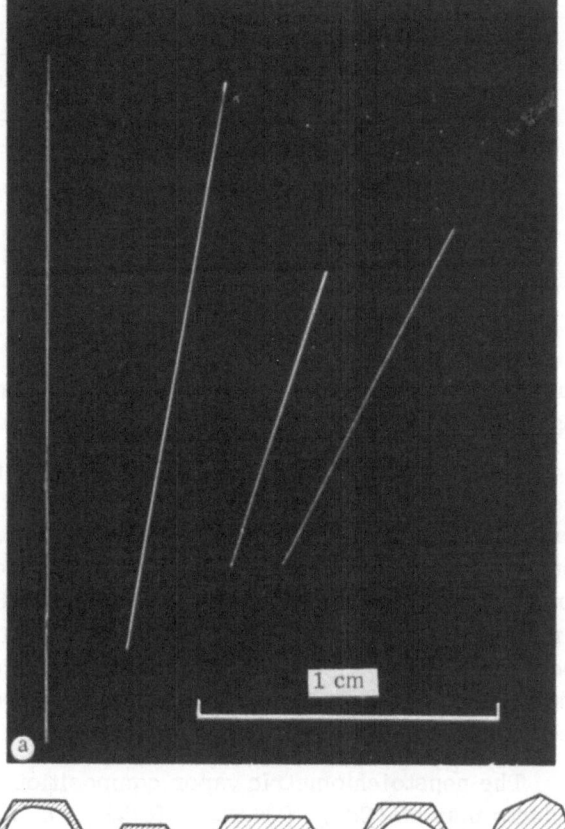

Fig. 7. CdS whiskers: a) photo; b) sections on (0001).

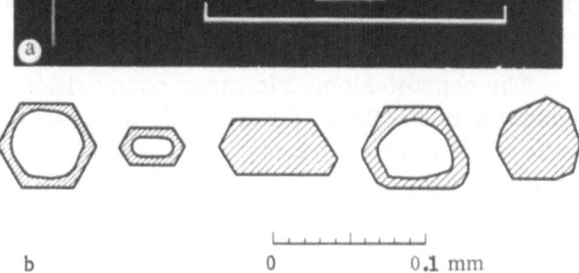

TABLE 1. Growth Rates of CdS Single Crystals

Shape	Growth rate (cm/sec) in direction	
	[0001]	[1$\bar{2}$10] or [10$\bar{1}$0]
Prisms	(3—8)·10^{-5}	(1,5—3)·10^{-6}
Needles	(2—6)·10^{-4}	(0,1—3)·10^{-6}
Whiskers	(3—6)·10^{-4}	(0,1—1)·10^{-7}
	[0001] or [10$\bar{1}$0]	[1$\bar{2}$10]
Plates	(3—8)·10^{-4}	(0,1—1,5)·10^{-6}
Strips	(3—5)·10^{-4}	(0,1—1,5)·10^{-7}

Our results show that cadmium sulfide crystals usually grow in a nonstoichiometric medium; on the other hand, it is familiar that the growth forms of cadmium sulfide crystals are dependent on the growth temperature, as has been described for crystals grown by sublimation, i.e., when the source of the vapor is of stoichiometric composition [16, 17]. The relationships are shown in Table 2, where the second column indicates the ratio between the elements in the growth region in synthesis at a constant crystallization temperature, while the third column gives the temperature ranges for production of the corresponding forms when the crystals are grown by sublimation [16, 17]. The values show that the growth forms at a given temperature are dependent on the element ratio, while the crystallization temperature is the controlling factor when a stoichiometric source is used. Consequently, the element ratio

TABLE 2. Shape of CdS Crystals in
Relation to Growth Conditions

Type	Element ratio	Growth temperature, °C, in sublimation [16, 17]
Prisms	~1.0	1100
Pyramids	1.0—1.3	(1050)
Needles	1.3—1.5	950—1050
Plates	1.5—2.0	880—980
Stripes	2.0—2.2	800—850
Needles (whiskers)	>2.2	750—800

above the growing crystal and the crystallization temperature are the basic factors determining the growth mechanism and form of the crystal.

There appears to be no reason to assume that the crystal growth on sublimation is different from that during synthesis, so the growth in the former case should occur in a medium of a definite composition, which means that the vapor composition above the source is different from that at the growth point. Of course, identical constructional elements are involved in the growth in both cases. It has been pointed out [18] that a real perfect lattice would appear to be more difficult to produce from single atoms than simultaneously from sets of atoms or blocks. From this viewpoint, the medium producing the crystal should provide appropriate conditions for nucleation and growth. This preparation of the medium can be provided by preliminary combining of the cadmium atoms with molecules or atoms of sulfur.

The nonstoichiometric vapor composition over the crystal indicates that the resulting products may be $[CdS]_n$, $[Cd_2S]_n$, or $[Cd_3S]_n$, or ones similar to these; however, the nonstoichiometric medium above the crystal is not the sole argument in favor of this theory. Another very important point is the role of the chemical reaction in the crystallization.

Chemical Reactions and Crystallization

The growth conditions in our experiments were investigated from the set concentrations of the parent elements in the crystallization region [11]. If the particles do not interact with one another, such an evaluation could give the true picture of the growth conditions, but in fact we have a system of reacting particles, which only indirectly give the final product. The chemical reactions during the crystallization are important in the synthesis of single crystals from vapor. Previously it had been assumed that the reaction accompanying the crystallization only provided an adequate amount of the material to be deposited [19], but now there is no doubt that the reaction is the factor that governs the crystallochemical, structural, and morphological features of the product [20]. On the other hand, it has been pointed out [20] that the crystal morphology is a very sensitive indicator of the actual composition for the vapor, and can assist in understanding the details of the reaction.

Reaction Kinetics

The observed reaction rates indicate that only some of the collisions between molecules result in reaction; for reaction to occur, the colliding molecules must have an energy exceeding a certain mean value, and molecules that have reacted have more energy than ordinary molecules, so it is kinematically essential for even a simple reaction to occur via several stages, such as the following [21]: normal molecules of the initial material ⇌ active molecules of the initial material ⇌ molecules of reaction products; i.e., any reaction occurs via an intermediate state, which requires activated molecules formed from normal molecules by collision. Only active molecules can participate in the reaction.

In the theory of the transition state [22], the activation energy takes the meaning of the energy of the transition state; for reaction to occur, the energy of the reacting system must be such as to allow the transition state to form. In any chemical reaction there is a continuous variation in the distance between the nuclei; the configuration corresponding to the initial state is transformed to some intermediate configuration representing the activated complex or transition state, which then gives way to the final configuration, i.e., the reaction can be represented as

$$A + B \rightleftarrows X^{++} \rightarrow C + D,$$

where A and B are the initial materials, X^{+} is the transition state or activated complex, and C and D are the reaction products.

Intermediate active particles are therefore necessary to the reaction; the basic conclusion for chemical kinetics is that only active particles can be involved in a reaction.

The Final Stage of a Reaction

For a reaction to occur, the system of atoms must be able to pass through the potential barrier separating the initial and final states; the effective activation energy is not a fundamental constant. While the energy difference between the initial and final states is not dependent on the path followed, the activation energy is very much dependent on that path, so the system selects the optimal transition path for passage to the final state in accordance with the external conditions [23].

The completion stage of a reaction is production of the final product; while this structure is not of decisive importance in chemistry, it is so in growing single crystals. In fact, in the latter case one selects such conditions that will enable the system to pass into the maximally ordered state as an infinitely extended structure with a small number of defects. Therefore, the path followed by the reaction, or the reaction mechanism, is bound to affect the structure and morphology of the crystal.

From the viewpoint of chemical kinetics, the role of the intermediate particles is clear, since these are the ones that are involved in the concluding stage of the crystallization; conditions can alter when a solid phase arises in a gas medium, since heterogeneous reactions must occur in that case. However, in kinetic theory it has been shown that particles in a vapor interacting with solid surfaces must also possess sufficient energy reserve to overcome potential barriers [24], so here again active intermediate particles should be considered as most suitably prepared for heterogeneous reaction.

If one considers the crystallization as the closing stage of a chemical reaction and assumes that mainly active intermediate particles are involved in nucleation and growth, one has a good basis for understanding many features of crystal growth, such as, for instance, the growth of cadmium sulfide crystals arising from intermediate complexes of cadmium with sulfur, which come together, while the participation of individual cadmium or sulfur atoms, and especially molecules, is very unlikely, since these are not in suitable energy states for reaction. This also means that the actual supersaturation in the vapor is determined by the concentrations of the complexes, which may be substantially less than those of the initial materials.

The vapor composition indicates that a variety of intermediate particles may be formed under different conditions; when the element ratio is 1, one may get [CdS] complexes, which are not neutral molecules of cadmium sulfide, but active complexes with unsatisfied bonds. When the ratio is 2, with two atoms of cadmium to one of sulfur, one gets $[Cd_2S]$. And finally, when the ratio is 3, one gets $[Cd_3S]$. At intermediate values, one expects mixtures of these; for instance, at 1.5, we have two atoms of sulfur to three of cadmium, so one expects [CdS] and [Cd$_2$S].

Vapor Composition for Nonstoichiometry

In general, one has the following expression for complexing:

$$aCd + bS = c\,[CdS] + d\,[Cd_2S] + e[Cd_3S]. \tag{1}$$

Then if n_{Cd} is the number of gram-atoms of cadmium and n_S is the same for sulfur, which react to form N_1 moles of [CdS], N_2 moles of [Cd$_2$S], and N_3 moles of [Cd$_3$S], the equations for the material balance are

$$N_1 + 2N_2 + 3N_3 = n_{Cd}, \tag{2}$$
$$N_1 + N_2 + N_3 = n_S = N, \tag{3}$$

where N is the total number of moles of the mixture of complexes. Then as we have $\alpha = n_{Cd}/n_S$, we get for 1 mole of the mixture that

$$N_1' = \frac{(3-\alpha) - N_2'}{2}, \tag{4}$$

$$N_3' = \frac{(\alpha - 1) - N_2'}{2}, \tag{5}$$

$$N_2' = (3-\alpha) - 2\,N_1', \tag{6}$$

$$N_2' = (\alpha - 1) - 2\,N_3'. \tag{7}$$

These equations should be obeyed throughout the range of variation in element ratio, i.e., $1 \le \alpha \le 3$; we see from (4) and (5) that N_2' can be zero throughout this range, its maximum value occurring, as shown by (6) and (7), for $N_1' = 0$ and $N_3' = 0$. As $N_2' \le 1$, (6) applies for $\alpha \ge 2$, and (7) for $\alpha \le 2$, so N_1' can be zero for $\alpha \ge 2$, while N_3' may be zero for $\alpha \le 2$.

We then have the following expressions for the limiting maximum values of N_1', N_2', and N_3':

$$N_{1max}' = \frac{3-\alpha}{2}, \quad N_{3max}' = \frac{\alpha-1}{2},$$
$$N_{2max}' = \alpha - 1 \quad \text{for} \quad \alpha \le 2,$$
$$N_{2max}' = 3 - \alpha \quad \text{for} \quad \alpha \ge 2. \tag{8}$$

For the minimum possible values we have

$$N_{1min}' = 2 - \alpha, \quad N_{2min}' = 0, \quad N_{smin}' = \alpha - 2. \tag{9}$$

The optimum relation between the complexes in the mixture should correspond to minimum mixture formation energy; if ε_1 is the energy of formation for 1 mole of [CdS], while ε_2 and ε_3 are the same for [Cd$_2$S], then the energy of formation of a mole of the mixture ε is

$$\varepsilon = \varepsilon_1 N_1' + \varepsilon_2 N_2' + \varepsilon_3 N_3'.$$

We express N_2' and N_3' in terms of N_1' to get

$$\varepsilon = (\varepsilon_1 - 2\varepsilon_2 + \varepsilon_3)\,N_1' + \varepsilon_2\,(3-\alpha) + \varepsilon_3\,(\alpha - 2). \tag{10}$$

Similarly,

$$2\varepsilon = (2\varepsilon_2 - \varepsilon_3 - \varepsilon_1)\,N_2' + \varepsilon_1\,(3-\alpha) + \varepsilon_3\,(\alpha - 1), \tag{11}$$

$$\varepsilon = (\varepsilon_3 - 2\varepsilon_2 - \varepsilon_1)\,N_3' + \varepsilon_1\,(2-\alpha) + \varepsilon_2\,(\alpha - 1). \tag{12}$$

Consider now the factors to N_1', N_2', and N_3'; if these are zero, the energy of formation for 1 mole of mixture is not dependent on the composition; if the coefficients are positive, the formation energy increases with the contents of the corresponding complexes, while negative values mean that the energy decreases when the contents of the corresponding complexes increase.

Then if

$$\varepsilon_2 = \frac{\varepsilon_1 + \varepsilon_3}{2} \tag{13}$$

the composition of the mixture may be arbitrary.

If

$$\varepsilon_2 > \frac{\varepsilon_1 + \varepsilon_3}{2}, \tag{14}$$

then the contents of the complexes are

$$N_1' \rightarrow N_{1max}', \quad N_2' \rightarrow N_{2min}', \quad N_3' \rightarrow N_{3max}'.$$

If

$$\varepsilon_2 < \frac{\varepsilon_1 + \varepsilon_3}{2} \tag{15}$$

the contents of the complexes are

$$N_1' \rightarrow N_{1min}', \quad N_2' \rightarrow N_{2max}', \quad N_3' \rightarrow N_{3min}'.$$

Equation (13) reflects the linear variation in the formation energy as the number of atoms of cadmium increases, i.e., it denotes additivity in the bond energies; one should not expect additivity [25] in the formation of intermediate complexes with small numbers of atoms when the interactions between them are important.

An electrostatic consideration of complexing [26] shows that the formation of $[Me_2X]^+$ from singly charged ions releases more energy than does the formation of [MeX], i.e., the $[Me_2X]^+$ complex is more stable than [MeX]. Therefore, if the number of Me^+ ions is sufficient, the [MeX] should be transformed to $[Me_2X]^+$. When $[Me_3X]^{2+}$ is formed, less energy is released than that in the formation of $[Me_2X]^+$, but $[Me_3X]^{2+}$ is still more stable than [MeX]. One can show that in that case (15) is realized.

If (15) is realized in the interaction of cadmium with sulfur, and the morphology of the crystals confirms this (see below), then the composition of the vapor will tend to the lower limits to the possible contents of [CdS] and $[Cd_3S]$, and to the upper limit for the possible content for $[Cd_2S]$ (Fig. 8). Then when $1 \leq \alpha \leq 2$, the vapor contains mainly [CdS] and $[Cd_2S]$,

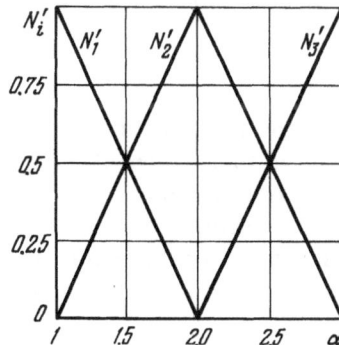

Fig. 8. Vapor phase composition for various ratios between the reacting elements.

while for $2 \le \alpha \le 3$ it contains mainly [Cd₂S] and [Cd₃S]. It is possible, of course, for [Cd₃S] to be formed in the first region and [CdS] in the second, but the concentrations should be small.

Vapor Composition for a Stoichiometric Ratio

We have seen from the results on cadmium sulfide that the crystals formed by sublimation grow in a medium of definite composition; evidence for this is that the difference between the source and crystallization temperatures varies with the growth form. For instance, prisms may be formed on the support, i.e., the crystals grow essentially at the source temperature, and the temperature difference $\Delta T \approx 0$. For needles T is 50-150°, as against 120-220° for platelets, 250-300° for strips, and 300-400° for whiskers (Table 2).

This evidence indicates that qualitative changes accompany the quantitative ones when the vapor temperature is reduced; we therefore have to consider how the vapor composition varies on congruent evaporation of the initial material. This is clearly possible if some of the atoms of one element (sulfur in the case of cadmium sulfide) do not participate in forming the intermediate complexes. Only intermediate active particles can participate in the reaction, and we know [27] that cadmium is present in the vapor state as atoms, while sulfur can form polyatomic molecules such as S_2, S_4, S_6, and S_8 [28]. From the reaction viewpoint, these molecules are less suitable for participation than are sulfur atoms, so these sulfur molecules should reduce the number of reactive sulfur atoms. Such formation of molecules is characteristic not only of $A^{II}B^{VI}$ compounds but also may be observed in the decomposition of many other inorganic compounds.

In general, the decomposition of a $A^{II}B^{VI}$ compound can be described by the equation

$$MeX \rightleftarrows Me + (1 - x) X + (x/2) X_2.$$

It has been stated [13] that this is the mechanism involved for zinc and cadmium tellurides, and the dissociation of Te_2 in the case of CdTe increases with temperature. Mass spectrometry shows that cadmium sulfide vapor contains Me^+, X_2^-, and X^-. The presence of the latter has been ascribed to dissociation of X_2 in response to the ionizing beam. The maximum evaporation temperature involved in the above experiments was 789°C, whereas cadmium sulfide crystals are grown by sublimation at a source temperature usually of 1050-1100°C, while the minimum source temperature that will give platelets is 900°C, i.e., these temperatures lie considerably above that covered by the above study.

We therefore assume that some of the sulfur is present in the atomic state in the temperature range usually employed in the evaporation of cadmium sulfide, so the decomposition of cadmium sulfide is described by the more accurate equation

$$CdS \rightleftarrows Cd + (1 - x) S + (x/2) S_2. \tag{16}$$

Then the following is the ratio of the elements capable of participating in the reaction:

$$\alpha = \frac{1}{1 - x}. \tag{17}$$

The concentration of S_2 molecules is inversely related to the temperature; the element ratio in platelet growth is close to 2 at about 1040°C, which shows that evaporation of cadmium sulfide above 1040°C produces sulfur mainly in the atomic state in the vapor. The growth of platelets by sublimation around 980°C indicates that here half the sulfur atoms are associated as S_2. The growth of whiskers at about 800°C indicates that some 2/3 of the sulfur atoms are here associated. Finally, virtually all the sulfur atoms are associated at lower temperatures.

It is clear that in single-crystal synthesis, association of sulfur molecules should lead, and actually does lead, to a reduction in the set element ratio in the growth region as the temperature is reduced.

Growth Mechanism of Cadmium Sulfide Crystals

Various views have been expressed in published papers on the growth mechanisms for crystals of cadmium sulfide and related substances; the two-dimensional nucleation mechanism and the displacement mechanism [29-33] may also be accompanied by a dislocation mechanism [17], which explains some features of the crystal growth.

Our results and the above assumptions about the vapor composition enable us to consider the growth mechanism of cadmium sulfide from a different viewpoint [34]. On the other hand, the morphology of the resulting crystals enables one to check the above assumptions.

Basic Structural Elements and Complex Association

Participation of intermediate active particles in constructing a crystal is a necessary condition for crystallization if this is considered as the final stage of a chemical reaction; one gets the intermediate complexes $[CdS]$, $[Cd_2S]$, and $[Cd_3S]$ in accordance with the vapor composition. We consider that these complexes are the basic building elements.

The second stage that needs to be considered is association of the main complexes, i.e.,

$$n\,[Cd_pS] \rightleftarrows [Cd_pSd]_n.$$

Association predominates in crystal growth, and this reaction goes from left to right; the final result of association is crystal formation and growth, so the association should tend to take up a geometrical configuration whose subsequent development leads to directional bonds in the crystal lattice. In other words, the complexes and simple associations of these are a form of nucleus, which will subsequently combine and develop into the structure characteristic of the compound. Of course, an association say of two complexes cannot be considered as a microcrystal of cadmium sulfide; however, as the number of complexes increases, the association gradually acquires the features of a crystal, and at a certain stage of development can be considered as a microcrystal or nucleus for a future crystal.

We can therefore consider associations with gradually increasing numbers of atoms to derive the structural element for the lattice; in that case, we can assign the spatial directions characteristic of the resulting lattice in constructing models for the complexes and associations of these, which we envisage as definite structural cells for the lattice. Figure 9 shows associations of complexes and the corresponding elements of the crystal lattice for cadmium suflide as used in our model constructions.

Growth of Cadmium Sulfide Crystals

We consider certain cases characteristic of various growth forms of these crystals.

a). $\alpha = 1$.   The vapor contains $[CdS]$ and $[CdS]_n$ associations.

The complexes may combine to give $[CdS]_6$ associations; such an association can grow at this vapor composition in the $[000\bar{1}]$ direction by addition of other such associations or of $[CdS]_3$ associations. Lateral growth can occur only by addition of $[CdS]$ complexes. The growth rate in these directions will be dependent on the concentration of $[CdS]$ in the vapor; the growth rate in the $[000\bar{1}]$ direction is determined by the probability of $[CdS]_6$ or $[CdS]_3$ formation, together with the probability of attachment to the growing surface. In fact, in this case we are dealing with two ordinary growth mechanisms (two-dimensional nucleation for the basal plane and displacement for the side faces), these being characteristic of prism growth [30].

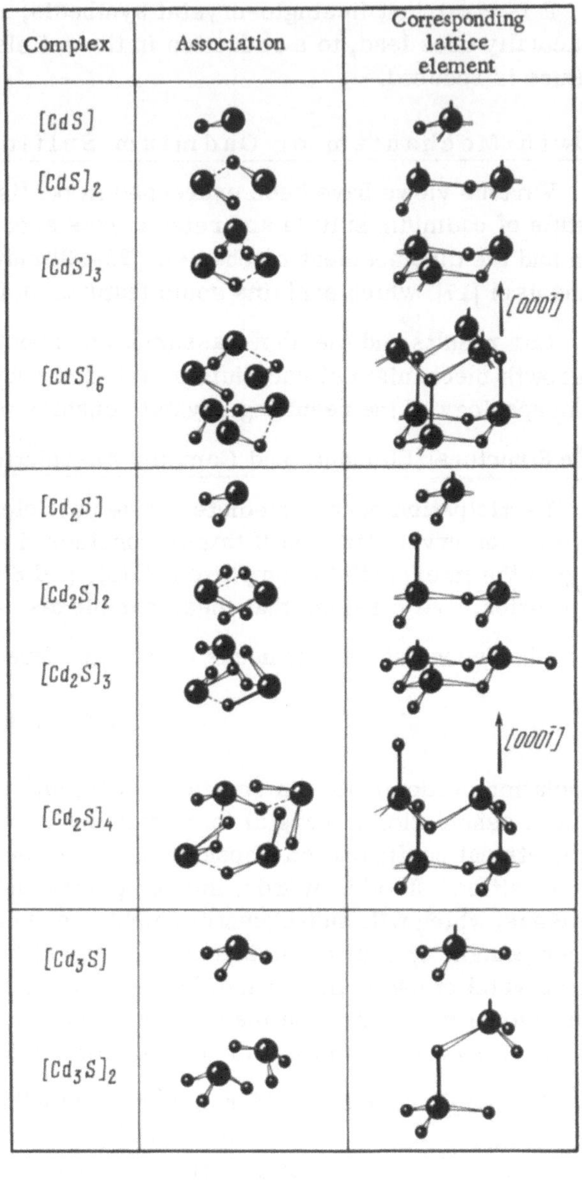

Complex	Association	Corresponding lattice element
[CdS]		
[CdS]₂		
[CdS]₃		[0001̄]
[CdS]₆		
[Cd₂S]		
[Cd₂S]₂		
[Cd₂S]₃		[0001̄]
[Cd₂S]₄		
[Cd₃S]		
[Cd₃S]₂		

Fig. 9. Complex associations and the corresponding lattice elements.

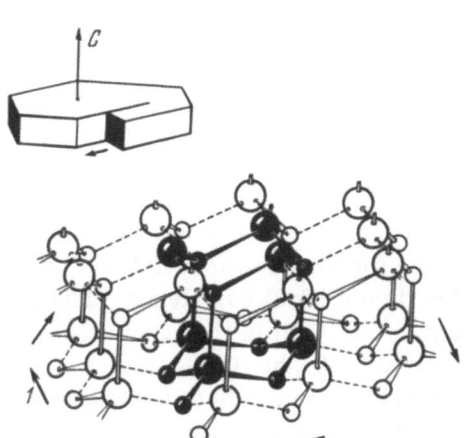

Fig. 10. Growth of [CdS]₆ association containing excess cadmium.

The probability of attachment of $[CdS]_6$ or $[CdS]_3$ to the surface may be dependent on the polarity of the latter; in a (0001) plane there are interaction forces between the cadmium atoms, while in a (000$\bar{1}$) plane there are the same between sulfur atoms. The preferential growth of CdS crystals in the [0001] direction goes with an equally probable direction for growth of CdSe crystals [35] to indicate that in CdS the bonding of an association on a (0001) surface is stronger, whereas the resulting bonds are comparable for CdSe.

b) $\alpha > 1$. One gets $[Cd_2S]$, complexes with the excess of cadmium, and these may participate in producing $[CdS]_6$, which results in associations with excess cadmium. Here spiral lateral growth in such an association is possible by addition of [CdS] or $[CdS]_2$ complexes (Fig. 10). Continuous growth can occur until a layer is completed. Such a process can involve also impurity atoms of other elements, which can have a substantial effect on the crystal growth [35, 36].

c) $\alpha = 2$. The gas contains mainly $[Cd_2S]$. If the conditions allow $[Cd_2S]_4$ to be formed, the latter can grow by addition of analogous associations, and, which is more likely, also by addition of $[Cd_2S]_2$ (Fig. 11). In that case, addition is possible and equally probable in the [000$\bar{1}$] and [10$\bar{1}$0] directions. Growth of the association results in a structure with a developed set of [1$\bar{2}$10] planes. The $[Cd_2S]$ complexes cannot be attached in the [1$\bar{2}$10] direction, but growth can occur in this direction by attachment of [CdS] (Fig. 12). Addition of these forms a new series analogous to that in the underlying substrate, so the growing $[Cd_2S]_4$ association forms a type of substrate on which one gets epitaxy of [CdS]. The growth rate in this direction will be dependent on the concentration; experience shows that increase in α, which should reduce the concentration of [CdS], causes the thickness of the platy crystals to fall, and needles are formed.

If the vapor contains concentrations such that $[Cd_2S]_3$ is preferred ($\alpha < 2$), then such associations can grow by combination in [000$\bar{1}$] directions. In lateral directions, one can attach a restricted number of $[Cd_2S]$ complexes, after which further growth can occur only via [CdS] (Fig. 13). These produce sites at which one can again attach $[Cd_2S]$. This alternation will continue until [10$\bar{1}$0] or [1$\bar{2}$10] side faces are formed, whose further growth is possible by epitaxy of [CdS], as for platelets. This process should result in needles, whose growth rate in the lateral directions is related to the [CdS] concentration.

d) $\alpha > 2$. When $[Cd_3S]$ complexes are formed in the vapor, the latter can associate only into chains; the growth mechanism for such associations is analogous to that considered for needles, but in this region the concentration of [CdS] is very small, so the lateral growth rate should also be very small. This may result in whiskers.

Fig. 11. Association of $[Cd_2S]$ complexes.

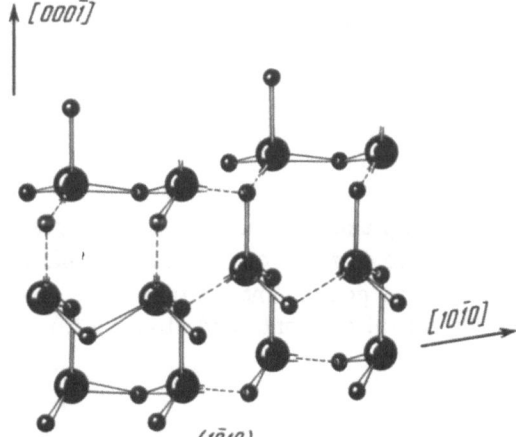

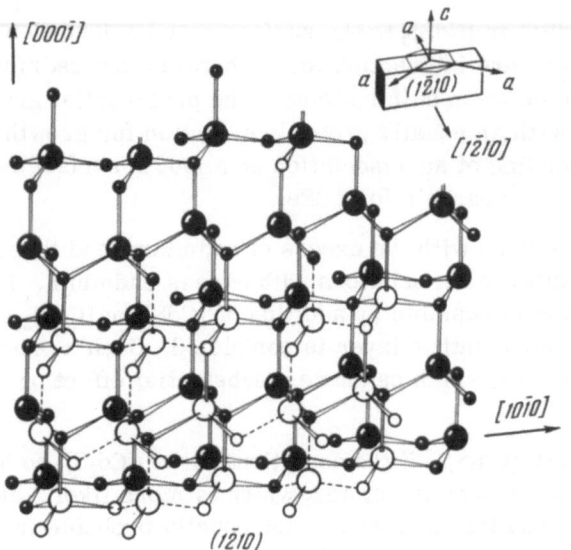

Fig. 12. Platelet growth.

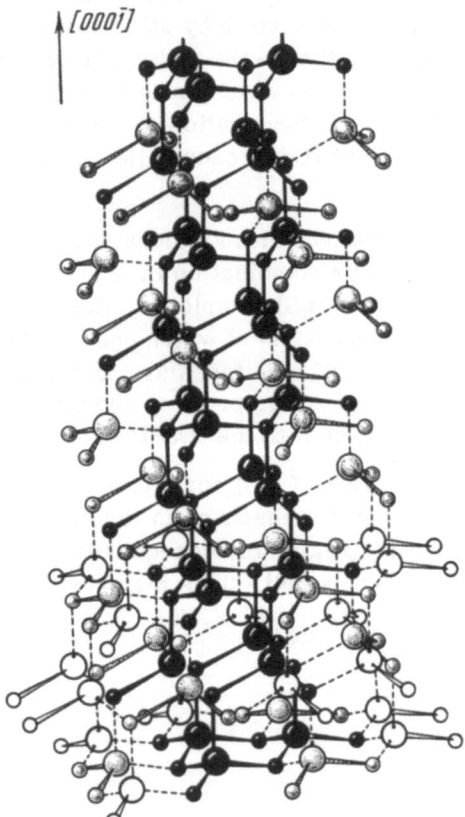

Fig. 13. Needle growth.

Then these examples enable us to explain the reasons for production of the various forms in relation to the growth conditions. The habit is dependent ultimately on the relations between the builder components in the vapor, and this relationship is dependent on the balance between the initial elements and also the temperature.

Conclusions

The real growth conditions have been considered for single crystals of cadmium sulfide in the vapor; it is shown that the ratio between the elements in the growth region goes with the crystallization temperature to control the growth form. The evidence also shows that crystallization cannot be considered in abstraction from the actual environment, particularly the composition and the chemical processes occurring around the crystal. This implies a careful study of the vapor composition, together with the effects of elements that participate in building the crystal. This is all the more necessary because it has been shown that even a stoichiometric composition may allow the crystals to grow from nonstoichiometric constructional elements.

The crystal growth is indissolubly linked to chemical reactions, and it is the final stage in the latter. This approach to the reaction in crystallization enables one to elucidate many features of the crystallization. This essential link to reactions in the vapor enables us to assume that the active intermediate particles produced in the reaction are the basic constructional elements for the growing crystal. Such particles for cadmium sulfide may be $[CdS]_n$, $[Cd_2S]_n$, and $[Cd_3S]_n$.

It is possible to explain the production of various growth forms and to control the occurrence of these, and this seems to be a strong argument in favor of the above theory.

I am indebted to N. N. Sheftal' for constant interest in the work, valuable discussions, and useful advice.

Literature Cited

1. K. Huml. Czechosl. J. Phys., 6A:535 (1961).
2. N. Kh. Abrikosov, V. F. Bankina, L. V. Poretskaya, E. V. Skudnova, and L. E. Shalimova. Production and Properties of Semiconducting Compounds [in Russian], Nauka, Moscow (1967).
3. M. R. Lorenz. II–VI Semiconducting Compounds, 1967 Internat. Conf., New York, (1967), p. 215.
4. N. N. Sheftal'. In: Growth of Crystals, Vol. 5A, Consultants Bureau, New York (1968), p. 25.
5. M. E. Bishop and S. H. Liebson. J. Appl. Phys., 5:24 (1954).
6. I. B. Mizetskaya, A. P. Trofimenko, and V. D. Fursenko. Zh. Neorg. Khim., 3:2276 (1958).
7. S. V. Svechnikov. Zh. Tekh. Fiz., 27:2492 (1957).
8. P. Hoschl and C. Konak. Czechosl. J. Phys., B13:364 (1965).
9. P. Flögel and R. Koepp. Phys. Status Solidi, 10:765 (1965).
10. B. M. Bulakh. Neorg. Mat., 5:25 (1969).
11. B. M. Bulakh and G. S. Pekar'. Neorg. Mat., 6:553 (1970).
12. A. D. Pogorelyi. Zh. Fiz. Khim., 22:721 (1948).
13. P. Goldfinger and D. A. Jeunehomme. Trans. Faraday Soc., 59:2851 (1963).
14. R. A. Burmeister and D. A. Stevenson. J. Electrochem. Soc., 114:394 (1967).
15. G. A. Samorjai and D. W. Jepson. J. Chem. Phys., 41:1389 (1964).
16. S. Ibuki. J. Phys. Soc. Japan, 14:1181 (1958).
17. J. Chikawa and T. Nakarama. J. Appl. Phys., 35:2493 (1964).
18. N. N. Sheftal'. In: Growth of Crystals, Vol. 1, Consultants Bureau, New York (1959), p. 5.
19. H. Buckely. Crystal Growth [Russian translation], IL, Moscow (1953).
20. Crystallization from the Gas Phase [in Russian] (N. N. Sheftal, ed.), Mir, Moscow (1965), p. 11.
21. E. A. Moelwyn-Hughes. Physical Chemistry (2nd ed.), Pergamon Press, New York (1964).
22. A. A. Zhukovitskii and L. A. Shvartsman. Physical Chemistry [in Russian], Metallurgizdat, Moscow (1963), 409.
23. N. N. Sheftal', E. I. Givargizov, B. V. Spitsyn, and A. M. Kevorkov. In: Growth of Crystals, Vol. 4, Consultants Bureau, New York (1966), p. 10.

24. N. M. Emanuel and D. G. Knorre. Textbook of Chemical Kinetics [in Russian], Vyssh. Shkola, Moscow (1962).

25. Ya. K. Syrkin and M. E. Dyatkina. The Structure of Molecules and the Chemical Bond [in Russian], Goskhimizdat, Moscow (1946).

26. E. Yu. Yanson. Coordination Compounds [in Russian], Vyssh. Shkola, Moscow (1963).

27. A. N. Nesmeyanov. Vapor Pressures of the Elements [in Russian], Izd. AN SSSR, Moscow (1961).

28. H. Braune, S. Peter, and V. Neveling. Z. Naturforsch., 6A:32 (1951).

29. J. M. Stanley. J. Chem. Phys., 24:1279 (1956).

30. D. C. Reynolds and I. C. Green. J. Appl. Phys., 29:559 (1958).

31. H. Samelson. J. Appl. Phys., 32:309 (1961).

32. K. Patek. Czechosl. J. Phys., B3:12, 216 (1963).

33. K. Hauptmanova and K. Patek. Phys. Status Solidi, 3:383 (1963).

34. B. M. Bulakh. J. Crystal Growth, 7:196 (1970).

35. R. V. Bakradze, L. A. Sysoev, E. K. Raiskin, and L. V. Konvisar. In: Growth of Crystals, Vol. 6B, Consultants Bureau, New York (1968), p. 72.

36. B. M. Bulakh and I. B. Mizetskaya. Ukr. Fiz. Zh., 7:1125 (1962).

GROWTH MECHANISM FOR VAPOR-DEPOSITED
CADMIUM SULFIDE

B. M. Bulakh and N. N. Sheftal'

The concepts of the molecular theory [1, 2] and of the theory of dislocation growth [3] provide the basis for understanding crystal formation, although the explanation is not complete. In particular, these theories do not incorporate growth from previously formed associated complexes and submicrocrystals. Little attention has also been given to the medium as regards its interaction with the growing crystal, although there is much experimental evidence to indicate that crystal morphology is very sensitive to the external conditions. Therefore, recently there have been attempts to take account of the state of the medium and to consider the growth of crystals not from single ions, atoms, or molecules, but from more complex formations, such as previously associated atoms [4-6], blocks, submicrocrystals [4], and molecular complexes in an intermediate state of aggregation [7]. However, the associative approach takes into account mainly the supersaturation; on the other hand, the morphology is also much dependent on the crystallization temperature and the medium composition.

One of us [8, 9] has discussed a growth mechanism in which the basic structural elements are intermediate complexes of various compositions arising by reaction; the balance between the complexes is dependent on the temperature and the composition of the medium. One of the special features of this approach is that it does not involve stoichiometry or the assumption that the crystal is built up from stoichiometric units, as in the case of cadmium sulfide, where the units are $[CdS]$, $[Cd_2S]$, and $[Cd_3S]$.

These concepts are used here to consider the growth mechanisms for cadmium sulfide crystals of various shapes: prisms, platelets, strips, needles, and whiskers.

Simulation of Crystal Growth

Structured spherical groups have been used [8, 9] to discuss the growth mechanism for cadmium sulfide crystals in vapor; such constructions are convenient for small groups of atoms, but their clarity is lost when the number of atoms becomes large, and it becomes difficult to establish the main trends. Therefore, in the present study we use model equivalents for the building elements to simulate the growth processes.

Model Equivalents

In choosing model equivalents for the complexes $[CdS]$, $[Cd_2S]$, and $[Cd_3S]$, we impose the requirement of maximum clarity and simplicity, with the scope for graphical representation of the growth. Figure 1 shows a sulfur atom surrounded by four cadmium atoms in wurtzite or sphalerite. Such an element can be replaced by a tetrahedron, as is often done in representing structural types; however, tetrahedra do not fill a space completely, and surplus lines are obtained on the figures, which complicate the pattern. The holes are eliminated if the

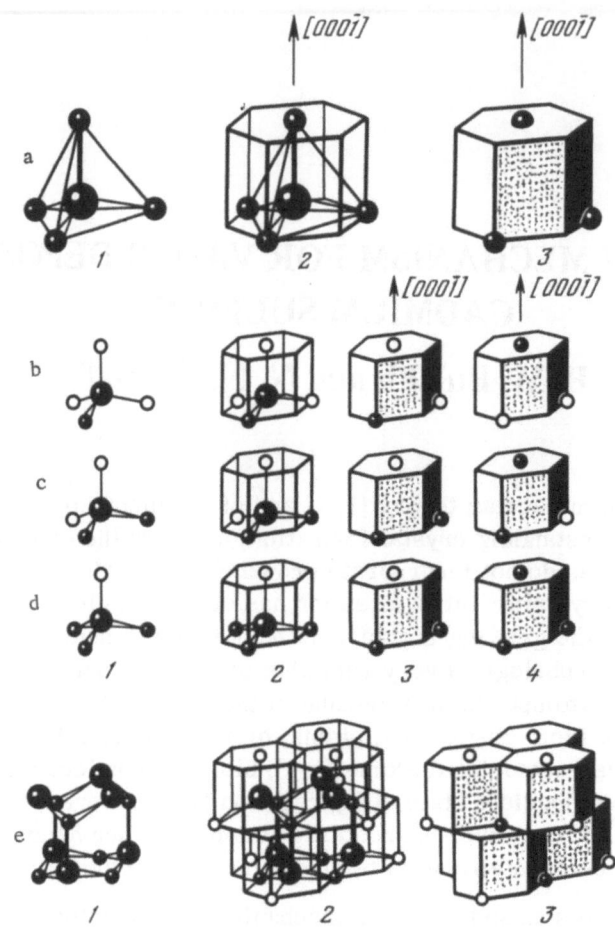

Fig. 1. Model equivalents of structural complexes.

tetrahedron is replaced by a six-faced prism (2 in Fig. 1a). If the positions of the cadmium
atoms are marked on the outer surface of the prism (filled circles), we get an element representing the model equivalent of a tetrahedral group (3 in Fig. 1a). The lacking cadmium
atoms are denoted by open circles.

The model equivalent for [CdS] (Fig. 1b) has one filled circle corresponding to a cadmium atom and three open circles, which denote cadmium vacancies (2 in Fig. 1b). The model
equivalent is spatially oriented, so it can be represented in two ways (3 and 4 in Fig. 1b). The
model equivalent for [Cd_2S] (Fig. 1c) has two filled circles and two open ones (2 in Fig. 1c).
The model equivalent is shown in 3 and 4 of Fig. 1c. Finally, the equivalent for [Cd_3S] (Fig. 1d)
consists of three filled circles (2-4 in Fig. 1d). An association of four [CdS] groups (Fig. 1e)
is shown as model equivalents in 2 and 3 of Fig. 1e. Figure 2 shows the associations considered in [8, 9] together with the model equivalents.

Figure 3 shows the external form of the various crystallographic planes in a wurtzite
structure as formed from model equivalents; the major planes characteristic of CdS single
crystals grown from the vapor are shown: the basal (000$\bar{1}$) plane, the ($\bar{1}$100) and (11$\bar{2}$0) planes
most prominent in prisms, plates, strips, needles, and whiskers, and finally the (10$\bar{1}$2) plane
frequently found at the vertices of needles and whiskers.

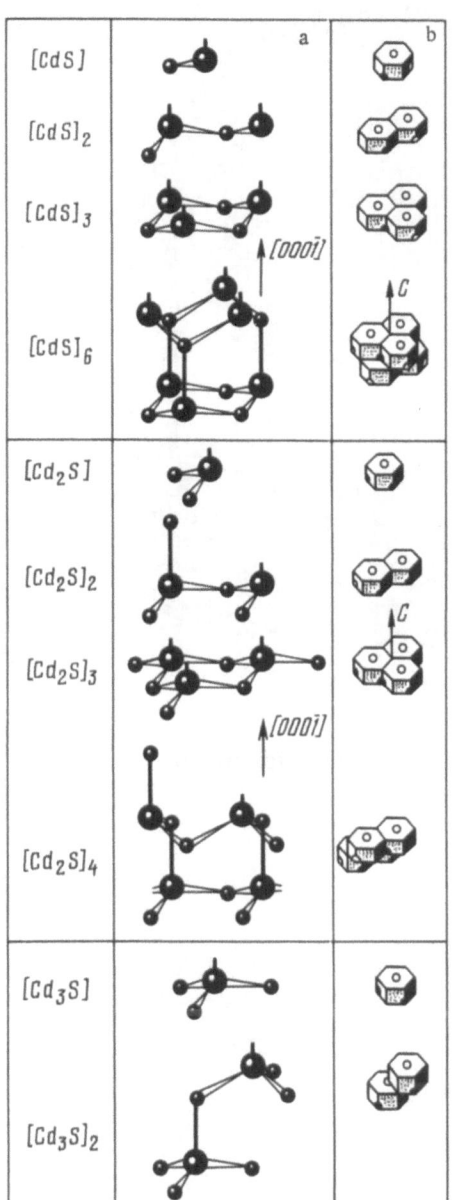

Fig. 2. a) Crystallographic planes of the wurtzite structure; b) the same represented by model equivalents.

Crystal Growth in a Medium with Varying Element Ratios

We consider several cases characteristic of formation of some cadmium sulfide crystal shapes from a vapor with various ratios α between the elements [8].

a) Crystal Growth from a Stoichiometric Medium ($\alpha = 1$): Prism Growth. The basic building elements here are [CdS] complexes; the supersaturation determines how far these are combined into $[CdS]_n$ associations. If one gets $[CdS]_6$, then the growth on [000$\bar{1}$] is possible by addition of associations or of $[CdS]_3$ associations (Fig. 4a). Lateral growth of the resulting association occurs by addition of [CdS] (the points of attachment in the lateral directions are shown by arrows). These points of attachment form niches in which the complexes are attached. The [CdS] groups taken up in these positions form two bonds with the growing association: one to the cadmium atom belonging to the core (at the corner of the niche), and the second via its own cadmium to the corresponding vacancy in the core. In all

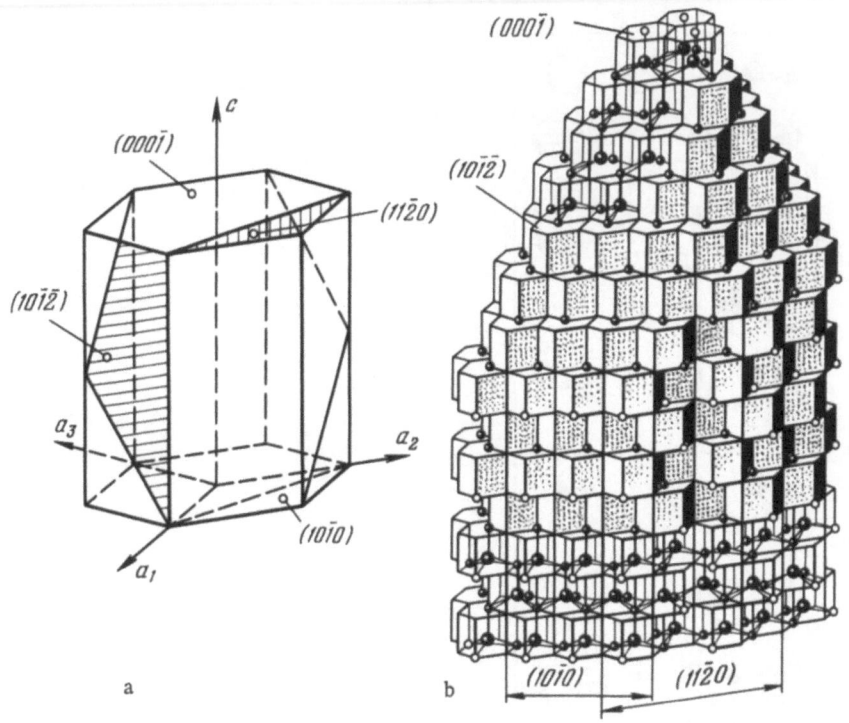

Fig. 3. Positions of crystallographic planes: a) in form; b)
in crystal structure.

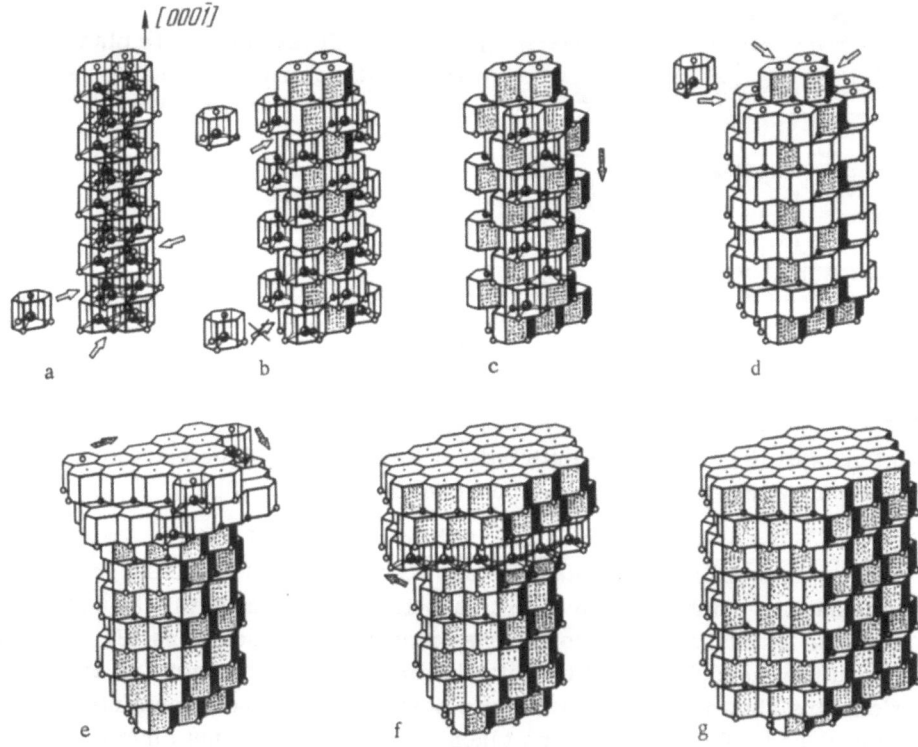

Fig. 4. Growth of prisms from stoichiometric media.

other positions, the group can form only one bond to the core, and such attachment appears unlikely and is neglected in the simulation.

If the association grows on [000$\bar{1}$], then the [CdS] groups are attached from below upwards as the growth proceeds; Fig. 4b shows the structure after attachment of the groups. The attached elements formed stepped balconies, between which there are places that would appear capable of being filled by [CdS] (these places are shown by bold arrows). Here an element also forms two bonds: to a cadmium atom and to a vacancy, but here the incorporation involves combining the vacancies of the center group and the element. The result is a very much weakened double bond, so these sites will not be filled.

Possible incorporation sites are shown by arrows (there are six); here the elements form two bonds: with the cadmium atom and with the cadmium vacancy in the core. The top incoming element combines its cadmium with the core vacancy, and produces under itself a site in which a [CdS] group can form three bonds to the core (Fig. 4c). The probability of such attachment should be larger than when there are only two bonds. The second incoming group produces an analogous place in the adjacent cell. These chain attachments continue from above downwards until all the cells have been filled (Fig. 4c). This stage should be rapid, and it can occur not only via [CdS] groups but also via [CdS]$_2$ ones. As a result, a structure as shown in Fig. 4d is formed, which resembles the initial one (Fig. 4a). Here there are analogous niches, which can be entered by [CdS], but the adjacent parallel niches cannot be filled simultaneously by [CdS] elements, since the latter should thereby combine their vacancy sites.

A second difference is the triple element at the vertex of the crystal; here also there are sites indicated by arrows for the attachment of [CdS] groups forming two bonds. A difference from the above lateral attachment is that here the incoming groups are linked to a (000$\bar{1}$) surface. The cleavage planes in cadmium sulfide crystals are parallel to $\{10\bar{1}0\}$ and $\{1\bar{2}10\}$, so one considers that a complex with (000$\bar{1}$) is stronger than ones found in the lateral directions, so the probability of attachment to the triple association is greater. There are six such attachment points. If a [CdS] group settles on one of these, it combines its cadmium atom with the crystal, and produces an attachment point in which elements to be attached can form three bonds. The sides of the triple element thus acquire overgrowths, and this in turn causes growth in the underlying layer, and a cap is formed on the vertex of the crystal, which consists of two layers of elements and which projects above the lower part of the crystal (Fig. 4e). The cadmium atoms belonging to the lower layer produce underneath points for attachment of elements involving three bonds; an incoming element creates in the same series a place in which three bonds can also be formed. The layer under the cap is sequentially filled along the perimeter (Fig. 4f); the resulting layer produces an analogous process in the next layer and so on, so the side faces grow at this stage by movement of steps over the whole perimeter of the crystal.

These processes produce a structure shown in Fig. 4g that is analogous to the initial structure (Fig. 4a); the lateral growth may now continue as described above, i.e., via the sequence a → b → c → d → e; however, if [CdS]$_3$ or [CdS]$_6$ groups settle on the (000$\bar{1}$) surface, a cap projecting above the crystal is again produced as a result of the rapid growth, and the process repeats via the sequence d → e → f.

We therefore have here two growth mechanisms: two-dimensional nucleation for the basal plane and a step motion for the lateral faces. It has been pointed out [10] that these mechanisms can occur in crystals with natural faces, and it is very interesting that step growth involves nucleation on the basal plane. Expansion of a two-dimensional nucleus causes the side faces to grow; a nucleus in its growth induces steps on the side faces, which move along the crystal. These conditions produce isometric crystals bounded by $\{10\bar{1}0\}$ lateral faces (Fig. 4g).

b) Crystal Growth in a Medium with a Small Cadmium Excess
($\alpha > 1$). If the vapor has excess cadmium, one begins to get [Cd_2S] groups; if such a group participates in producing [CdS]$_6$ groups, we get conditions for spiral lateral growth of such an association via an attachment of [CdS] or [CdS]$_2$ groups. Figure 5 uses model equivalents to show the growth stages of an association containing excess cadmium, which strengthens the binding of the group to the surface and produces sites for preferential attachment of [CdS] or [CdS]$_2$ (Fig. 5a). When the latter is attached, it forms three bonds to the crystal (Fig. 5b). The incoming group produces ahead of itself a site for preferential attachment in which five bonds can now be formed (Fig. 5c). The next incoming group is linked to the crystal by three bonds (Fig. 5d). The next attachment is accompanied by formation of four bonds (Fig. 5e). Subsequent attachments (Fig. 5, f and g) lead to a position analogous to the initial one (Fig. 5a).

Although attachment of [CdS]$_2$ results in nonequivalent sites for the next incoming groups, the sequence should be as described above, since the attachment conditions are less favorable at all other sites. When a group with excess cadmium grows, there is an effective messenger cadmium atom, which facilitates the next attachment; the growth continues until the layer is completed.

A nucleus expanding on a (000$\bar{1}$) surface induces growth steps on the side faces, so a small excess of cadmium in the vapor should improve the growth of isometric crystals, as has actually been found [11].

c) Needle Growth ($\alpha < 2$). If there is a considerable cadmium excess in the vapor, one gets [CdS] and [Cd_2S], and the concentration of the latter increases with α. If the supersaturation is appropriate, one will get [Cd_2S]$_3$, and such groups can grow by combination only along [000$\bar{1}$] and [0001] (Fig. 6a). The result is a core with attachment sites where an element forms two bonds; these sites are suitable for attaching [CdS] and also [Cd_2S]; however, while [Cd_2S] groups use all their bonds, [CdS] has one vacant bond unused. One therefore expects that preference will be given to [Cd_2S] (Fig. 6b); the resulting core appears screened by cadmium atoms, and then only [CdS] groups can be attached (Fig. 6c). The resulting structure again has sites allowing attachment of [Cd_2S]. There are six of these in each row; however, one does not get simultaneous filling of two adjacent sites by [Cd_2S], since then the two cadmium atoms of two adjacent groups would meet in one node. Therefore, only three sites are filled in each row, and these are symmetrically disposed (Fig. 6d). Subsequent growth is possible only via [CdS] (Fig. 6e), and these groups on attachment set up sites for attachment of [Cd_2S], and so on.

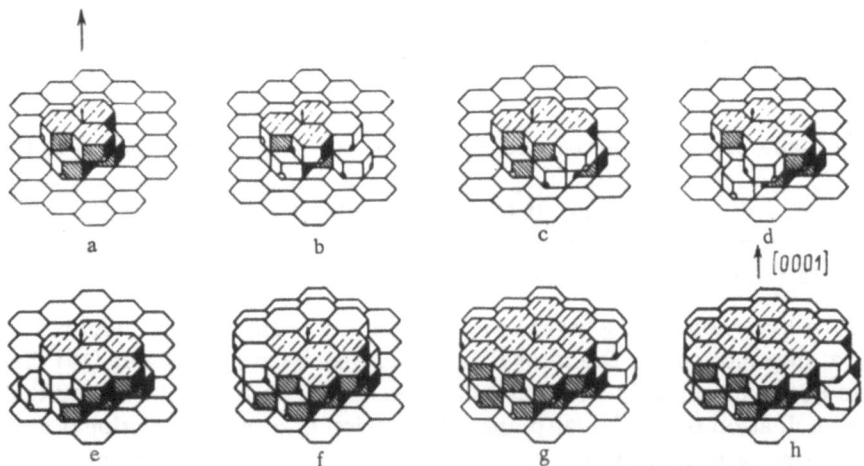

Fig. 5. Growth of prisms from a medium with a slight excess of cadmium.

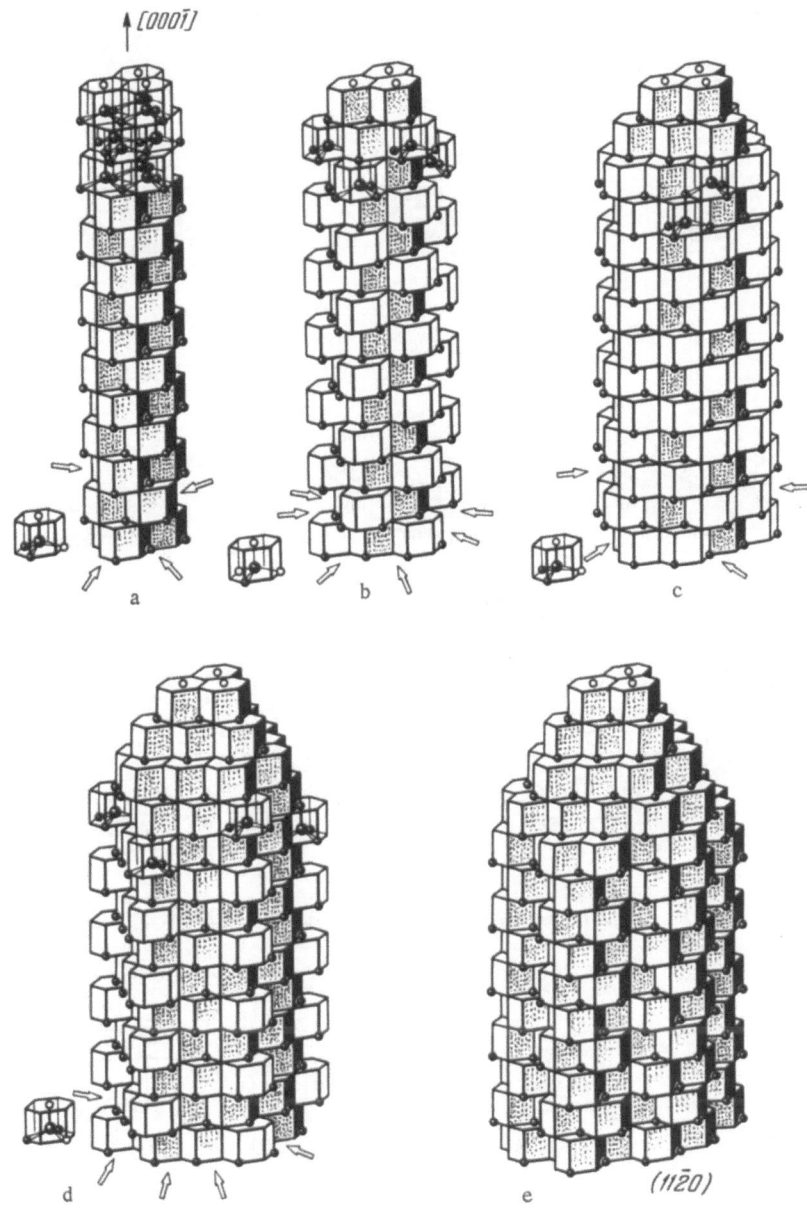

Fig. 6. Sequential addition of complexes in needle growth.

It is clear from the above figure that a growing needle is bounded by $\{1\bar{2}10\}$ planes; the [Cd$_2$S] groups preserve the $\{1\bar{2}10\}$ planes, as Fig. 7 shows, in terms of sections of a crystal consisting of three building layers at various stages of growth. We consider the top layer. Three [Cd$_2$S] elements are attached to the top layer (Fig. 7a and b), and then [CdS] elements (Fig. 7c), which create six sites for attachment of [Cd$_2$S]. It is impossible for two groups to be attached simultaneously, so three of the six cells are filled (Fig. 7d). The incoming elements in the lower layer are turned through 60°. After the [CdS] groups have been attached (Fig. 7e), one again gets attachment of [Cd$_2$S] groups (Fig. 7f). The gaps between the [Cd$_2$S] groups are filled by [CdS] (Fig. 7, g and h), and so on. The last figure clearly shows the structure of the side faces, which corresponds to $\{1\bar{2}10\}$ planes.

At the vertex of a crystal growing on [000$\bar{1}$] (Fig. 6) one gets sites with a lower attachment probability; therefore, the incoming groups gradually become less common downwards,

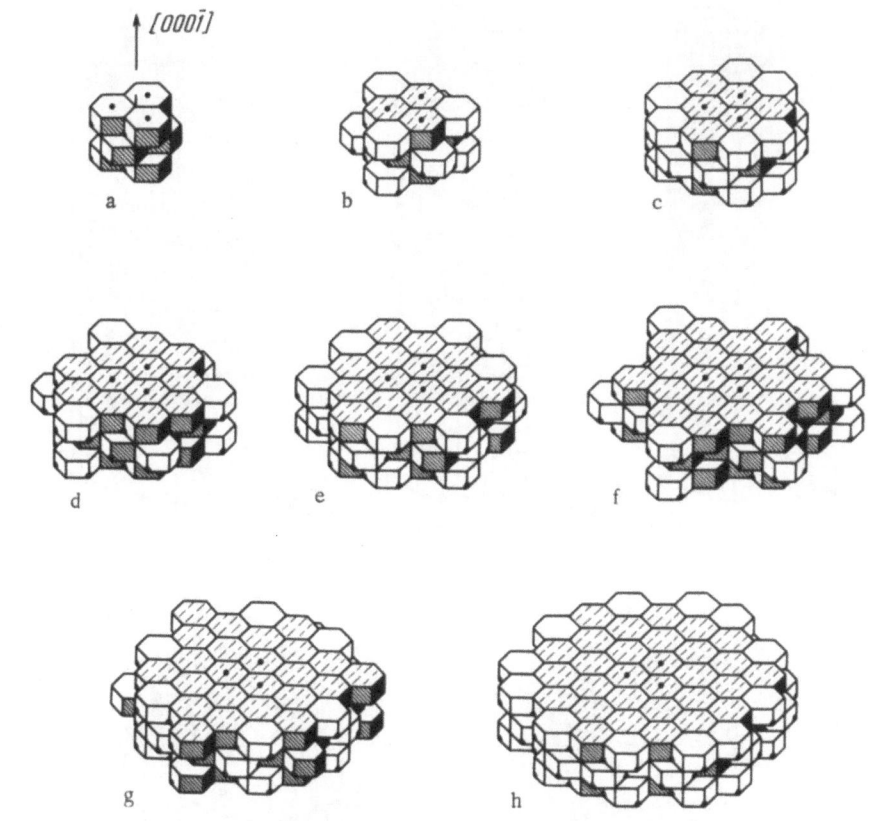

Fig. 7. Growth of needles along [0001].

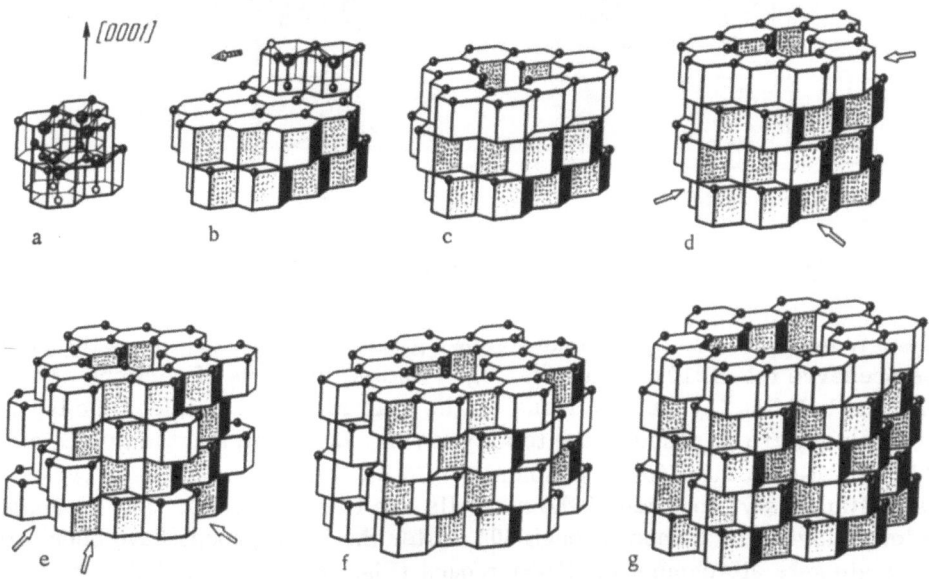

Fig. 8. Stages in needle growth: formation of a cavity.

which results in a conical vertex bounded by $\{10\bar{1}2\}$ planes, as have actually been observed at the tops of needles [12]. If the top of a crystal continues to grow, one will find different planes at the vertex on account of interaction between the lateral and longitudinal modes of growth. The base of the crystal is planar and corresponds to a (0001) face.

This last feature is responsible for the growth behavior of needles on [0001] (Fig. 8). In this case the initial association (Fig. 8f) expands laterally to form a planar surface, on which cadmium atoms emerge; this can grow on [0001] by attachment of $[Cd_2S]_3$ groups, but this is not the only possibility. A (0001) surface can accommodate also $[Cd_2S]_2$ groups, and a preferable site for attachment of these lies at the periphery of the crystal (Fig. 8b). An incoming $[Cd_2S]_2$ group produces a site more favorable for attachment of $[Cd_2S]$ or $[Cd_2S]_2$; this is a messenger-type process and terminates with closure of the chain (Fig. 8c). Within the chain there remains a cavity, which cannot be filled by the building elements; in fact, the substrate on which the $[Cd_2S]_2$ group deposits exerts an orienting influence on the incoming element, and it forces the groups to form ring chains repeating the shape of the substrate. Subsequent attachments on [0001] for $[Cd_2S]$ groups accentuate the resulting cavity (Fig. 8d).

The lateral growth of a needle occurs as for the growth of $[000\bar{1}]$, i.e., the incoming $[Cd_2S]$ groups (Fig. 8e) alternate with [CdS] ones (Fig. 8f). As the crystal becomes thicker, the $[Cd_2S]$ groups attached on [0001] form closed rings around the periphery and cause the internal cavity to expand (Fig. 8g). Then the growth of a $[Cd_2S]_2$ association on [0001] results in hollow needles; the growth on [0001] should be faster than that on $[000\bar{1}]$, since the change on (0001) can be built up from different forms of $[Cd_2S]$. This conclusion agrees with our experimental results.

d) Platelet Growth ($\alpha \approx 2$). When the ratio between the atoms is close to 2, the vapor contains mainly $[Cd_2S]$, with few [CdS]; if the supersaturation is appropriate, one gets $[Cd_2S]_4$, and these groups may expand by attachment of analogous ones, and also, which is more likely, by attachment of $[Cd_2S]_2$ in $[000\bar{1}]$ and $[10\bar{1}0]$ directions. The resulting structure has an extensive $(1\bar{2}10)$ plane (Fig. 9a); the arrows indicate the sites for possible attachment of building elements; the sole element that can be attached here is [CdS], and attachment of this gives the structure shown in Fig. 9b, which has sites that can also accommodate only [CdS]. The incoming elements form three bonds to the core, and they combine their vacancies with the cadmium atoms of the core and present their own cadmium atoms at the surface of the crystal. Then the growing $[Cd_2S]_4$ association forms a type of substrate providing for epitaxy of [CdS] groups.

Apart from these sites at the edges of the crystal, there are sites (shown by arrows) in which a [CdS] group will form only two bonds to the crystal (Fig. 9b). The probability of attachment at such points is therefore less, which might appear to imply that the more rapidly growing $(1\bar{2}10)$ plane would be displaced by the slowly growing $(0\bar{1}10)$ and $(10\bar{1}0)$ planes; however, this does not occur because these sites are taken up by $[Cd_2S]$ groups, whose concentration is high (Fig. 9c); these groups even out the edge of the crystal, and, as in needle growth, prevent the displacement of the $(1\bar{2}10)$ planes. It seems particularly interesting that the $[Cd_2S]$ groups should play such a role in the growth of the crystal.

Further attachment of [CdS] groups reproduces the structure of the base (Fig. 9d); the next layer of groups deposits (Fig. 9e) to produce along the edges sites that are filled by $[Cd_2S]$; the growth rate on $[1\bar{2}10]$ should be dependent on the [CdS] concentration; if the growth rate is proportional to the concentration of the corresponding building elements, then the typical size of $10 \times 10 \times 0.1$ mm enables one to estimate the relative concentration of [CdS], which is less by a factor 10^2 than that of $[Cd_2S]$. The concentration of [CdS] falls as α increases, and the plates become thinner, and turn into strips.

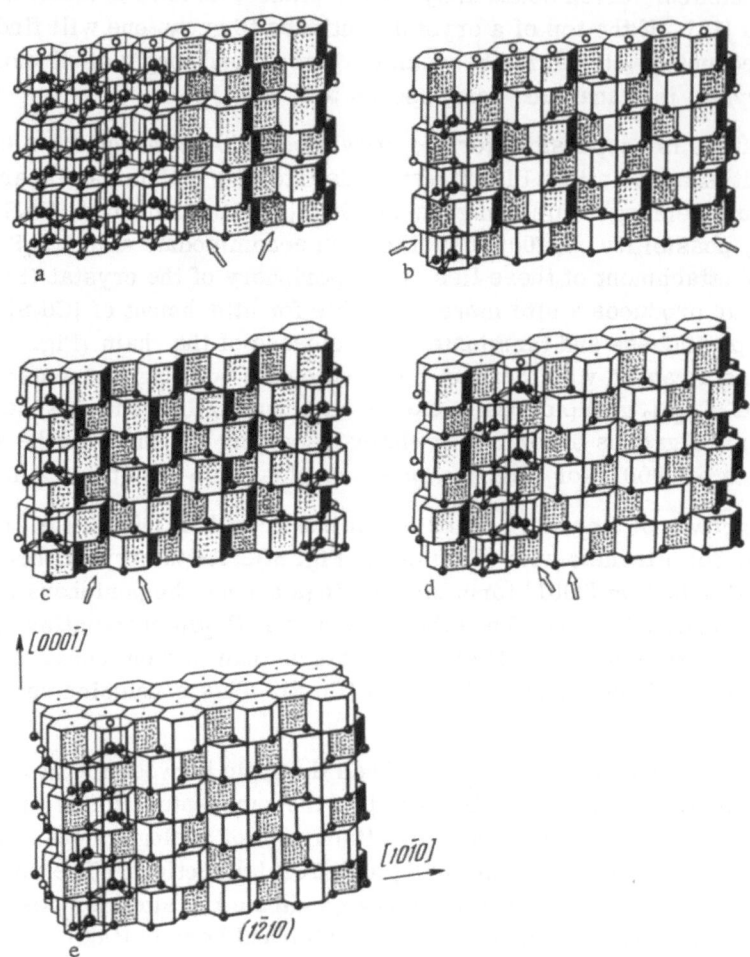

Fig. 9. Stages in platelet growth.

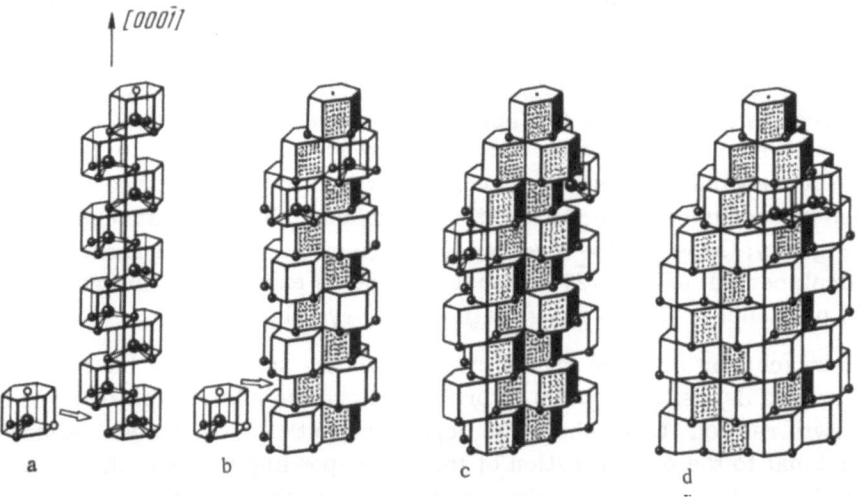

Fig. 10. Whisker growth.

e) Whisker Growth ($\alpha > 2$). If the ratio is greater than 2, one gets [Cd$_3$S], which at adequate concentrations can produce only chain associations; the growth mechanism for such an association is analogous to that considered above for needles. The successive stages are shown in Fig. 10. The growing [Cd$_3$S] association (Fig. 10a) has sites for attachment of [Cd$_2$S] groups (Fig. 10b), which in turn provide conditions for attachment of groups of the same composition (Fig. 10c). Then only [CdS] groups can be attached to the core (Fig. 10d); these create sites that will take [Cd$_2$S], and so on. The process takes the sequence described for needle growth, but the initial configuration of the growing association results in some distinctive features. The incoming [CdS] groups lie not at angles of 120° (as in needles), but at 139 and 82°, so the cross section of the crystal is a deformed hexagon with two parallel planes more extensively developed, and these crystals are very common (see Fig. 8a in [9]).

The [CdS] concentration is low in this region, so the lateral growth rate is extremely small, and hence one gets very thin needles (whiskers).

Nonstoichiometry of Cadmium Sulfide Crystals

With the exception of growth in a stoichiometric vapor, all other instances relate to incorporation of nonstoichiometric complexes such as [Cd$_2$S] and [Cd$_3$S], so the resulting crystal should have excess cadmium (be nonstoichiometric); the deviation from stoichiometry in cadmium sulfide crystals is considered in the next section.

Nonstoichiometry of Needles

To determine the deviation from stoichiometry in needles grown on [000$\bar{1}$] one needs to follow in sequence the attachment of the elements during growth. Figure 11 shows the sequential stages in the growth of a layer of elements for a needle, where the filled hexagons denote [Cd$_2$S] groups. The filled hexagons with white circles indicate [Cd$_2$S], while the hatched hexagons denote [CdS]. It is clear from Fig. 11 that the [Cd$_2$S]$_3$ association, which consists of three [Cd$_2$S] groups, (shown as +3" in the figure), will attach three [Cd$_2$S] groups (+3"), and then six [CdS] (+6), after which we get attachment of three [Cd$_2$S] (+3"), and so on.

We denote [Cd$_2$S] by a and [CdS] by b; then the growth sequence of an element layer may be represented as

$$3a + 3a + 6b + 3a + 6b + 3a + 6b + 12b + 3a + 12b + 3a + 12b + 18b + 3a + 18b + $$
$$+ 3a + 18b + 24b + 3a + 24b + 3a + 24b + 30b + \ldots$$

There are repeating periods or cycles in this sum, which equals

$$N(n) = 6a + (6a + 18b) + (6a + 18 \cdot 2b) + (6a + 18 \cdot 3b) + \ldots + \left[6a + 18b \frac{n(n+1)}{2}\right], \qquad n = 0, 1, 2, 3, \ldots,$$

whence

$$N(n) = 3(n+1)(2a + 3nb),$$

where n is the number of cycles; $N(0) = 6$ for n = 0, and we have the initial association with the [Cd$_2$S] elements attached to it (2 in Fig. 11); when one is determining the total number of groups in a layer, one needs to bear in mind that a denotes one [Cd$_2$S] element, while b denotes one [CdS], i.e., in that case $a = b = 1$; then the total numbers of the two forms of elements in one layer n is given by

$$N_e(n) = 3(n+1)(2 + 3n).$$

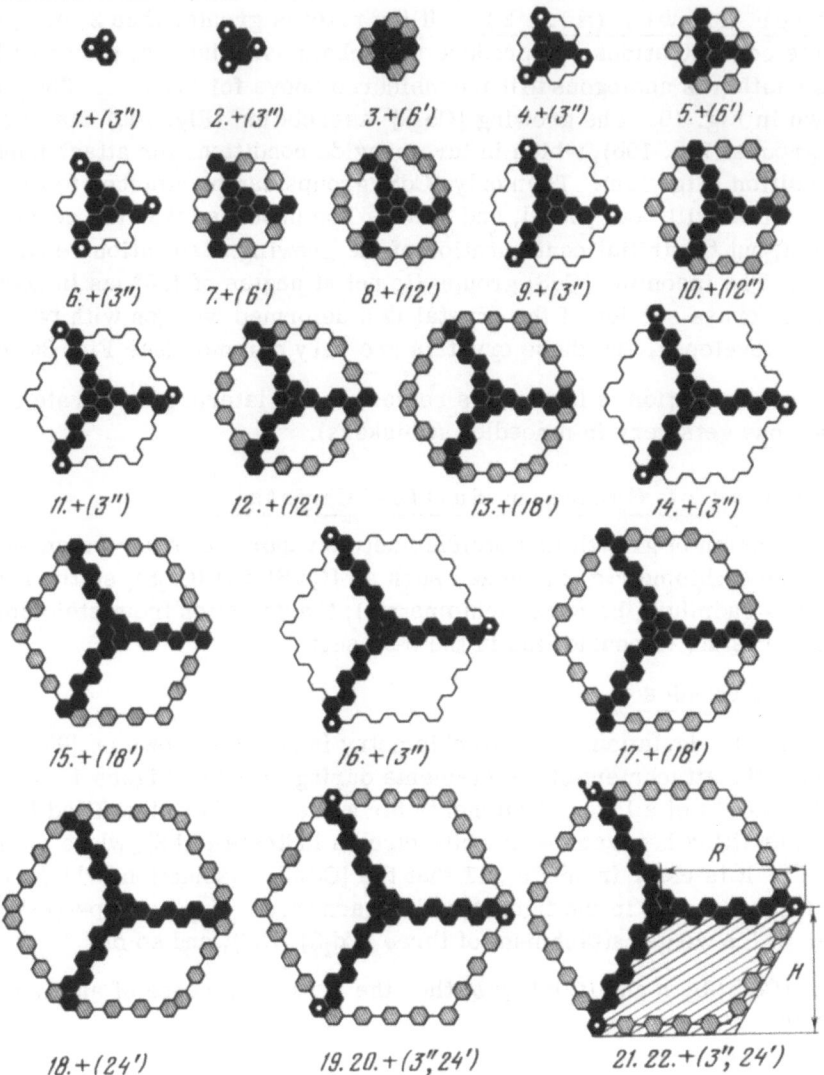

Fig. 11. Nonstoichiometry of needles.

The number of sulfur atoms is equal to the number of building elements:

$$N_B(n) = N_e(n) = 3(n+1)(2+3n).$$

The number of cadmium sulfide molecules is

$$N_{AB}(n) = N_B(n) = N_e(n) = 3(n+1)(2+3n).$$

Each [Cd₂S] group contains two cadmium atoms, so in calculating the number of cadmium atoms in a layer we must put $a = 2$ and $b = 1$; the number of cadmium atoms in a layer is

$$N_A(n) = 3(n+1)(4+3n).$$

The element ratio in the crystal is

$$\varkappa = \frac{N_A(n)}{N_B(n)} = \frac{3n+4}{3n+2}.$$

The absolute excess of cadmium is

$$\Delta N_A (n) = N_A (n) - N_B (n)$$

and the deviation from stoichiometry in a crystal is

$$\eta = \frac{\Delta N_A (n)}{N_{AB} (n)} = \frac{N_A (n) - N_B (n)}{N_{AB} (n)} , \qquad \eta = \frac{2}{3n + 2} .$$

If n = 0, then η = 1, i.e., for the $[Cd_2S]_3$ group after the first attachment of three $[Cd_2S]$ groups, and there is one excess cadmium atom for each CdS molecule; $\eta \to 0$ if $n \to \infty$, i.e., stoichiometry is a limiting state to which the crystal tends.

The parameter n (number of cycles) reflects the thickness of the crystal; each successive cycle denotes a thickness increase by a definite amount. For practical purposes, it is more convenient to relate the nonstoichiometry to the thickness. The base area of a hexagonal building element is

$$S_e = \frac{4a^2 \sqrt{3}}{3} ,$$

where a is the interatomic distance. The cross-sectional area of a six-faced crystal is

$$S_c = \frac{3D^2}{2 \sqrt{3}} ,$$

where D is the diameter of the circle inscribed in the cross section (thickness of the crystal). If the cross section contains N_e building elements, we have

$$S_c = \Sigma S_e = N_e S_e,$$

whence

$$9n^2 + 15n + 6 = \frac{3D^2}{8a^2} ,$$

and

$$n = \frac{\sqrt{1 + 3D^2/2a^2} - 5}{6} .$$

We substitute for n in the equation for η to get

$$\eta = \frac{4}{\sqrt{1 + 3D^2/2a^2} - 1} .$$

As $3D^2/2a^2 \gg 1$ even for D > 0.1 μ, we can put

$$\eta = \frac{8a}{D \sqrt{6} - 2a} = \frac{4a}{1.22D - a} .$$

If a needle has D = 10^{-2} cm, then η = 0.82 $\cdot$ 10^{-5} atoms of Cd per CdS molecule; if D = 10^{-1} cm, η = 0.8 $\cdot$ 10^{-6} Cd atoms per CdS molecule and so on. The deviation from stoichiometry in needles is therefore dependent on the crystal thickness, and for typical dimensions lies in the range 1–10 $\cdot$ 10^{-6} atoms of Cd per CdS molecule.

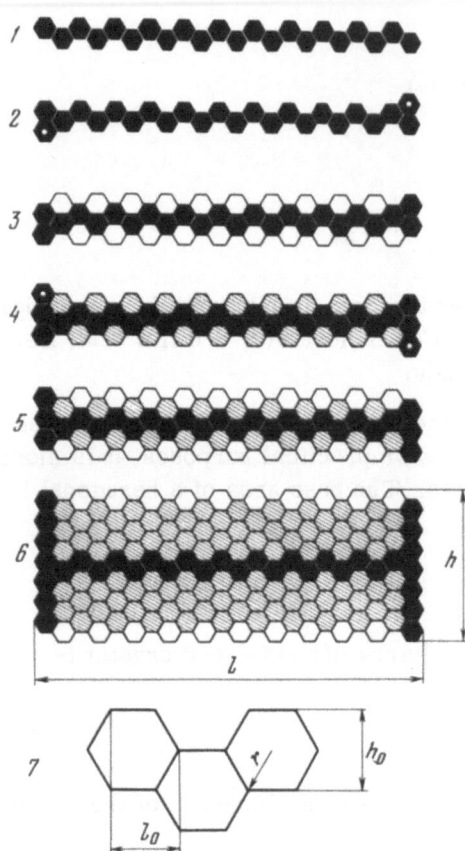

Fig. 12. Addition of building elements in platelet growth.

Platelet Nonstoichiometry

Figure 12 shows the comparatively simple sequence in which the building elements are attached in platelet growth.

The basis of the crystal is layed by growth on a $[Cd_2S]_4$ association, and for one layer of elements is represented by a chain (1 in Fig. 12), to the edges of which $[Cd_2S]$ groups are attached (2 and 4 in Fig. 12), while the internal regions between these are filled by $[CdS]$ (3 and 5 in Fig. 12).

Consider the number of building units contained in the cross section of the crystal, the length being l and the thickness h (6 in Fig. 12); 7 in Fig. 12 shows the necessary dimensions of the bases of the building elements, and we have for the radius of an element

$$r = \frac{2\sqrt{2}}{3}a,$$

and for the length of an element

$$l_0 = \frac{3}{2}r = a\sqrt{2},$$

and finally the width

$$h_0 = \frac{2a\sqrt{6}}{3}.$$

Then the number of elements in a series l is

$$N_l = \frac{l}{l_0} = \frac{l}{a\sqrt{2}}.$$

The number of elements in an h series is

$$N_h = \frac{h}{h_0} = \frac{3h}{2a\sqrt{6}}.$$

The number of excess cadmium atoms in an l series is

$$\Delta N_A(l) = N_l,$$

and that in an h series is

$$\Delta N_A(h) = 2N_h,$$

since the crystal is bounded by $[Cd_2S]$ elements on two sides. The total number of excess cadmium atoms in a cross section is

$$\Delta N_A = \Delta N_A(l) + \Delta N_A(h) = N_l + 2N_h,$$

$$\Delta N_A = \frac{l + \sqrt{3h}}{a\sqrt{2}}.$$

The total number of elements in a layer is

$$N_e = N_l N_h = \frac{3hl}{4a^2\sqrt{3}}.$$

The total number of CdS molecules in a layer N_{AB} equals the number of elements N_e, i.e.,

$$N_{AB} = N_e = \frac{3hl}{4a^2\sqrt{3}}.$$

The nonstoichiometry of the platelet is

$$\eta = \frac{\Delta N_A}{N_{AB}} \approx 2.45a\left(\frac{1}{h} + \frac{1.73}{l}\right).$$

As $h \ll l$ for platelets, the thickness is the main factor in determining η; typical platelets have $l = 1$ cm, $h = 10^{-2}$ cm, and then $\eta = 4.2 \cdot 10^{-6}$ atoms of Cd per CdS molecule; if $l = 1$ cm and $h = 2 \cdot 10^{-2}$ cm, then $\eta = 2.1 \cdot 10^{-6}$ atoms of Cd per CdS molecule.

It is of interest to compare these results with the measured nonstoichiometry for CdS crystals [13]; all cadmium sulfide crystals grown from the vapor, no matter what technique was employed, contain excess cadmium to the extent of $(1-10) \cdot 10^{-6}$ atoms of Cd per CdS molecule. CdS crystals grown in an H_2-H_2S atmosphere at 900°C gave values of $(1.1-2.7) \cdot 10^{-6}$ Cd atoms per CdS molecule. Unfortunately, no statement is made of the size of the crystals used in these measurements. Our results however appear to agree well with experiment.

In conclusion we give some forms of growth model for cadmium sulfide crystals to illustrate the scope for representing growth mechanisms using these groups.

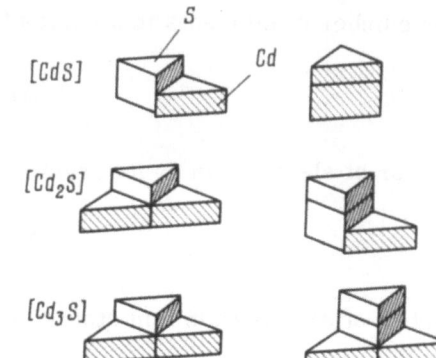

Fig. 13. Single CdS, Cd$_2$S, and Cd$_3$S complexes represented by trigonal prisms .

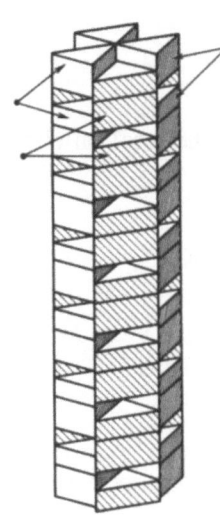

Fig. 14. The first stage in prism growth.

First Form.   We represent the cadmium and sulfur as trigonal prisms to take into account the structural features of wurtzite; the prism representing the sulfur atom has twice the height of the cadmium prism. Figure 13 shows the single groups CdS, Cd$_2$S, and Cd$_3$S in terms of such prisms. Figures 14-17 use these equivalents to indicate the growth stages in a prism. The points of attachment of the [Cd$_n$S] groups are shown by arrows.

Second Form.*   This is illustrated by reference to platelets and needles of cadmium sulfide, since these require fewer groups for construction; these show quite clearly how the shape of the groups in the medium dictates the growth mechanism, and hence the growth shape of the crystal. It is clear from the figure that the incorporation sequence is determined by the number of bonds to be completed; the attachment probability increases with the number of such bonds.

Needle Formation when $\alpha < 2$ (Fig. 18). Combination of [CdS]$_3$ groups in the [0001] direction gives rise to a core, where there are sites for attachment with two free bonds each, which will take [Cd$_2$S] groups using all their bonds. These are therefore the ones to be preferred. The [CdS] do not use one bond to the core, so attachment of these is less likely (Fig. 18a).

After the [Cd$_2$S] groups have been attached, the core is screened by Cd atoms, and now only [CdS] groups can be taken up (Fig. 18b), which is the second stage of growth.

───────

* Devised by A. N. Buzynin.

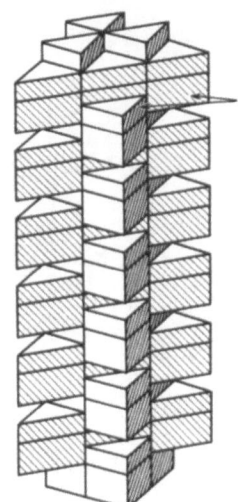

Fig. 15. The second stage in prism growth.

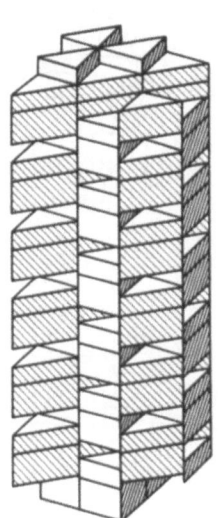

Fig. 16. The third stage in prism growth.

Fig. 17. The fourth stage in prism growth.

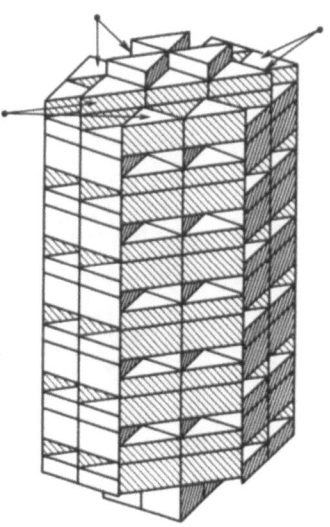

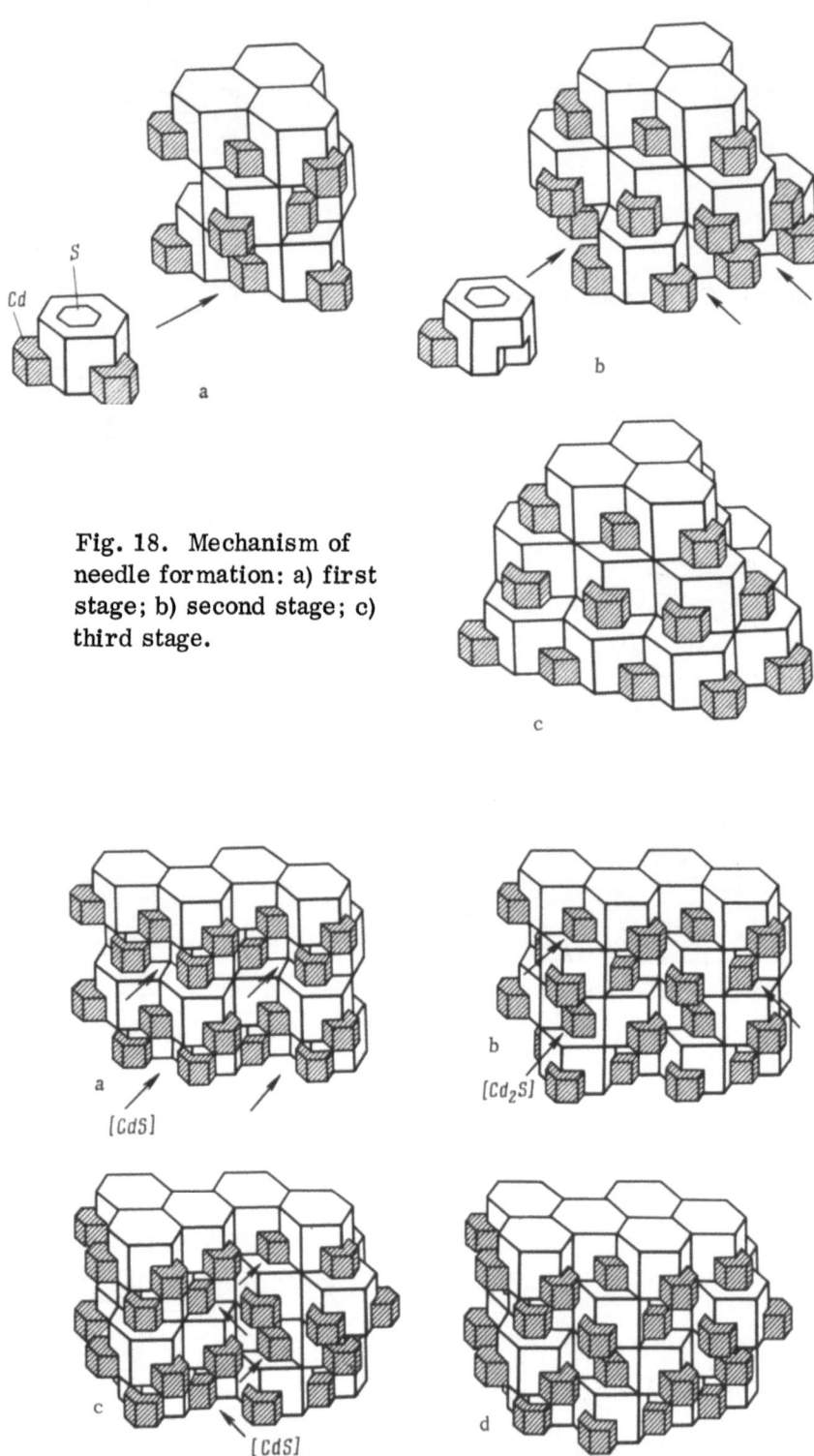

Fig. 18. Mechanism of needle formation: a) first stage; b) second stage; c) third stage.

Fig. 19. Mechanism of platelet formation: a) first stage; b) second stage; c) third stage; d) fourth stage.

Fig. 20. Growth model for the wurtzite
structure with atoms as tetrahedra.

The resulting structure again has sites that will accommodate $[Cd_2S]$, which provide for retention of the $(1\bar{2}10)$ planes.

Sites with lower attachment probabilities arise at the top of the core, so that one gets a conical vertex bounded by $(10\bar{1}2)$ planes (Fig. 18c).

Platelet Formation when $\alpha = 2$ (Fig. 19). The basis of a platelet is a $[Cd_2S]_n$ association; the sole element that can attach to the niches in this core is $[CdS]_n$ (Fig. 19a).

Incoming elements form three bonds to the core, and they combine their vacancies with the Cd atoms of the base and project their own Cd atoms to the surface; now the niches with sites for attachment providing three bonds are accompanied by new sites at the edges of the crystal where there are two free bonds, these being suitable for $[Cd_2S]$ groups, whose concentration is high (Fig. 19b).

The [CdS] groups thus even out the edge of the crystal and prevent displacement of the $(1\bar{2}10)$ planes (Fig. 19c).

The subsequent epitaxy of $[Cd_2S]$ groups reproduces the structure of the core (Fig. 19d).

Third Form (Fig. 20). In this form we have discarded the concept of continuous filling of the available space, this being a basic concept in the previous forms, since we assumed that such filling should provide greater clarity in simulating the growth; however, various difficulties are encountered in realizing this idea, in particular difficulties in making models.

In the third form, which is a discontinuous model, each atom is assigned a facetted form obtained by drawing planes perpendicular to the bond directions to the nearest neighbors. Then the shape of the atom reflects the number and disposition of the nearest neighbors, but only the nearest ones in the lattice. This agrees with the concept that the final growth form of a perfect crystal is determined by the interaction of the atoms only with their nearest neighbors [14].

The dimensions of the facetted atoms correspond approximately with the radii; $A:B = 1:2$ for crystals with the wurtzite structure.

The polyhedra are linked one to another by short rigid bonds (wires). The atoms and groups are linked by these bonds, which enables one to assemble and dismantle the structure in a definite sequence controlled by the morphology of the core and the groups.

Conclusions

A study of the real growth conditions for cadmium sulfide in vapor indicates that the supersaturation is not the sole factor; important effects come also from the ratio between the elements in the growth region and the crystallization temperature, which may have the decisive effect on the growth form. The results show that the crystallization cannot be considered in isolation from the medium, especially the composition and the chemical processes occurring there.

Crystal growth is indissolubly related to chemical reactions; consideration of the latter opens the way to understanding many features of crystal growth. It is possible to explain from a single viewpoint many trends in growth shapes as influenced by external conditions by discussing the nucleation and growth mechanism from the viewpoint of the intermediate complexes formed in the vapor.

Growth process simulation via intermediate complexes enables one to consider in detail the various stages of crystallization, together with the details of production for various crystal shapes; on the other hand, even the simplest examples, where the behavior of the groups is most clear, would indicate a fairly complex and very detailed interaction between the growing crystal and the medium. Even slight changes in the medium are rapidly reflected in the growing crystal. Crystal morphology can therefore act as a very sensitive indicator for the processes in the medium.

Literature Cited

1.   W. Kossel. Nachr. Ges. Wiss. Göttingen. Math. Phys. Kl. (1927), p. 135.
2.   I. N. Stranski. Z. Phys. Chem., 136:259 (1928).
3.   W. K. Burton, N. Cabrera, and F. C. Frank. Nature, 163:398 (1949).
4.   N. N. Sheftal'. In: Growth of Crystals, Vol. 1, Consultants Bureau, New York (1959), p. 5.
5.   S. A. Stroitelev. Izv. AN SSSR, Neorg. Mat., 4:1411 (1968).
6.   S. A. Stroitelev. In: Mechanism and Kinetics of Crystallization [in Russian], Izd. AN BSSR, Minsk (1969), p. 183.
7.   R. O. Grisdale. In: Theory and Practice of Crystal Growth [Russian translation], Metallurgiya, Moscow (1962), p. 176.
8.   B. M. Bulakh. J. Crystal Growth, 5:243 (1969).
9.   B. M. Bulakh. This volume, p. 88.
10.   D. C. Reynolds and L. C. Green. J. Appl. Phys., 29:559 (1958).
11.   N. Hemmat and W. Weinstein. J. Electrochem. Soc., 114:851 (1967).
12.   H. Samelson. J. Appl. Phys., 32:309 (1961).
13.   L. Hildisch. J. Crystal Growth, 3(4):130 (1968).
14.   N. N. Sheftal'. Z. Kristall and Technik, 8(1-3):149 (1973).

PINACOID GROWTH PYRAMIDS IN
ARTIFICIAL QUARTZ CRYSTALS

G. V. Kleshchev, A. N. Bryzgalov, P. P. Butorin, L. N. Chernyi, V. N. Turlakov, A. F. Kuznetsov, and L. V. Skobeleva

Artificial quartz crystals differ from natural ones in having pinacoid growth pyramids, i.e., planes that do not occur in natural crystals as faces and which are essentially created by the shape of the seed.

Such pyramids differ morphologically from the adjacent growth pyramids for the S trigonal dipyramid and the X trigonal prism, being separated by the L_1 and L_2 boundaries from these (Fig. 1). These are formed at the stage of regeneration of the steady-state growth form [1-7], and they taper off completely when this process is completed.

No adequate explanations have been given for the nature of these pyramids, the formation mechanism, and the characteristic features of their structure.

Here we use new evidence to analyze the structure of the pinacoid growth pyramids described in [1-7].

Autonomous Growth Regions in a Pinacoid
Growth Pyramid and Nature of the Boundaries

The most characteristic feature of the pinacoid growth pyramids is their dissection into regions of autonomous growth (growth cones in [8], regions of cellular growth in [9]). These regions differ in zoning or banding and are separated by boundaries well seen in various methods [4, 10-16] (see Figs. 1-3 and 7 in [1], part 1 of Fig. 4 in [10], and in [14]).

The shadow projection method (Fig. 2) [14] gives the clearest picture that is not complicated by secondary details, and one that enables us to establish the main features of the boundaries. The boundaries of an autonomous growth region are seen in this case as lines of dark-light contrast, which indicates that the image is of diffraction nature.

If the plate is cut parallel to an end face of the crystal (Y-cut), then one sees two types of such lines: lines of the first type λ_1 parallel to the L_1 boundary and lines of the second type λ_2 parallel to the L_2 boundary. This relation in orientation between λ_i and L_i is of organic character; if the growth conditions are altered, and hence the relation between the growth rates in the corresponding faces, then the orientation of the L_i boundaries will alter also, and with it the orientation of the λ_i boundaries, each line remaining parallel to the corresponding L_i boundary. This indicates that the boundaries of the λ_1 and λ_2 growth regions and the boundaries between the pyramids L_1 and L_2 are of identical nature. We now consider the origin of this.

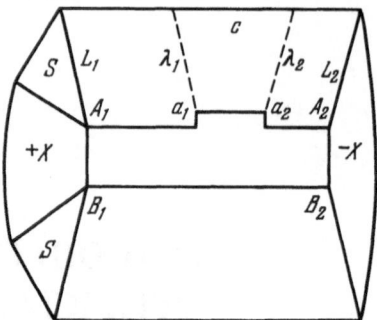

Fig. 1. λ defects in a pinacoid overgrowth pyramid and L boundaries between overgrowth pyramids (L_1 between S and C, L_2 between C and −X).

The L_1 and L_2 boundaries (Fig. 1) are formed on the A_1 edges (and B_1) and on the A_2 ones (and B_2), these edges being perpendicular to the plane of the figure, and each denoted by one letter; consequently, these edges may be considered as active sources for the L_1 and L_2 boundaries.

Let a step $a_1 a_2$ be formed on the growing surface of the crystal; edge a_1 of this step resembles A_1 edge in being an active source of a λ_1 boundary analogous to L_1, while edge a_2 is a source of a λ_2 boundary analogous to L_2. On the photographs, this means that we get lines of defects λ_1 and λ_2.

These defects are inherited from the seed. Let the seed contain a defect of type λ_2, which continues into the growing crystal, and which in the upper and lower parts remains parallel to the L_2 boundaries (Fig. 3), being refracted at the lower boundary of the seed. A λ_1 defect behaves similarly.

This property of the λ_1 and λ_2 defects is readily explained by the above formation mechanism; as the autoclave is heated from room temperature to the set point, the surface of the seed plate becomes etched, and the region containing the λ defect is etched more rapidly, so one gets a pit with a projecting edge (Fig. 3). If the seed plate contains a λ_2 defect, this will be an edge of A_2 type on the upper and lower sides of the plate, and so one would get a λ_2 defect on both sides; a defect of λ_1 type will behave analogously.

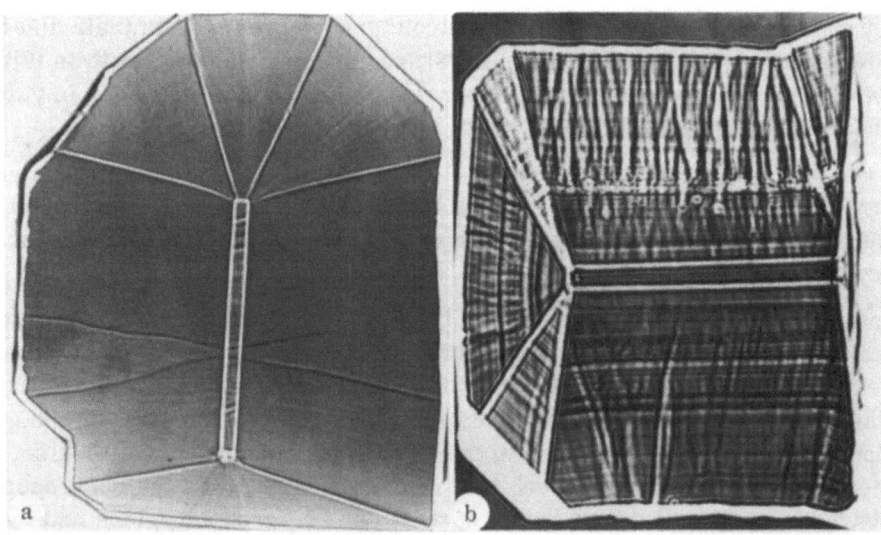

Fig. 2. V defects in an artificial quartz crystal, (10$\bar{1}$0) section, schlieren projection (×1.5): a) unheated crystal; b) macroscopic particles as sources of V defects (heated crystal).

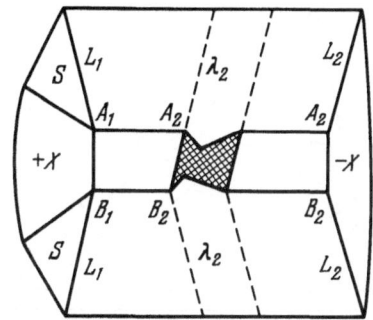

Fig. 3. Inheritance of a seed λ_2 defect. Etching gives type A_2 pits with projecting edges. Sizes of step greatly exaggerated.

The following conclusions can be drawn from this:

1. Any factor that initiates defects involving formation of macroscopic steps on the growing surface leads to the production of defects of λ_1 and λ_2 types; examples are deposition and incorporation of macroscopic particles, composition differences, structure differences in the seed plate, in particular defects of types λ_1 and λ_2, coarse scratches from mechanical processing, cracks, twins, and so on. The numbers of these defects in the grown part of the crystal may increase because the existing ones are inherited and new defects may appear.

2. The structure of the newly grown part of the crystal is substantially dependent not only on the conditions and mechanism of the growth but also on the structure of the seed, i.e., on the previous history of the crystal.

We have considered a one-dimensional pattern for the distribution of the λ_1 and λ_2 defects; when we extend it to the three-dimensional case, one must bear in mind that edges of type A_1 can take up three possible dispositions at 60° in any (0001) plane, and also three A_2; therefore, a region of autonomous growth is a spatial formation enclosed between these differently oriented and statistically distributed boundaries of types A_1 and A_2. This is clearly visible from sections of such boundaries by a (0001) plane (Fig. 5c in [10], Fig. 15c in [3], and Fig. 5c in [9]).

Then: (a) a region of autonomous growth is the most characteristic element in the pinacoid growth pyramid structure; (b) the boundaries of such a region are analogous in nature to boundaries of types L_1 and L_2 between the corresponding growth pyramids; and (c) these boundaries arise from macroscopic steps on the growing surface of the crystal.

As regards their constant reproduction, these steps may be considered as the macroscopic analog of screw dislocations; these features of the macroscopic steps mean that the dislocation mechanism for growth of a quartz crystal is hindered in a pinacoid pyramid, and often is not realized.

Banding in Quartz Crystals and Growth

on Pinacoid Pyramids

It has been shown [6] that quartz crystals in general have three forms of banding or zoning, which differ in nature; the boundaries can also differ substantially in orientation.

Banding of the third type (secondary zoning in [3]) appears when the cut of the seed plate is not parallel, but nearly so, to a face with simple indices on which the crystal can grow (Fig. 4); this has been considered in [6], and we will not stop over it.

Banding of the first type arises from changes in the thermodynamic growth conditions; the boundaries correspond to equal growth times, i.e., positions of the growth front at times of more or less sharp change in the growth conditions. This type of banding in general is

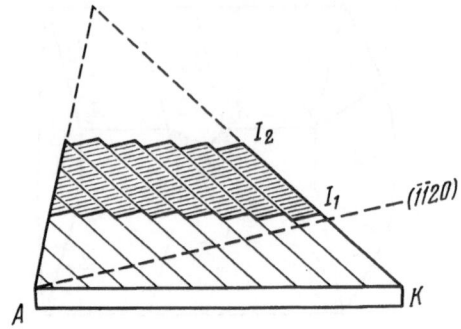

Fig. 4. Banding of the first and second types. Growth layers (plane of the second type) parallel to ($\bar{1}\bar{1}20$); growth isochrons I_1 and I_2 (plane of the first type) at times T_1 and T_2.

completely unrelated to the growth layers and bears no information about the planes involved in deposition of the material.

Banding of the second type is closely related to the growth layers; the boundaries are parallel to these layers and retain impressions of them. This banding is related not so much to change in the growth conditions as to features of the growth itself, especially the rhythmic nature of the growth [17]. As a consequence, the boundaries in banding of the second type are less sharp, and they may fail to be detected if coarser zoning of the first type is superimposed upon them. This is especially so when the cut of the seed plate is parallel to a face on which growth occurs; it is then practically impossible to distinguish the two types of boundary.

Each autonomous growth region in a pinacoid growth pyramid has banding whose boundaries have a distinct orientation, which frequently differs from that in the adjacent regions.

The surface of a pinacoid is never smooth; it is formed by accessories taking the form of rounded three-faced pyramids based on trigonal prisms. The faces of the accessories have stepped relief.

The above boundaries from banding precisely reproduce the outside surface of the crystal at a small depth.

The features of the boundaries in this type of banding in a pinacoid pyramid lead us to assign the banding to the first type: different orientation in the autonomous regions, sufficiently precise reproduction of the outer surface.

For this reason we may ask how banding of the second type makes itself felt in a pinacoid growth pyramid. The boundaries of the accessories have a stepped structure, and one naturally supposes that these steps are the points of emergence of banding boundaries of the second type (Fig. 5). We now examine which faces of the crystal are parallel to the growth layers in the accessories, i.e., on which faces one gets accessory growth.

Accessories are analogous to pyramids as regards structure and disposition, particularly ones formed by the S trigonal dipyramid, except that the faces of them form angles with (0001) less than do the corresponding S planes. The faces of the trigonal dipyramids are unique and have an orientation close to that of the accessory faces. Also, in some cases the pinacoid growth pyramids show layers almost parallel to the planes of the trigonal dipyramid (Fig. 61c [3]).

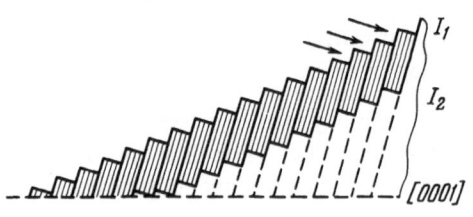

Fig. 5. Overgrowth of an accessory. Growth layers (shown hatched) parallel to the face of a trigonal dipyramid. The arrows show the deposition direction. The outer surface of the accessory (I_1 and I_2) is stepped.

Then the accessories on the surface of a pinacoid grow approximately in accordance with the faces of a trigonal dipyramid, and such growth results in the formation of autonomous growth regions.

Causes of Banding of the First Type in
Quartz Crystals

Usually, banding in crystals is related to instability in the thermodynamic growth conditions; we now examine what should be understood by such conditions.

In practice, one understands the following by such conditions: the pressure and temperature in the crystallization chamber, the temperature difference ΔT between the dissolution and crystallization chambers, and the solution concentration C. These parameters characterize the state of the solution in macroscopic volumes, but it is not these parameters that are responsible for the banding. All the growth features of the crystal, including variations in impurity trapping, are dependent not only on these parameters but particularly on the state of the solution directly adjoining the growing crystal. Variation in the local growth conditions is the principal cause of banding. There is no doubt that marked change in the macroscopic parameters should undoubtedly result in banding, but the local growth conditions can vary substantially and lead to banding even when the external parameters are relatively constant.

In fact, when crystals are grown hydrothermally, the supersaturated solution is brought up to the crystal by convection, and the approach of deposition material to the surface is provided not by ordinary diffusion but by convected diffusion.

The convected motion arises because of the temperature difference between the solution and crystallization chambers. To set up this difference, a diaphragm is placed between the chambers [18], with gaps for the rising hot and descending cool flows to pass. The area of the gaps is 3-5% of the total area of the diaphragm, so the proportion of hot solution in the crystallization chamber is about the same proportion of the total volume of solution there. Three zones can be distinguished in the crystallization chamber: the rising zone, the descending zone, and the stagnant zone. The growth conditions vary between these zones.

The convected flow is itself inhomogeneous; it expands as it rises upwards, and the temperature falls. At the same time, a radial temperature gradient is set up, since the outer part is in contact with cooler solution.

A characteristic feature of convection is its instability, so the crystal is flushed at different times by convected solutions whose state tends to vary. In particular, the crystal may be at one time in a rising convected flow and at another in a descending one.

The temperature difference ΔT is maintained by heaters; the convected currents arise near these, and are substantially dependent on the heater temperatures. Temperature instability in the heating elements is the most important cause of instability in the convected currents and hence in banding of the first type.

We now consider in somewhat more detail the reasons for the differences in impurity trapping by the crystal at different stages of growth, which leads to banding.

The amount of material q deposited in unit time is defined, for heterogeneous reactions, by the equation $q = D[(C_H - C_0)/\delta]S$, where S is the area of the crystal available for deposition, C_H is the solution concentration far from the crystal, C_0 is the solution concentration near the surface, D is the diffusion coefficient, and δ is the thickness of the diffusion layer. If one has convected diffusion, this δ is constant, being substantially dependent on the speed V of the incident flow. To a first approximation, $\delta \sim D^{1/3} V^{-1/2}$. These relationships show that the numbers of impurity atoms diffusing to the crystal and depositing there are substantially

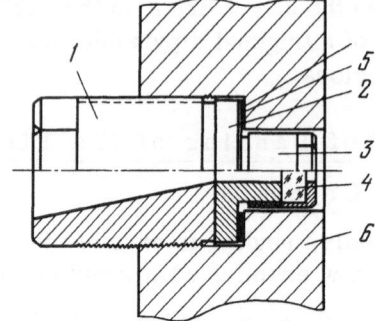

Fig. 6. Observation unit in the autoclave: 1) pressure nut; 2) tube; 3) thrust nut; 4) white sapphire cylinder; 5) copper seal; 6) vessel wall.

dependent on C_H and δ. The convected flow is inhomogeneous, so C_H varies, since the temperature varies from one part to another. At different instants, the crystal will receive convected flows differing in temperature and velocity, and the differences in C_H and δ will cause the amounts of impurity to vary, i.e., the crystal will have a banded structure.

Direct Observation of the Solution
in the Autoclave

We have observed the state of a solution in the autoclave and the movements of heating; for this purpose we made a special autoclave with two transparent windows opposite one another. A powerful light source shone through one window into the autoclave, while observation was made through the other. Figure 6 shows the system.

The tests showed that transparent plates made of white sapphire crystals grown by Verneuil's method only fairly slowly become cloudy from contact with the hot soda solution at the working pressures and temperatures.

The autoclave had heaters for the upper and lower chambers, which were outside; the upper and lower parts were separated by a diaphragm. The height of the autoclave was 75 cm, while the internal volume was 1.5 liters.

We now give a brief description of the series of experiments directly related to the present study.

First Series. The autoclave was filled to 100% with distilled water.

Above 30°C, we saw an unusual motion of the solution, which became more evident at higher temperatures; at 50°C (pressure 5 atm), the field of view revealed clearly chaotic mo-

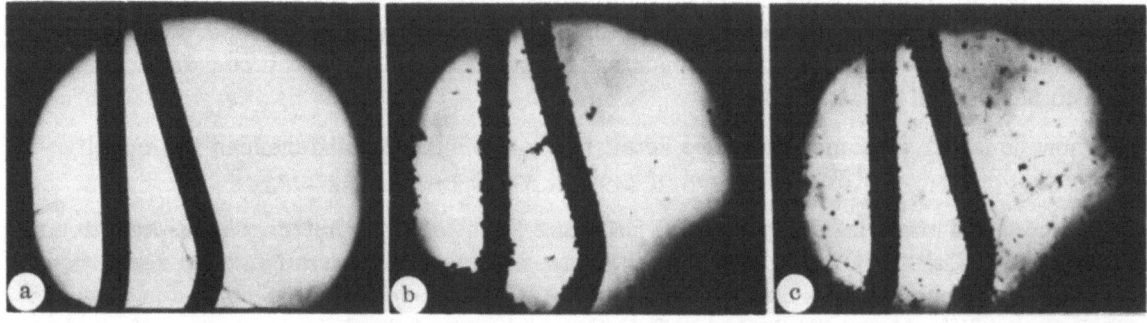

Fig. 7. Observations on solution through window, ×15: a) initial state, solution transparent; b) first heating period, particles moving rapidly; c) second period, flow stopped, particles deposited.

tion of pulsed density waves. The frequency of these waves increased with the temperature. The solution remained transparent up to 300°C (pressure 500 atm).

Second Series. The autoclave was filled with 5% soda solution to 80% of the volume; a temperature difference ΔT of about 20° was maintained between the two chambers. Vein quartz was placed in the dissolution chamber.

As in the previous experiments, we saw density waves in the solution; when the lower heaters were switched off, the solution at once ceased to move, while it started to move again when the heaters were restarted. When the temperature of the upper chamber reached 340°C, while that in the lower chamber was 360°C (pressure 430 atm), flocculent particles appeared; when the experiment was completed, the autoclave contained a large quantity of amorphous silica. The solution during the run therefore had a fairly high concentration of silica. The exact value could not be estimated, but it would seem that the solution was saturated at the instant of formation of the flocculation.

Third Series. The conditions were analogous to the previous series.

The initial solution was transparent (Fig. 7a). Then, as the temperature was raised, we began to see particles trapped by the flow (Fig. 7b, temperature 249°C, pressure 34 atm). The speed of these particles increased with the temperature and reached 10-20 m/sec. Then the motion of the solution became slower. The start of this corresponded to filling of the entire volume by the solution, after which the pressure rose rapidly. The flocculent particles began to deposit as the solution moved more slowly. Many small particles appeared in the solution, which subsequently enlarged and precipitated (Fig. 7c, 270°C, 60 atm).

Fourth Series. The conditions were analogous to the preceding except that the vein quartz was not added.

In this case, as in the previous series, we found flocculent particles at 180°C, which then coagulated.

These results show as follows: (a) The convected motion is very sensitive to the switching of the autoclave heaters; (b) flocculent particles are produced in a pure soda solution and in a solution containing silica in certain temperature ranges, these subsequently being deposited.

The first fact provides a basis for the above views on the reasons for banding in artificial quartz crystals.

The second fact needs to be borne in mind in considering the nature of macroscopic inclusions and their causes; as we have seen above, macroscopic inclusions are efficient centers for producing V defects.

Literature Cited

1. G. V. Kleshchev and A. N. Bryzgalov. This volume, p. 133.
2. I. V. Kabanovich, G. V. Kleshchev. L. N. Chernyi, and K. F. Kashkurov, Min. Sborn., Izd. L'vov. Univ., No. 4, p. 571 (1967).
3. A. N. Bryzgalov. Thesis: Some Trends in the Internal Structure and Growth of Quartz Crystals. Sverdlovsk (1969).
4. G. V. Kleshchev, A. N. Kashkurov, K. F. Kashkurov, A. V. Simonov, G. M. Safronov, and P. P. Butorin. Neorg. Mat., 4(3):362 (1968).
5. G. V. Kleshchev, K. F. Kashkurov, and L. N. Chernyi. Dokl. AN SSSR, 174:585 (1967).
6. A. N. Bryzgalov and G. V. Kleshchev. ZVMO, 98:760 (1969).
7. G. V. Kleshchev, A. V. Bryzgalov, L. N. Chernyi, A. F. Kuznetsov, and P. I. Nikitichev. This volume, p. 147.

8.   G. M. Safronov and V. E. Khadzhi. Trudy VNIIP, 2(6):101 (1958).
9.   V. E. Khadzhi. Trudy VNIIP, 6:31 (1962).
10.   V. F. Miuskov and A. Lang. Appl. Phys., 38(6):2377 (1967).
11.   L. I. Tsinober, V. E. Khadzhi, A. A. Gordienko, and M. I. Samoilovich. In: Growth of Crystals, Vol. 6A, Consultants Bureau, New York (1968), p. 25.
12.   C. S. Brown and L. A. Thomas. J. Phys. Chem. Solids, 13:337 (1960).
13.   G. G. Lemmlein and L. I. Tsinober. Trudy VNIIP, 6:13 (1962).
14.   V. S. Doladugina. In: Growth of Crystals, Vol. 3, Consultants Bureau, New York (1962), p. 340.
15.   A. Lang. Acta Crystallogr., 12:249 (1959).
16.   I. Smid. In: Growth of Crystals, Vol. 6A, Consultants Bureau, New York (1968), p. 53.
17.   N. N. Sheftal'. Vestnik MGU, No. 6, p. 28 (1966).
18.   R. A. Laudise, A. A. Bolman, and G. C. King. J. Phys. Chem. Solids, 26:8 (1965).

GROWTH CONDITIONS AND THE STRUCTURE OF ARTIFICIAL QUARTZ CRYSTALS

G. V. Kleshchev and A. N. Bryzgalov

Studies have been made on artificial quartz crystals by hydrothermal etching [1, 2]. It is found that the internal morphology is closely related to the growth conditions and to the composition of the medium incorporated into the crystal [3].

Methods

We have examined quartz crystals grown by transport in aqueous solutions of Na_2CO_3, K_2CO_3, and NH_4F, and also in aqueous Na_2CO_3 at various supersaturations. The last was controlled via the temperature difference between the dissolution and growth chambers. The growth rate serves as a measure of the supersaturation, and it varied from 0.2 to 1 mm/day.

An important point was to record the position of the growth front at different times, for which purpose the temperature difference was altered about half-way through the growth cycle, the crystal being grown briefly at a different rate, which left a mark from the growth front.

The internal structure was revealed by etching in a soda solution at high temperatures and pressures; thin plates were cut parallel to the (0001), (11$\bar{2}$0), and (10$\bar{1}$0) faces. The etching conditions were selected by trial. Lower temperatures and shorter times were required for the less perfect crystals; the temperatures varied from 200 to 400°C and the etching times from some minutes to hours, while the pressures varied from 400 to 1000 atm. The surfaces were carefully polished before etching. The etching was done in the autoclave, after which the plates were washed with dilute hydrochloric acid. The surfaces were coated with evaporated aluminum, and then examined under the microscope at magnifications from 10 to 300.

Structure Features Revealed by

Hydrothermal Etching

The etching begins at points of greatest structural and other nonuniformity related to accumulation of impurities, where the bonds between the particles are weakened and the chemical activity is high. The surface of the crystal then reveals relief and etch figures, from which one can judge the internal structure and the defect distribution. A soda solution is more sensitive to concentration nonuniformities than to structural ones (dislocations, block boundaries, etc). The latter are revealed only in so far as they contain elevated amounts of impurities. The following are some features of the structure and defects of these quartz crystals.

A typical etch pattern on a (10$\bar{1}$0) section reveals the seed plate cut on (0001) (below), as well as the crystal—seed boundary and the zone around the seed (Fig. 1). The seed plate is more perfect in structure, while the crystal—seed boundary is seen as a set of deep pits on

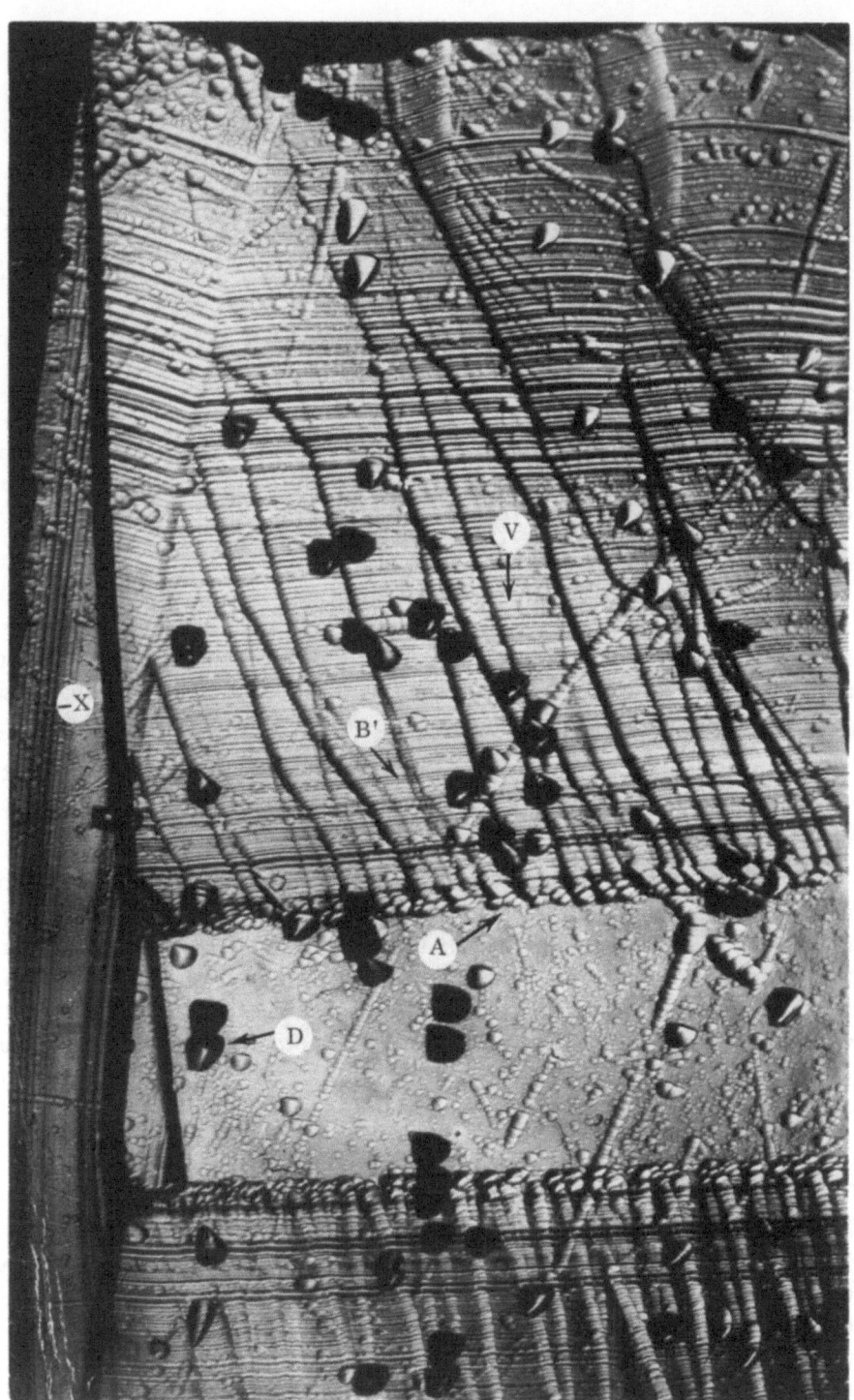

Fig. 1. Structure of a quartz crystal revealed by hydrothermal/etch-
C pinacoid, +X positive trigonal prism, −X negative trigonal prism,
crystal and seed, G between growth cones, K between overgrowth
ridges, F cells, D emergence of hollow dislocations.

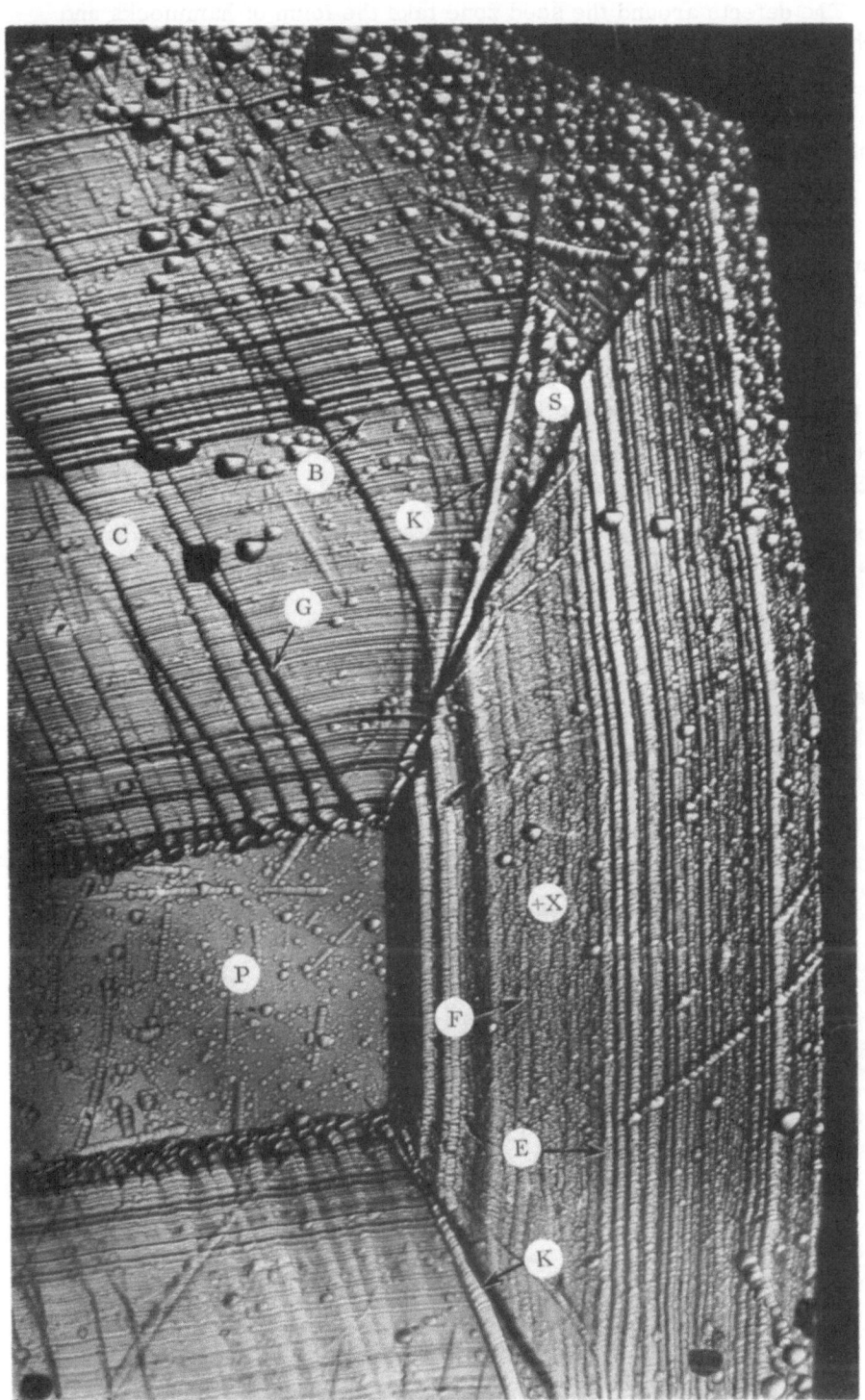

ing; growth rate 0.5 mm/day, (10$\bar{1}$0) section, ×10. Growth pyramids: S dipyramid, P seed plate, AB' seed zone. Boundaries: A between pyramids, V layer of first type, B layer of second type, E transverse the surface of the seed plate, which indicate uneven etching during heating of the autoclave to the growth temperature. The defects around the seed zone take the form of hummocks and growth cones, etc., which are related to transients before the steady state is reached.

The higher perfection in the seed plate indicates that this was grown under more favorable conditions; the difference in structure is due in part to imperfect matching of the seed plate to the later layer of crystal.

The hydrothermal etching reveals also sector and zonal structures in the crystals; a large part of the volume is taken up by growth pyramids of the C pinacoid (Figs. 1 and 2). There are also prominent pyramids from the +X and −X trigonal prisms. The first takes up more volume than the second. Between the C and +X pyramids there lies the pyramid of the S trigonal dipyramid.

All these growth pyramids subsequently taper off, and the steady-state growth form is formed by faces of the m hexagonal prism and the R and r rhombohedra.

Growth on the trigonal prism faces leads to regeneration of the m hexagonal prism faces, on which no growth occurs, and so these are left behind.

When the steady form of the crystal has been regenerated, the subsequent growth occurs on the rhombohedron faces, which also produce an increase in the tangential extent of the hexagonal prism faces.

Fig. 2. Layered structure of crystal growth at 0.6 mm/day, (10$\bar{1}$0) section, × 10. Deposition of crystal as in Fig. 1.

Fig. 3. Layers and boundary of growth cone in
pinacoid overgrowth pyramid, ×300.

Crystals grown at high rates show growth layers (Figs. 1-3). Etching starts at the boundaries between the layers (Fig. 3). These are seen on brief etching as small narrow grooves, while the layers themselves form projections above these. Figures 1 and 2 show the grooves as dark, while the projections are light. The grooves are almost triangular in cross section (Fig. 3).

The depth and width increase with the etching time, while the cross section becomes trapezoidal.

High magnifications show that each of the layers itself consists of thinner layers (Fig. 4, a and b), the boundaries between which are only weakly etched.

The layers can be classified into two types for the pinacoid pyramid (Fig. 1). The first type forms during stabilized growth, and the layers are separated by fine grooves. A layer of width up to 0.02 mm grows in 40 min and consists of thinner layers of width 0.003 mm (Fig. 4, a and b). The layers of the second type are much broader (up to 0.1-0.2 mm). These arise on marked change in temperature and pressure during the transient state, and the boundaries between them are etched as broader and deeper grooves. Such layers are observed also near the seed (Fig. 1).

The trigonal prism growth pyramids (Fig. 1) have less coarse relief; they are covered with oval etch pits or cells. Crystals of good quality show cells also in the pinacoid pyramid. The dimensions of the layers and cells are given in Table 1.

Fig. 4. Structures of crystals grown from solutions:
a) Na$_2$CO$_3$; b) K$_2$CO$_3$; c) NH$_4$F. (10$\bar{1}$0) section, $\times$ 300.

TABLE 1

Solvent	Layer width, mm		Cell diameter, mm	Transition temperature, °C
Na$_2$CO$_3$	0,02—0,03	0,002—0,003	0,01—0,1	560
K$_2$CO$_3$	0,02—0,03	0,002—0,003	0,05	560
NH$_4$F	Not examined		0,002	530

Layers of the first type are not seen in the −X growth pyramid, while they are only weakly seen in the +X one.

Layers of the second type are visible in the growth pyramids of the trigonal prisms, but there is no exact correspondence with the layers in the pinacoid pyramid.

The etching reveals clearly the boundaries between the growth pyramids and also the structure of these (Figs. 1 and 2); the widths are considerable (up to 0.5 mm) and the layer structure is quite specific.

Growth cones (Figs. 1–3, 5b, and 5c) are an important element in the crystal structure [4–7]; the boundaries between adjacent cones are seen on etching as steps (Figs. 1 and 3). The layered structure is clear at the boundary of the cones.

The dispositions of the layers in the cones vary; usually they are convex, and the convexity faces toward the growing surface.

The growth cones begin on the surface of the seed plate; the sources of them are deeply etched pits on the surface of the seed. As the growth proceeds, the cones tend to displace one another.

The surface has deep conical etch pits (Figs. 1, 2, and 5); on the surface of each pit there are steps linked to the layers. Two types of pit can be distinguished via the structure and distribution on the various faces and cross sections.

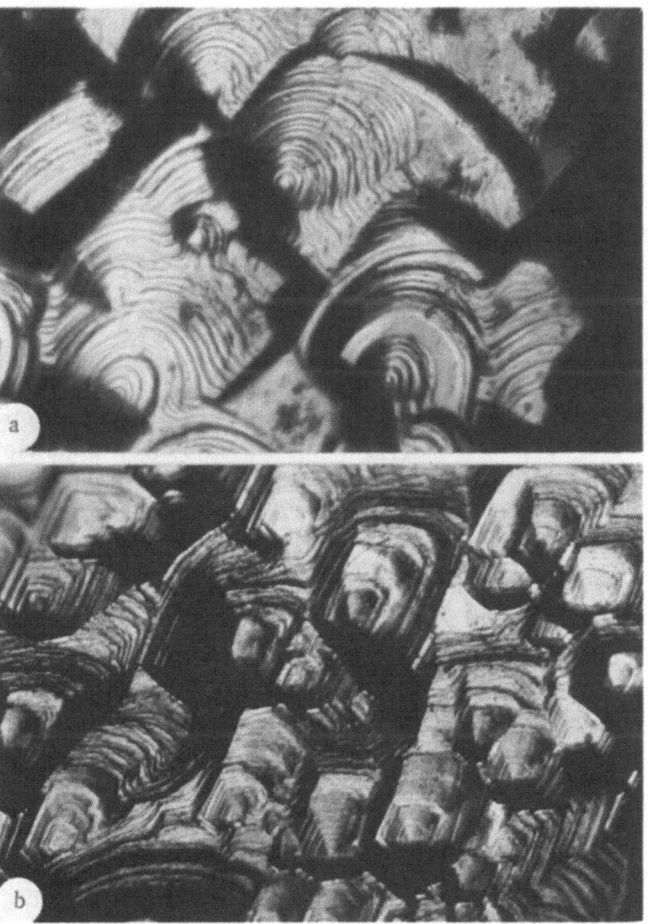

Fig. 5. Etch pits on faces: a) positive rhombohedron; b) hexagonal prism. ×20.

Fig. 6. Region of a hollow dislocation in the plane of section;
$(11\bar{2}0)$ section, × 300.

The pits on a $(10\bar{1}0)$ section (Figs. 1 and 2) are randomly disposed, with wide exposures on the surface; they are deeply etched. The number of pits increases with the number of macroscopic inclusions in the crystal.* Shadow patterns [4, 8] show that they are related to the emergence of striations on the surface, and the lines of these are usually curved, with the axes of the conical pits intersecting the etched surface at acute angles.

Etch pits of the first type are related to linear defects arising mainly from macroscopic inclusions; these are hollow channels or hollow dislocations [4, 9, 10] filled with impurity atoms and therefore readily etched. Such defects are incorporated into the crystal during the growth. Figure 6 shows a case of etching of a hollow dislocation lying in the etch plane. To the right and left one sees clearly growth cones with their own layer orientation. The layers are differently oriented near the hollow dislocation.

Pits of the second type (Fig. 5a) are seen on R faces; the face is completely covered with these pits, which are shallow and etch slowly, with characteristic transitions of layers into adjacent pits. Our results and published ones [11, 12, 15] on the etch-pit distribution indicate that the pits are related to growth dislocations; therefore, growth on the rhombohedron faces occurs by a dislocation mechanism. The faces of the hexagonal prisms have analogous etch pits (Fig. 5b), the only differences being that the outlines tend to be polygonal and the axes are inclined to the surface of the face.

Traditional methods do not reveal growth layers and other structural features in more perfect crystals.† Hydrothermal etching, however, reveals rounded or approximately polygonal pits similar to the cells of Fig. 7a.

Crystal Structure in Relation to Supersaturation

The internal structure is substantially dependent on the supersaturation [13]; crystals grown at low supersaturations, with rates of 0.2-0.3 mm/day, have a cellular structure, the

* Inclusions of the heavy phase.
† Shadow patterns, firing, and exposure to ionizing radiation.

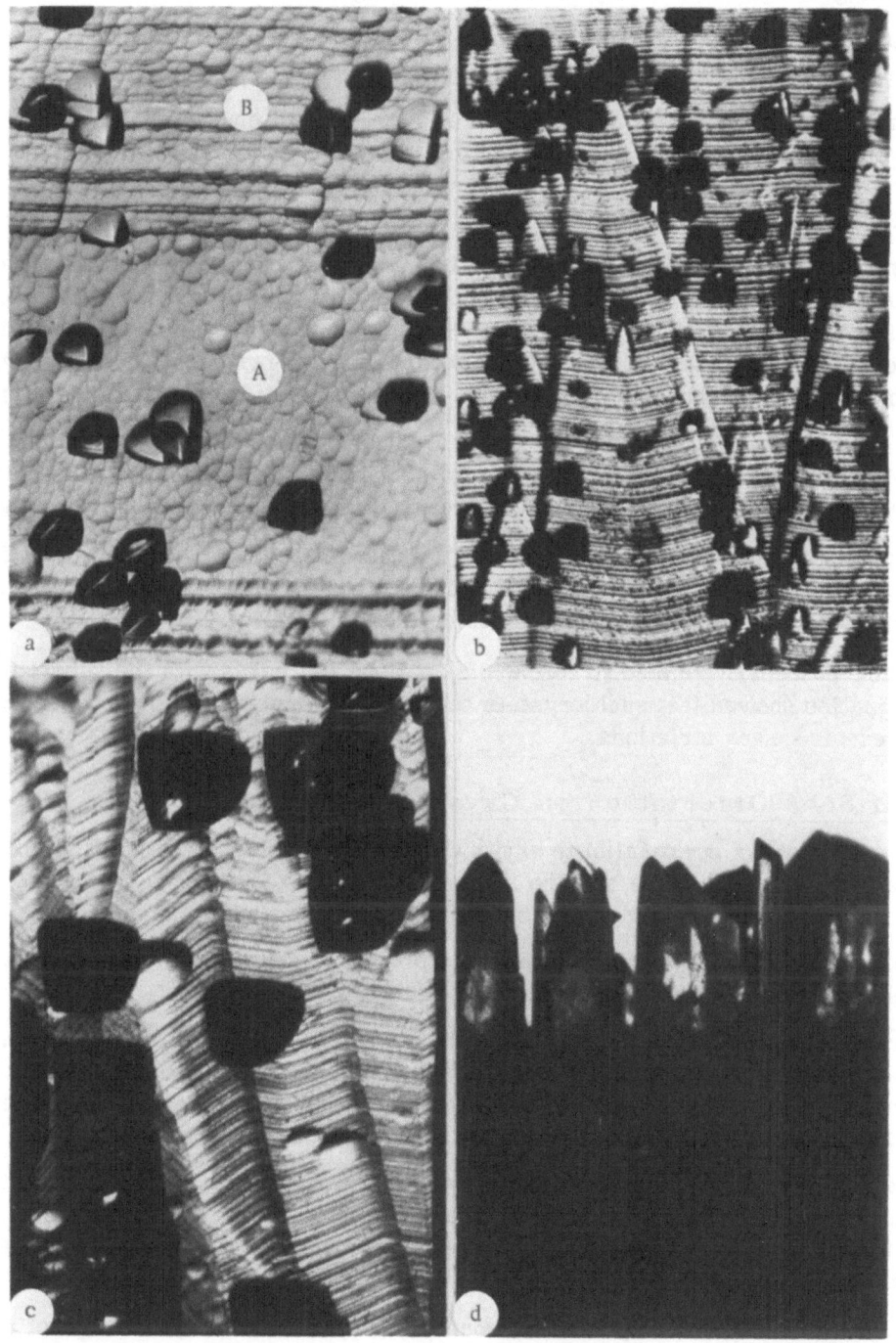

Fig. 7. Structure of crystals grown at different rates: a) 0.3 mm/day (A) and 0.5 mm/day (B); b) 0.6 mm/day; c) 0.8 mm/day; d) 1 mm/day. Separate crystals grew on a single seed plate at 1 mm/day. $(10\bar{1}0)$ section, ×10.

dimensions of the cells varying between the growth pyramids. The cells become larger in size as the growth rate is reduced.

Crystals grown at 0.6 mm/day or more have a layered structure (Fig. 2).

Crystals grown at 0.3-0.6 mm/day have a mixed structure, with the cellular component becoming less prominent as the growth rate increases.

The transition from the cellular structure to the layered one is related to the supersaturation; a crystal grown at 0.3 mm/day followed by a raise to 0.5 mm/day, and then a return to the initial rate at the end shows a cellular structure in the parts grown at the low rates and a mixed one in the parts grown at the high rates (Fig. 7a, A, and B).

The growth cones are also dependent on the supersaturation; they are not seen at low supersaturations, but they appear in crystals of mixed structure (Fig. 7, a and b); at 0.7-0.9 mm/day (Fig. 7c), the boundaries between the cones become more prominent, and the layers approximate in position those in the rhombohedron growth pyramids. Some cones are separated from their neighbors by deep cracks. The number of such cracks increases with the growth rate.

If the growth rate is raised further, the crystal splits up into independent small crystals growing on a common seed (Fig. 7d), and the growth of the parts becomes largely independent.

As the growth rate increases, there are also changes in the angles at the vertices of the growth cones; these angles are 20-30° at 0.6 mm/day, but they become almost zero at 0.7-0.8 mm/day, and the cones degenerate into growth cylinders (Fig. 7c).

The etch-pit density related to hollow dislocations increases with the supersaturation; visual examination showed that such crystals contain many macroscopic particles, while shadow projection revealed many striations.

Effects of Seed Orientation on Crystal Structure

If the seed plate is cut parallel to any of the principal faces, the stratification in the growth pyramid is parallel to the surface (Fig. 8, a and b); exception must be made for the the m face, on which there is no appreciable growth.

If the surface makes a certain angle with a principal face, the internal structure becomes much more complicated; we consider some examples below.

Figure 8b shows a crystal grown on a plate cut at 20° to a trigonal prism face. The crystal has grown well towards the positive direction of the X axis, and has formed a growth pyramid on the +X face, which tapers out, and faces of the m hexagonal prism are produced. No growth occurs on the m faces, and the growth layers terminate there.

There are deep stepped grooves related to sharp changes in the growth conditions, which indicate that the growth front ran parallel to the seed plate. There are also visible layers parallel to the faces of the +X trigonal prism, which are analogous to the layers of Fig. 8a and are not parallel to the growth front.

The layers are separated by internal boundaries parallel to the m hexagonal prism face, and as a result the crystal becomes zoned in what is called a secondary fashion. The layers of the first type terminate at each internal boundary, and the extent of these boundaries in the tangential direction increases as a result of growth on parts of the trigonal prism face; here again one sees the above feature of the m face.

Figure 8d shows the prominent secondary zoning involving a face of the R positive rhombohedron; here, as in the case above, the deposition layers were not parallel to the growth front, and they terminate at the boundary parallel to the rhombohedron face.

Fig. 8. Structures of crystals grown on variously oriented seed plates, ×10. Sections a and b) trigonal prism; c and d) positive rhombohedron. The secondary-zoning boundaries are visible parallel to the faces of the hexagonal prism (b) and to those of the positive rhombohedron (d).

Then if the seed surface is not parallel to principal faces, one gets secondary zoning, which is not directly related to changes in the growth conditions (to the conditions giving the primary zoning), and slowly growing faces are involved.

This feature must be borne in mind when the zoning in natural crystals is used to establish what faces were involved in the growth, since the growth layers may not coincide with the zoning boundaries.

Effects of Solution Composition on

Crystal Structure

Figure 4 shows etched sections of crystals grown by transport of SiO_2 in aqueous solutions of Na_2CO_3, K_2CO_3, and NH_4F. Crystals grown in K_2CO_3 and NH_4F are more readily etched; to get comparable relief on crystals grown in K_2CO_3, the etching time must be reduced by a factor 2, while for NH_4F the factor is 10, these values being relative to crystals grown in Na_2CO_3.

Crystals grown in K_2CO_3 have structural elements and defects as described above, but the boundaries between layers are more deeply etched and the cells are smaller, while the transverse grooves are more readily visible in the pinacoid growth pyramid, and the hollow-dislocation density is higher by an order of magnitude.

Crystals grown from NH_4F have a defective structure; the surfaces reveal very deep relief, with numerous small cells and deep grooves, which show that the crystal consists of individual blocks. Therefore, crystals grown from NH_4F have a loose structure. The phase transition point in such a crystal is lower by 30°C than that for crystals grown in soda solution (Table 1). We assume that one of the reasons for the open structure and reduced transition point in the case of NH_4F is that the F ions enter the crystal and form Si−F or Si−O−F bonds, which disrupt the Si−O−Si bonds in the crystal.

Discussion

The first process in growth of a quartz crystal on a seed plate of arbitrary orientation is regeneration, which is followed by growth proper.

During the regeneration period, the pinacoid and trigonal-prism faces at first grow rapidly and then taper off; the steady-state growth form is formed by faces of the rhombohedra and hexagonal prism. The latter, as in natural crystals [5, 14, 15], do not grow appreciably.

The hexagonal prism faces in natural crystals are often covered with a layer with a highly defective structure, which has an elevated capacity for adsorbing passivating impurities [3, 5, 16].

Direct crystal growth occurs on rhombohedron faces; numerous studies of etch figures [11-13] indicate a dislocation mechanism for growth of these, which is confirmed by the etch figures we have obtained on rhomohedron and hexagonal-prism faces.

Quartz crystals show a tendency to produce growth cones and secondary zoning, especially when grown at high supersaturations; these effects have a common cause: macroscopic unevenness of the growth surface, with growth in some parts running ahead of that in others, which constantly reproduces the unevenness. The rapid etching of the boundaries between cones indicates an increased impurity content there relative to the rest of the crystal. The impurity atoms are rejected at the growth cone boundaries during the growth.

The secondary zoning is similar in nature to the growth cones, but it has a somewhat unusual formation mechanism; here the decisive part in producing the unevenness is played by the fact that the parts of the faces involved in the layering are not parallel to the growth front, so the growing surface has macroscopic steps, which on the sides are terminated by slowly growing m or R faces, on which the growth layers terminate. The surfaces of these faces increase tangentially during growth, and they form internal boundaries (secondary zoning), which are revealed by the hydrothermal etching, so they must have an elevated impurity content.

The stratification in quartz crystals deserves particular attention. The macroscopic layers of the second type revealed by etching represent primary zoning; they are produced by marked variations in temperature and pressure during the transient growth, and the macroscopic layers correspond to periods of more or less similar growth conditions, acting as indicators for the growth front. The layers of this type are not dependent on the growth direction and are always parallel to the seed surface.

Macroscopic stratification arises from entry of foreign atoms and groups; we have examined crystals after exposure to ionizing radiation [4] and studies by shadow projection and firing [4, 8] also indicate that the impurity contents in these layers vary. It has reliably been established that the layers differ in contents of Na and Al. Milkiness* occurs on firing, and

---

* Milkiness in the crystals is due to OH groups; it has been found [21, 22] that the milky color of natural crystals is due to large amounts of submicroscopic regions filled with water molecules, e.g., 10^{14} per cubic centimeter of dimensions from 200 to 1000 Å. The milkiness of artificial crystals is related to the formation of water molecules during firing by detachment of OH groups from the Si ions and collection of these into submicroscopic formations.

the OH content of the crystal is very much dependent on the growth rate, which shows that the layers also differ in OH content.

Silica exists in hydrothermal solutions not as SiO_2 or $Si(OH)_4$ [17-19], but in more complicated forms, some of which contain OH groups; during normal growth, these complexes break up near the growing surface and release SiO_2 or SiO_4^{2-}. If the growth conditions change, or if the supersaturation is high, the process may not go to completion, and the crystal may grow via these complexes [20], so it would be forced to take up OH, as well as impurity H, Na, Al, Fe, etc.

The nature of the layering is to be explained from this viewpoint via differences in the compositions of these various complexes that are deposited at the various growth stages, since the composition of these is very sensitive to the details of the growth conditions.

However, this cannot explain the origin of layers of the first type, which grow in a few minutes; this layering is due to a cyclic process in the crystal growth itself [23]. Impurities rejected by the crystal tend to accumulate near the growing faces, and this hinders the access of fresh material to the faces, whose growth rate is thereby reduced, which in turn tends to reduce the amount of heat of crystallization released, and hence the temperature falls near the growing face, which increases the supersaturation, and this ultimately leads to a marked increase in growth rate and trapping of the adjacent impurity atoms. This process becomes cyclic, and the proportion of impurities captured by the crystal periodically varies, which results in a layered structure.

The substructure of the layers is seen on etching as transverse bands, which is most likely due to a secondary process occurring after the growth of a given layer is completed. The layers differ in impurity content, so they differ also in lattice parameters [14, 24]. Dimensional and orientational discrepancies arise between the lattices in successive layers, which produce internal stresses, and this has the following consequences.

The internal stresses may relax by dislocation formation; dislocation redistribution can itself result in small-angle boundaries, and hence in submicro blocks. Crystals grown in NH_4F are extensitvely split into blocks for this reason. The stress relaxation will be accompanied by redistribution of the impurity atoms, with diffusion to the block boundaries; these processes are facilitated by the reasonably high temperatures at which the crystals grow, together with the small thickness of the layers on the crystal. The stress relaxation and the impurity redistribution are responsible for the layer substructure.

The cellular structure is superficially similar to the substructure of the layers and is also a secondary effect; the relaxation processes near the growing surface are accompanied by structural modification in the completed zone of the crystal, which may substantially influence the crystal growth; dislocations and other defects generated in this process may emerge at the surface, which is bound to affect the growth, which is determined not only by the activity of the face but also by the scope for relaxation in the underlying material.

Then in studying crystal structure one has to bear in mind the scope for formation of a structure unrelated directly to the growth mechanism, but arising afterwards in the underlying crystal; one also has to take account of processes in the underlying crystal in so far as they affect the growth.

Finally we consider a particular topic of practical importance. The shadow method reveals striations in artificial quartz crystals, and it has been shown [4, 16] that these striations are due to hollow dislocations surrounded by impurity atom atmospheres. However, this is not the sole cause of the optically observed striations, since etching shows that the boundaries between growth cones have elevated impurity contents, and therefore the refractive index at such a boundary is different from that of the rest of the crystal, so the cone boundaries may be seen as striations in specimens prepared for optical investigation.

I am indebted to N. N. Sheftal' for many valuable comments on the paper, and also to K. F. Kashkurov and A. V. Semenov for providing facilities for the experimental part of the work, in which assistance was received from V. N. Mironov.

Literature Cited

1. I. Smid. in: Growth of Crystals, Vol. 6A, Consultants Bureau, New York (1968), p. 53.
2. A. N. Bryzgalov, L. N. Chernyi, I. V. Kabanovich, and G. V. Kleshchev. In: Optics and Molecular Spectroscopy [in Russian], part 1, Izd. Chelyab. Ped. Inst. (1968), p. 78.
3. N. N. Sheftal'. In: Growth of Crystals, Vol. 3, Consultants Bureau, New York (1962), p. 3.
4. L. I. Tsinober et al. In: Growth of Crystals, Vol. 6A, Consultants Bureau, New York (1968), p. 25.
5. G. V. Kleshchev, I. V. Kabanovich, et al. Min. Sborn., No. 4, Izd. L'vov. Univ. (1967), p. 575.
6. C. S. Brown and L. A. Thomas. J. Phys. Chem. Solids, 13:377 (1960).
7. G. M. Safronov and V. E. Khodzhi. Trudy VNIIP, 2(6):127 (1958).
8. V. S. Doladugina. In: Growth of Crystals, Vol. 3, Consultants Bureau, New York (1962), p. 340.
9. E. V. Tsinzerling and Z. A. Mironova. Kristallografiya, 8:117 (1960).
10. A. Varma. Crystal Growth and Dislocations [Russian translation], IL, Moscow (1958).
11. I. Augustine and D. R. Hale. J. Phys. Chem. Solids, 13:344 (1960).
12. K. I. Chepizhnyi. Dokl. AN SSSR, 166:486 (1966).
13. R. A. Laudise and J. W. Nielsen. J. Solid Phys., 12:149 (1961).
14. N. N. Sheftal'. In: Growth of Crystals, Vol. 1, Consultants Bureau, New York (1959), p. 5; Vol. 3 (1962), p. 3.
15. K. V. Kashkurov, G. V. Kleshchev, L. N. Chernyi, and V. T. Ushakovskii. In: Optics and Molecular Spectroscopy [in Russian], Vol. 1 Izd. Chelyab. Ped. Inst. (1968), p. 86.
16. G. V. Kleshchev, I. V. Kabanovich, and I. V. Rabanovich. Dokl. AN SSSR, 174:585 (1967).
17. R. Ailer. Colloid Chemistry of Silica and Silicates [Russian translation], IL, Moscow (1959).
18. I. I. Ganeev. Sov. Geol., No. 12, p. 25 (1963).
19. Chang Yuan-lung and Chang Kuei-fen. In: Growth of Crystals, Vol. 6A, Consultants Bureau, New York (1968), p. 37.
20. I. Haven and A. Kats. Silicates Industr., 27:137 (1962).
21. A. C. McLaren and A. Phakey. J. Appl. Phys., 36(10):3244 (1965).
22. L. A. Bursill and A. C. McLaren. J. Appl. Phys., 36(10):2084 (1965).
23. N. N. Sheftal'. Vestnik LGU, ser. geol., 6:28 (1966).
24. U. Bonse. Z. Phys., 13(5):184 (1965).

SOME TRENDS IN SHAPE PRODUCTION FOR
ARTIFICIAL QUARTZ CRYSTALS

G. V. Kleshchev, A. N. Bryzgalov, L. N. Chernyi, A. F. Kuznetsov, and P. I. Nikitichev

It has been shown [1-3] that the supposed growth pyramids of the m hexagonal prism in natural quartz crystals have growth layers parallel to rhombohedron faces, from which it was concluded that the m faces are passive, with no growth occurring on them, and that the growth pyramids of m faces do not exist in the usual sense. This does not agree with the scheme of [4] for the sector structure of quartz crystals. This meant that it was necessary to examine the trends in form production and sector structure in quartz crystals grown under artificial control conditions. Particular interest attaches to crystals grown on small spherical seeds, since in that case we approximate to the growth conditions for point seeds [4].

Here we give a comparison of crystals grown from spherical and platy seeds of (0001) cut; the results also enable one to relate the shape to the growth conditions.

First of all we consider the results with spherical seeds, which were cut from artificial crystals and had initial diameters of 25 mm. The surfaces were ground with fine power. These seeds were placed in the autoclave in such a way that the threefold axis was horizontal. The growth occurred under two conditions: a) with a small temperature difference between the dissolution and crystallization chambers (low growth rates and low supersaturation) for crystals A; and b) with a large temperature difference (correspondingly, large growth rates and large supersaturations) for B crystals.

The internal structures of specimens cut from the crystals in various orientations were examined by hydrothermal etching [3, 5, 6] in soda solution, and also by x-ray diffraction topography [1, 7]. The topograms were recorded with Mo radiation in a KRS camera.

The A and B crystals acquired a facetted two-headed growth form (Fig. 1). At each head of the crystal there were faces turned upwards and downwards. The growth conditions of these were different: the lower faces were screened by the upper ones from macroscopic inclusions deposited under the action of gravity, while the upper faces were open to contamination.

The growth shape of the B crystals included traces of the m hexagonal prism and the R positive rhombohedron; the A crystals had not only these faces but also faces of the r negative rhombohedron. The extent of the r faces in this case was much less than that of the R faces which indicates a higher growth rate on the r faces relative to R.

The faces of the R rhombohedron turned upwards (Fig. 1) were uneven, with many accessories (Fig. 2); the B crystals had the faces on a given head unequally developed (Fig. 3), so the crest of the head was displaced relative to the C_3 geometrical axis of the crystal, the

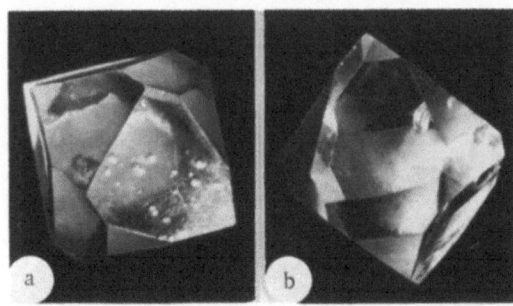

Fig. 1. Quartz crystals (×2) grown at: a) low rate; b) high rate. Macroscopic inclusions are visible on the upper face of the positive rhombohedron R.

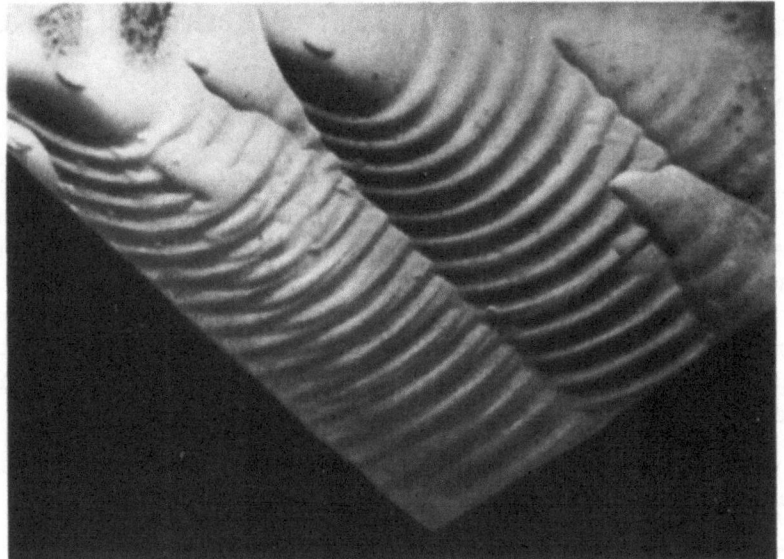

Fig. 2. Accessories on the upper face of the R rhombohedron, ×5.

displacement being up to 5 mm. The R faces from A crystals developed identically in a given head, and the above displacement did not exceed 1 mm.

The diffraction patterns (Figs. 4 and 5) and the etch patterns (Fig. 6) show the following trends in the regeneration and growth on the spherical seeds:

1. Regeneration is rapid and occurs on planes of the trigonal prism (the six-faced form is restored) and on the pinacoid; intermediate growth pyramids taper off, and the crystal acquires its steady-state shape, which consists of a hexagonal prism and two rhombohedra.

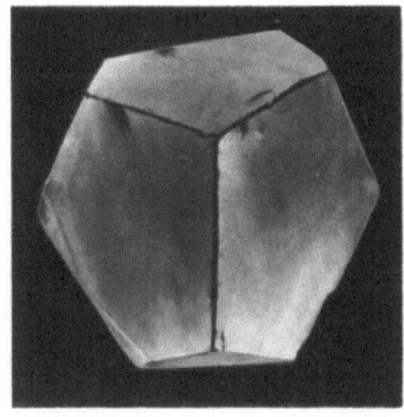

Fig. 3. Uneven development of R faces, ×5.

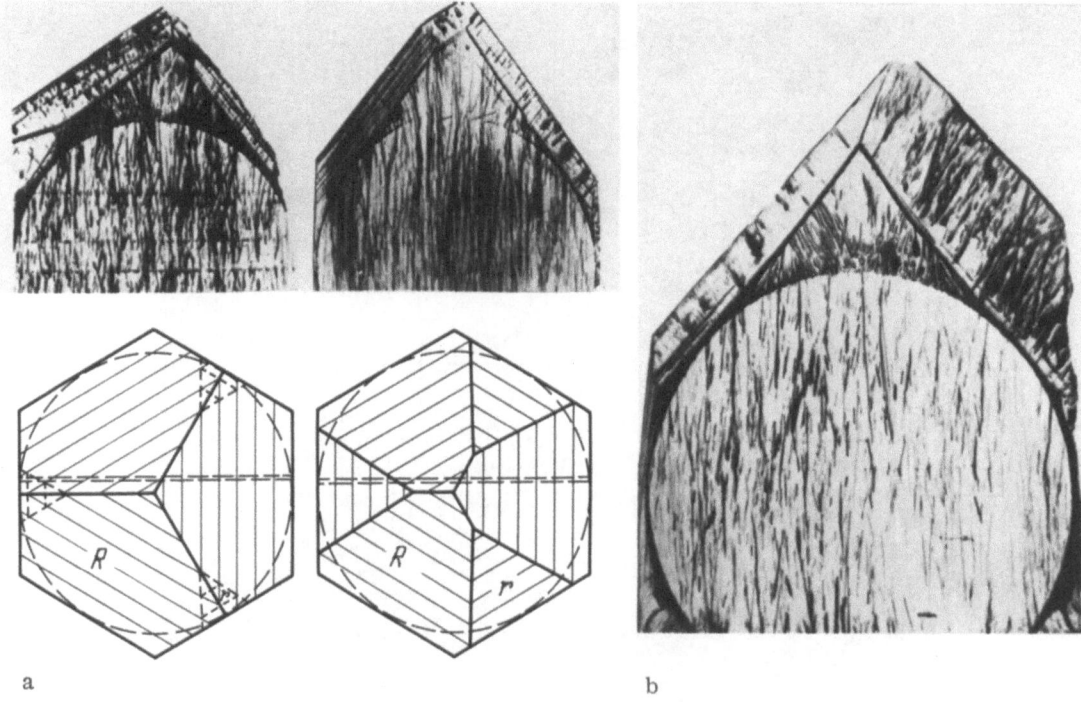

a   b

Fig. 4. X-ray topograms (×5) of type B crystals. The position of the section in the crystal is shown by a broken line. The topogram of crystal B (a and b) shows different growth rates on the upper and lower rhombohedra, and also tapering of the r pyramid.

The subsequent growth occurs only on the rhombohedron faces (growth period proper). As for natural quartz crystals [1-3], there was no appreciable growth on the m faces, and the extent of these in the tangential direction increased only by growth on the rhombohedra.

2. The growth rate on the R faces was much less than that on r; in the B crystals, the r faces rapidly disappeared even during the regeneration (Fig. 4a), while in A crystals they persisted in the growth form much longer. Consequently, the relative growth rates on the R and r faces are substantially dependent on the supersaturation.

3. The growth rates on the upper R faces were almost four times those on the lower ones (Fig. 4b), since the growth pyramids on the upper faces contained far more macroscopic inclusions and defects (Fig. 1).

4. The growth pyramids of the rhombohedra showed layering; this was more prominent in the A crystals, and it clearly arises from the smaller number of defects in these crystals, which facilitates diffraction contrast from the stratification. The R growth pyramids have the layers in the form of bands up to 200 μm wide, while the r pyramids have ones only up to 20 μm, which indicates that the growth on the r faces is more sensitive to variations in the conditions. Layers were observed also in the pyramids for the trigonal prisms (Fig. 5). The $-$X pyramid had these layers parallel to the spherical seed (Fig. 5a), and the layers were about 10 μm thick and evenly separated, which indicates that this type of layering is due to a cyclic feature of the growth mechanism, and not to changes in the growth conditions [7].

We now consider the results for plates cut parallel to the pinacoid. Figure 7 shows an example of a hydrothermal etch pattern for a crystal grown in mode B. The above features are clear: a) tapering in the pinacoid growth pyramids; b) appearance and extension of the m faces

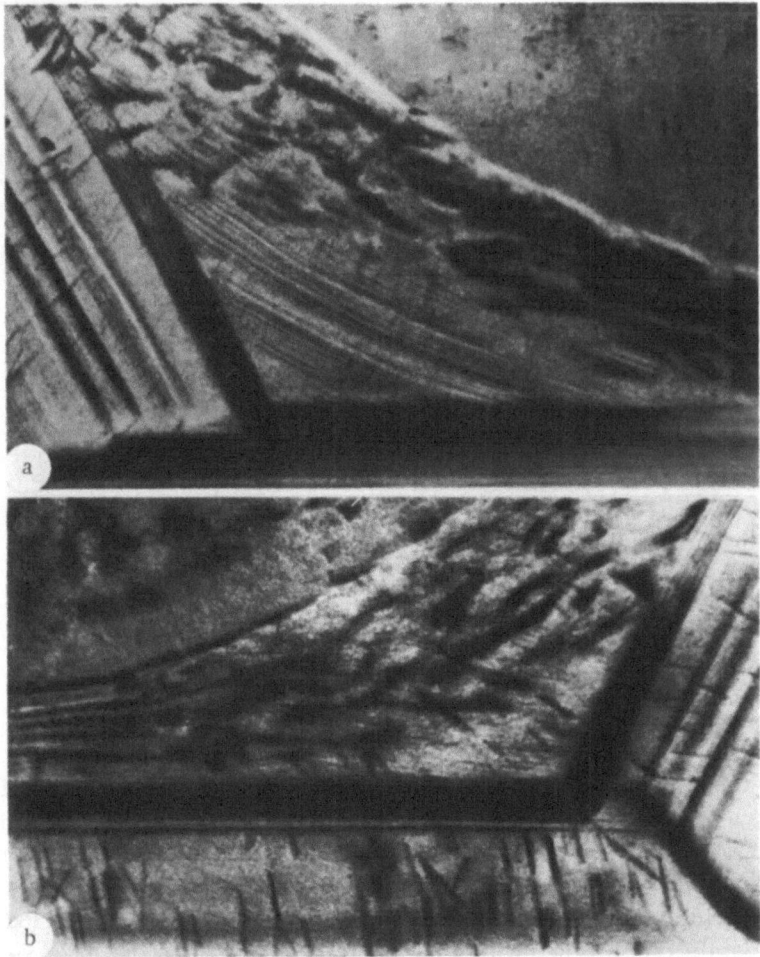

Fig. 5. Growth sectors of trigonal prisms: a) −X; b) +X.

by growth on the rhombohedron faces; c) growth rates on the r negative rhombohedron faces greater by a factor 4–5 than those on the R positive rhombohedron ones.

Figure 8 shows the crystal habit in relation to the shape of the seed plate; it is clear that a plate formed by straight lines not parallel to the faces of the m prism causes growth to start with production of the hexagonal prism faces, growth of the crystal proper occurring only after this. In all cases, there is no growth on the m faces, and the extent of these in a tangential direction increases by a growth on the rhombohedra.

Figure 9 shows schemes for crystal production in relation to growth rate, i.e., supersaturation with silica. It is clear that the relative growth rates on the R and r faces substantially vary with the supersaturation; when the latter is high, the r growth rate is considerably larger than the R one, and the r pyramid rapidly disappears. The difference is less at low saturations, and the r pyramid is slower to disappear.

In all these cases the m faces appear as passive; there is hardly any growth in a direction perpendicular to such a face. There is an appreciable difference in growth rates for the upper and lower rhombohedra, which is responsible for the difference in contents of macroscopic impurities and defects in the growth pyramids, which leads us to expect that these impurities and defects could stimulate growth on the m faces also. This is actually so, for Figure 10 shows a diffraction pattern from a natural quartz crystal indicating that in this case

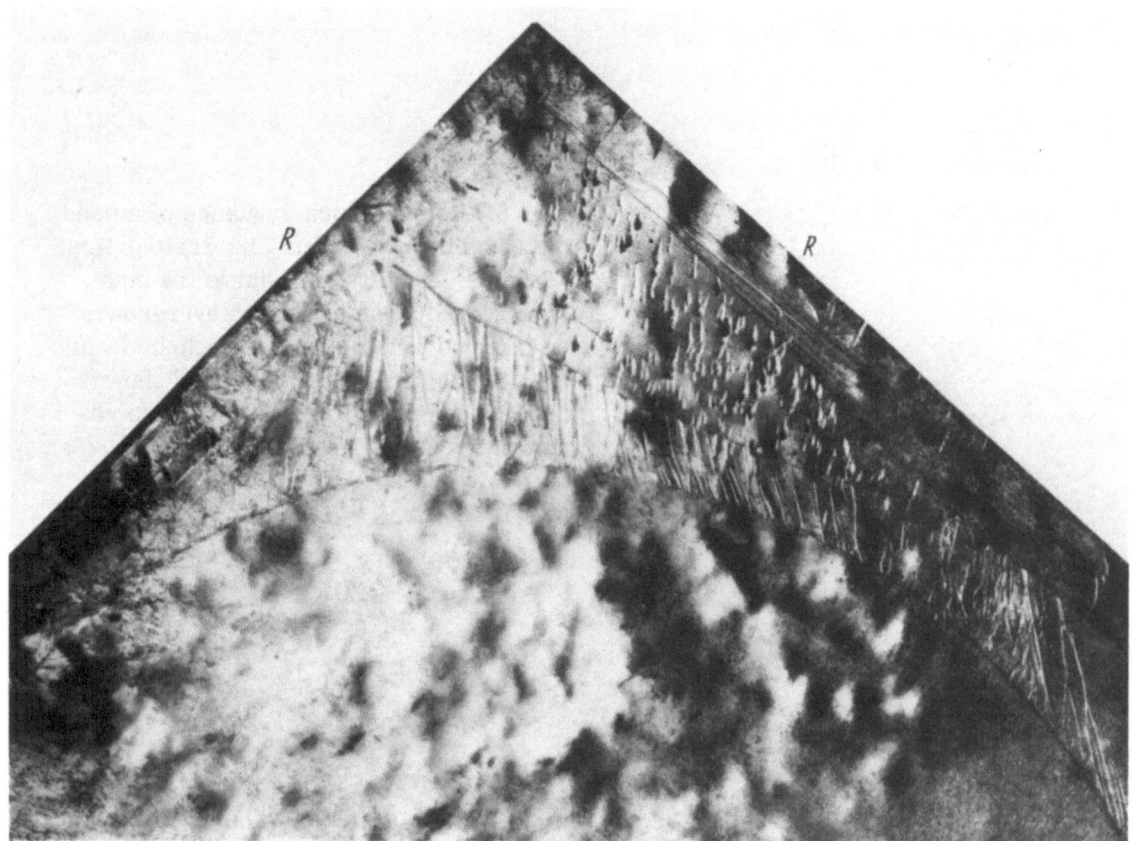

Fig. 6. Etched (11$\bar{2}$0) section of a crystal grown on a spherical seed, ×10.

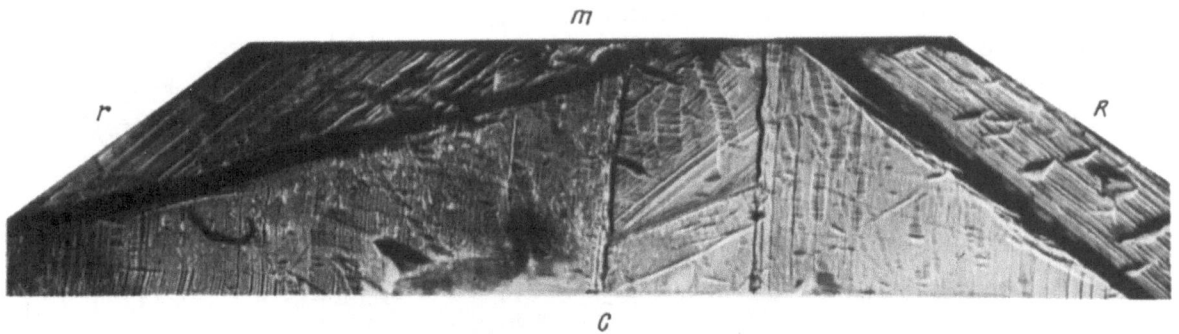

Fig. 7. Etched (11$\bar{2}$0) section of a crystal grown from a (0001) plate at a high rate, ×10. Overgrowth layers are seen on the faces of the C pinacoid and R and r rhombohedra. The faces of the m hexagonal prism are produced by overgrowth on R and r.

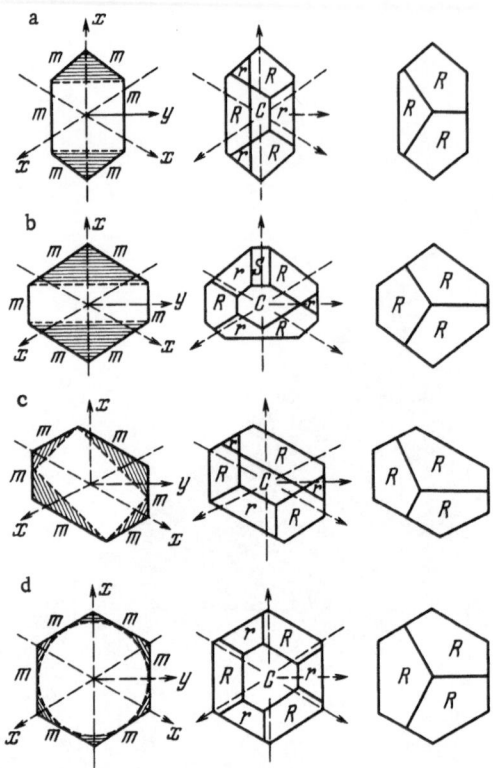

Fig. 8. Regeneration sequence of a seed plate (view on [0001]). The first vertical series in the left (a-d) shows the plate shapes and regeneration by overgrowth on trigonal prisms; the second shows intermediate forms, and the third shows the final forms of the crystal heads resulting from the first series.

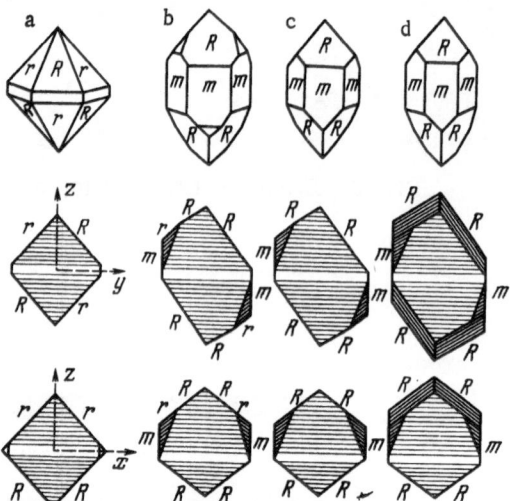

Fig. 9. Effects of growth rate on the structure and shape of quartz crystals. The first vertical series (a) represents low rates, with no appreciable growth on the rhombohedron faces; the last series (d) represents high rates, with considerable growth on the rhombohedra. The R faces exposed after regeneration. The second and third series (b) and (c) represent intermediate growth stages. The face of the m prism expands tangentially only from growth on the rhombohedra.

Fig. 10. Topogram (×5) of a natural crystal, and layering on the rhombohedron faces and defective growth on the m hexagonal prism.

there was growth on the m faces also. However, the illustrations show that the layer is extremely defective, as is clear from the uniform blackening of the diffraction pattern from this part of the crystal, which contains macroscopic inclusions and twins.

Literature Cited

1. K. F. Kashkurov, V. T. Ushakovskii, L. N. Chernyi, I. V. Kabanovich, and G. V. Kleshchev. Zap. Vses. Min. Obshch., ser. 2, 96:25 (1967).
2. V. T. Ushakovskii, K. F. Kashkurov, L. N. Chernyi, I. V. Kabanovich, and G. V. Kleshchev. In: Optics and Molecular Spectroscopy [in Russian], Vol. 1, Izd. Chelyab. Ped. Inst., (1968), 86.
3. A. N. Bryzgalov, L. N. Chernyi, I. V. Kabanovich, and G. V. Kleshchev. In: Optics and Molecular Spectroscopy [in Russian], Vol. 1, Izd. Chelyab. Ped. Inst., (1968), 86.
4. G. G. Lammlein. Sector Structure in Crystals [in Russian], Izd. AN SSSR, Moscow (1948).
5. G. V. Kleshchev, A. N. Bryzgalov, K. F. Kashkarov, G. M. Safronov, A. V. Simonov, and P. P. Butorin. Zh. Neorg. Mat., 4(3):362 (1968).
6. G. V. Kleshchev and A. N. Bryzgalov. This volume, p. 133.
7. I. V. Kabanovich, L. N. Chernyi, G. V. Kleshchev, V. T. Ushakovskii, and K. F. Kashkarov. Min. Sborn., No. 4, Izd. L'vov Univ. (1966), p. 575.
8. N. N. Sheftal'. Vestnik LGU, No. 6, p. 22 (1966).

GROWTH ANISOTROPY OF GALLIUM ARSENIDE AND GERMANIUM IN GAS-TRANSPORT SYSTEMS

L. G. Lavrent'eva

Interest has recently increased in crystallization in gas-transport systems, because the process is of technological importance and also because one needs to elucidate the growth conditions. An advantage of the method is that one can carry out the crystallization at temperatures below the melting point while maintaining conditions close to equilibrium. There are many papers on the morphology of crystals obtained by deposition from the vapor.

Films with (111) and (100) crystallographic indices have a layer growth mechanism, and studies have been made on the basic types of growth defects, with the causes and trends in formation of these [1-5]. There are difficulties in studying the kinetic aspects by varying the temperature and supersaturation because these factors alter the composition of the gas phase and also the rate-limiting step of the process as a whole [6]. However, the crystal structure is such as to provide a means of examining the crystallization under set conditions. This involves examining the growth anisotropy and doping on smoothly varying the angular deviation θ from a certain singular plane $(h_0 k_0 l_0)$ toward the direction (hkl); if the $(h_0 k_0 l_0)$ plane grows by two-dimensional nucleation and layer expansion, then one sets the deviation to specify the relief on the growth surface (the distance and shape of the growth steps). If the crystal has the sphalerite or diamond structure (gallium arsenide or germanium), it is assumed that a deviation from (111) toward (100) should result in a series of parallel steps bounded by the nearest singular planes, i.e., (111) and (100) [7]. If the heights of these steps are constant and independent of θ, one can calculate the growth-rate anisotropy due to the change in step density on varying θ, as has been performed within the framework of the model due to Barton et al. [8, 9], and the results are given in Fig. 1 for certain fixed values of λ_s / y_0, where y_0 is the distance between steps in a series and λ is the surface diffusion length for an adsorbed atom. Growth by nucleation has also to be borne in mind near the singular and adjoining vicinal faces.

Crystallization of spheres [10, 11] is one of the methods of examining growth anisotropy, since it enables one to observe simultaneously all the growth planes and to reveal the faces that persist under given conditions. However, this method also has some disadvantages: it is difficult to maintain the shape of the sphere in chemical or gas etching, and it is almost impossible to obtain information on the doping anisotropy under the same conditions. Also, some of the vicinal orientations are overlapped by the expanding singular face. When chemical crystallization from the vapor is being used, it is more convenient and informative to examine the anisotropy in growth and doping using planar substrates with any necessary orientation, since one can then set accuracy and monitor the orientation.

Figure 2 shows the observed dependence of the growth rate on the orientation for GaAs as obtained by this method [8, 9, 12]. The curves relate to two different gas-transport sys-

154

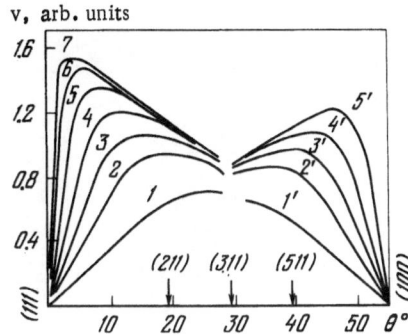

Fig. 1. Growth anisotropy due to motion of
straight steps, values of λ_s/y_{311} of: 1) 0.6;
2) 1.2; 3) 1.5; 4) 3.0; 5) 6.0; 6) 12.0; 7) 29.0;
1') 0.5; 2') 1.0; 3') 1.5; 4') 2.5; 5') 5.0.

tems and somewhat differing crystallization conditions, but there is no doubt as to their similarity. The results show that the growth mechanism for GaAs films has much in common for
the different systems, which is clearly due to the anisotropy in the GaAs crystal, especially
the surfaces. The calculated and experimental relationships have been compared for curve 1
of Fig. 2 in [9], and it is clear that the best correlation between the shapes of the curves occurs for the range (111)A–(311)A, while the deviation is largest near (100). Electron microscopy shows that this deviation is due to steps of increased height (up to 100-200 Å) on surfaces
vicinal relative to (100) [13]. The shape of the growth steps alters somewhat as one deviates
from (100) toward (111)A and (111)B. In the second case, the steps were more nearly rectilinear. The macroscopic relief also reveals this difference in growth step shape. Figure 3
shows characteristic growth figures on (100) and on layers deviating from (100) toward (111)A
and (111)B (photograph recorded by V. A. Moskovkin). The shapes of the figures show that
the step speed in the [011] direction is less than that along [0$\bar{1}$1], and in the second case the
surface anisotropy is less pronounced. The basic surface of the layers (between the figures)
is extremely flat and free from relief or else has very slight wavy relief, this being reproducible on varying the growth time.

These results on the growth-rate anisotropy show that this model correctly reflects the
trends; the coalescence of steps (which means that the condition h = constant is not met) only
slightly influences the result under these conditions; however, it will be incorrect to suppose
that this is always so. Experiment shows that in certain instances the coalescence occurs very
extensively, and the growth steps enlarge up to macroscopic dimensions and acquire facets.
In that case, the layer growth rates fall appreciably. The process is initiated by local nonuniformities in the relief and by growth defects, which produce groups of growth steps [14, 15],
which then develop. One expects that a step interaction and reconstruction should be most
prominent when there are small differences between steps and the crystallization temperature
is low. In fact, in studying the growth rate anisotropy of iodide germanium (crystallization
temperature 400°C), it was found that there were marked changes in the height and shape of
the steps as the deviations from (111) and (100) toward the center of the range increased.
The surfaces of (211), (311), and (511) films showed macroscopic steps, the facets consisting

Fig. 2. Growth anisotropy of gallium arsenide in the following systems: 1) GaAs$-I_2H_2$
[8]; 2) Ga$-AsCl_3-H_2$ [10]; 3) Ga$-AsCl_3-$
Zn$-H_2$.

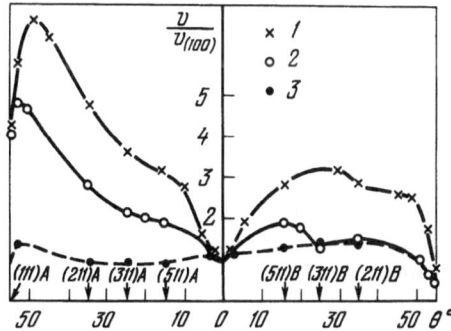

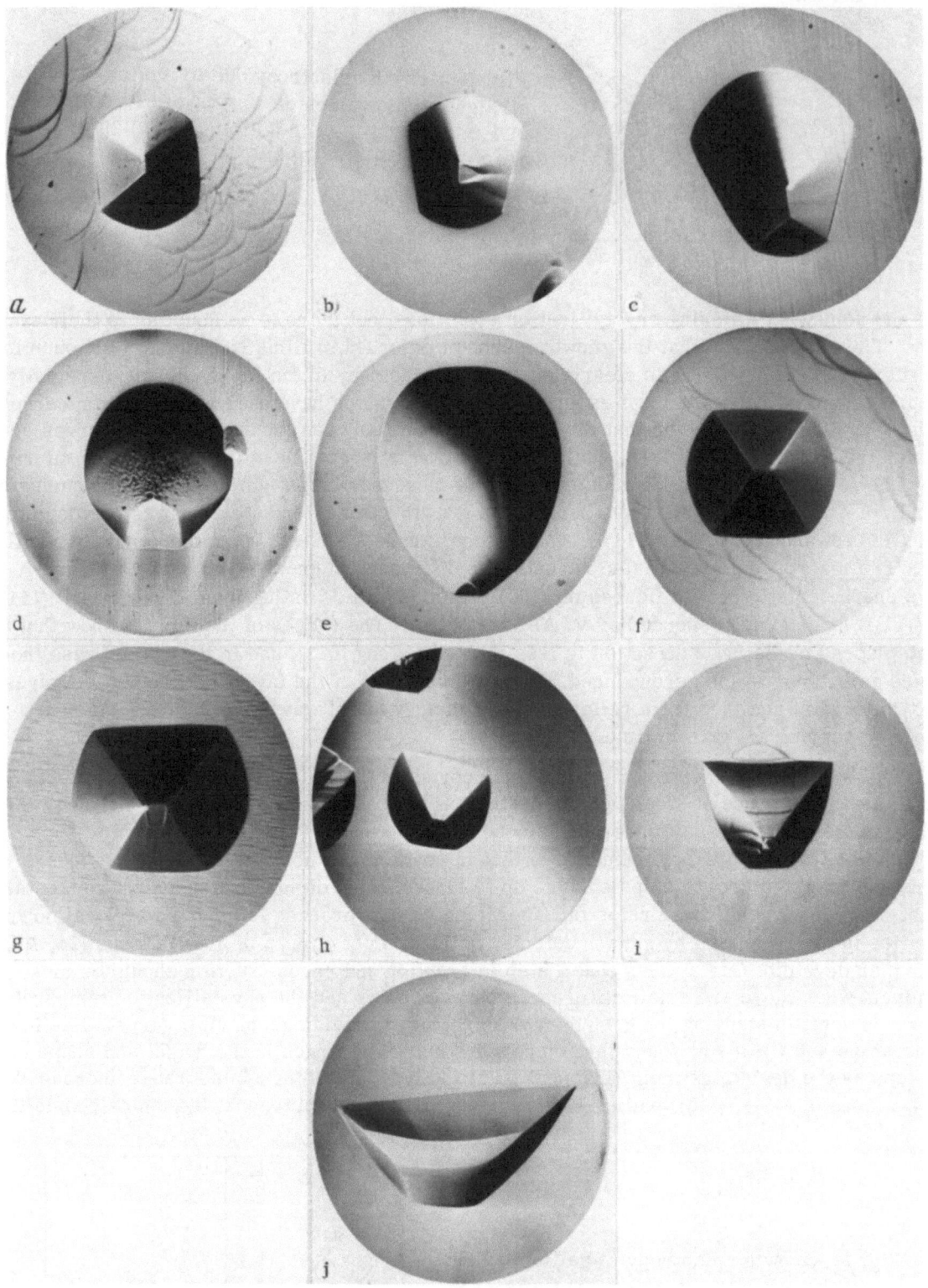

Fig. 3. Characteristic growth figures on iodide gallium arsenide. Angular deviations: a–e) from (100) to (111)A; f–j) from (100) to (111)B in degrees. a) 0; b) 2; c) 6; d) 10; e) 16; f) 0; g) 2; h) 6; i) 10; j) 16; a, b, f, g, h, and i, ×60; c, d, e, and j, ×100.

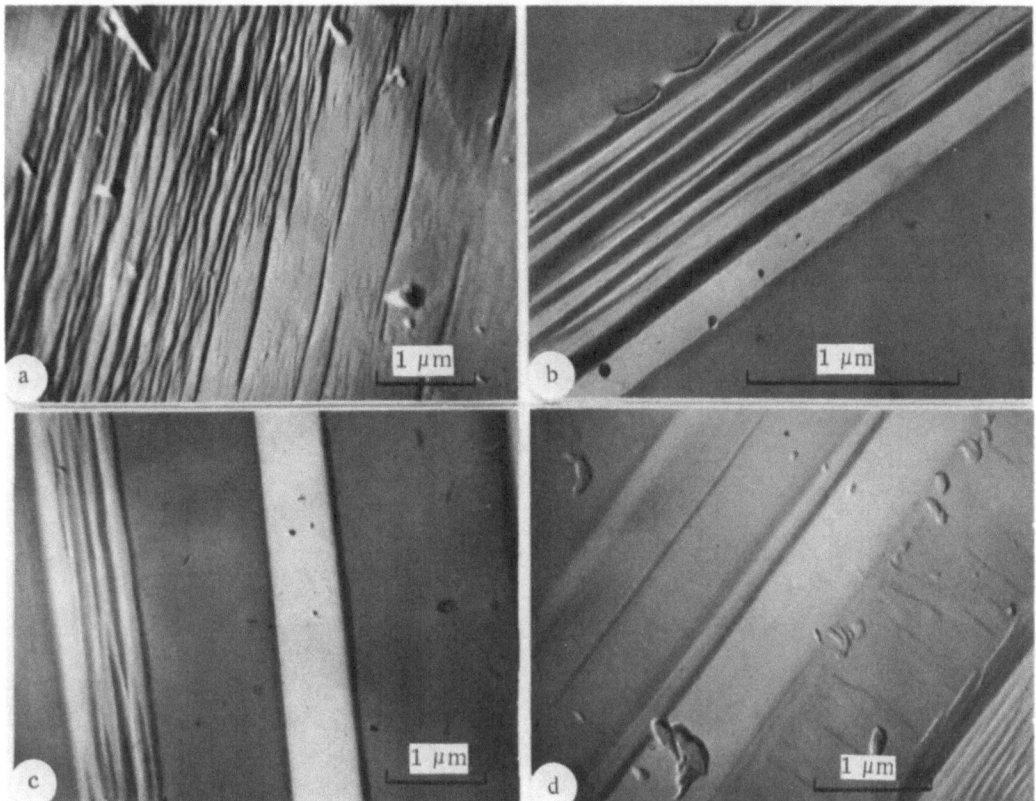

Fig. 4.   Morphology of germanium films, Ge −I₂ system.   Angular deviations: in
degrees: a) 4; b) 10; c) 19.5; d) 29.5.

of (111) and (100) of height up to 1-2 μm (Fig. 4). Here in essence we have the coarser relief
whose mechanism has been described in [16]. This coarsening of the relief is accompanied by
the occurrence of a smooth and deep minimum in the growth rate at the center of the range
(Fig. 5) [13]. The relief alters fairly rapidly with time, as is clear from the photomicro-
graphs of Fig. 6 for (311) plane. In the same time, no appreciable changes were seen in the
relief on a plane deviating 2° from (111). Consequently, the growth-rate anisotropy curve
should vary with time under these conditions. The driving force of the relief modification in
this case is the change in surface energy at the end of a step, since the step coalescence has
become much less pronounced if the lateral deviation prevents the formation of a singular
face on the end.

Gallium arsenide shows a similar smooth and broad minimum in the growth rate near
the (311) orientation; Fig. 7 shows data for this in the GaAs −ZnCl₂ closed system at 790°C [17],

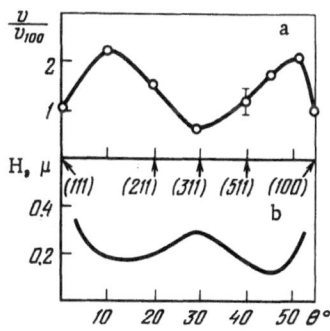

Fig. 5. Growth of germanium in the Ge −I₂ system [11];
a) anisotropy in the growth rate; b) mean height of macroscopic steps in relation to orientation.

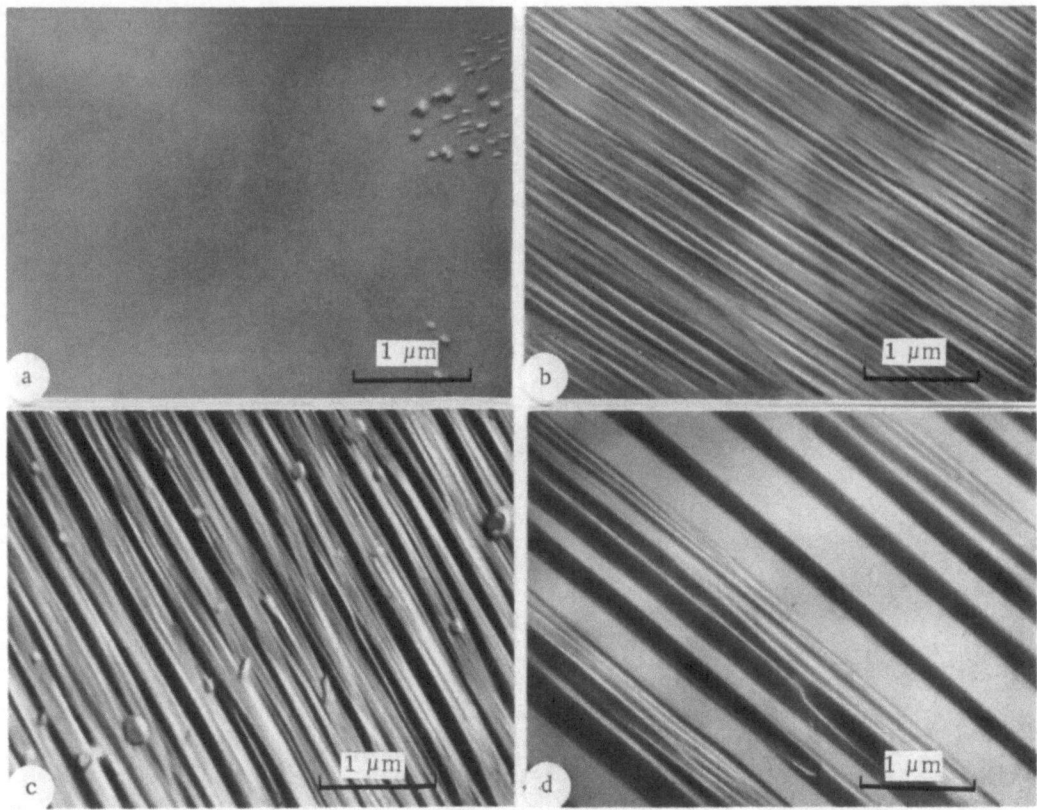

Fig. 6. Time-dependent increase in step height for (311) orientation in the Ge−HI−H$_2$ system. Growth times (min): a) 0; b) 1; c) 15; d) 240.

with (111)B-(100) orientation. It is clear that the relationships are similar to those for germanium: the minimum in the growth rate corresponds to macroscopically rough relief. Figures 2 and 6 together show that the minimum in the growth rate, hardly appreciable for undoped GaAs films (curves 1 and 2) here fully develops; it may be that the altered structure is facilitated by the high zinc concentration in the vapor and at the growth surface, especially since there is scope for nonuniform distribution of the zinc over the surface at high concentrations. When gallium arsenide doped with zinc crystallizes in the Ga−AsCl$_3$−Zn−H$_2$ open system (experiments by L. P. Porokhovnichenko and P. N. Tymchishin), one does not find this

Fig. 7. Substrate orientation effects for GaAs in the GaAs−ZnCl$_2$ system [14] as regards: a) growth rate; b) height of relief.

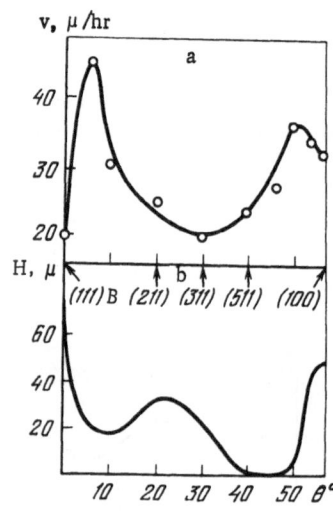

coarsening of the surface, although appreciable roughness occurs on certain surfaces, special-
ly (311)A and (311)B. The basic effect of the zinc is seen in this case in a considerable reduction of the growth rate for films of all orientations relative to undoped films, especially
near the A orientations (Fig. 2, curve 3), where there is a slight minimum in the growth rates
in the middle of the range also. It follows from this that the adsorbed zinc is very anisotropi-
cally distributed. The growth rate reduction in the presence of zinc may be ascribed to poison-
ing of kinks on the growth step [18].

The above results indicate that orientations of (h11) type can have two effects under
growth conditions; they may grow reproductibly, retaining the initial macroscopic very smooth
relief [19], and in that case they enable one to obtain epitaxial films of high perfection, which
are necessary for technical purposes [20]. However, in certain conditions even small perturbations in the relief can be sufficient to produce instability, which is seen initially as waviness
[19, 21] and then as coarsening in the relief, with production of macroscopic steps and reduction in the growth rate. Defects that cause relief instability may be related to initial processing of the substrate or to the growth process itself (high supersaturation, doping). The stability of the surface against such perturbations improves as the distances between the steps increase: then to obtain films with smooth surfaces it is preferable to use small-angle vicinal
planes, on which the growth steps are reasonably resistant to modification, while the outward
growth of defects from the substrate (which is characteristic of singular faces) is greatly
suppressed [9]. In conclusion we note that relief alteration on the growth surface during
growth is accompanied by change in the doping level of the films, which may result in very
undesirable nonuniformity in the electron distribution in the epitaxial films, the extent of this
being dependent on the orientation of a substrate and on the crystallization conditions [15, 22].

Literature Cited

1. N. N. Sheftal' and E. S. Givargizov. Kristallografiya, 9(5):686 (1964).
2. E. S. Givargizov. Kristallografiya, 9(6):902 (1964).
3. Kh. A. Magomedov and N. N. Sheftal'. Kristallografiya, 9(6):902 (1964).
4. Kh. A. Magomedov and N. N. Sheftal'. Izv. AN SSSR, Neorg. Mat., 1(12):2113 (1965).
5. S. Mendelson. Single-Crystal Films, Pergamon Press (1964), 251.
6. F. A. Kuznetsov. In: Growth and Structure of Single-Crystal Semiconductor Films
   [in Russian], part 1, Nauka, Novosibirsk (1968), p. 50.
7. R. C. Sangster. Compound Semiconductors, Vol. 1, Preparation of III–V compounds,
   New York (1963).
8. L. G. Lavrent'eva, V. A. Moskovkin, and M. P. Yakubenya. Izv. VUZ, Fizika, No. 3,
   p. 69 (1970).
9. L. G. Lavrent'eva, Yu. Kataev, V. A. Moskovkin, and M. P. Yakubenya. Kristall und
   Technik, 6(5):607 (1971).
10. A. V. Belyustin, A. V. Kolina, and N. S. Stepanova. In: Growth of Crystals, Vol. 3,
    Consultants Bureau, New York (1962), p. 111.
11. A. N. Stepanova. Proceedings of the Fourth All-Union Conference on Crystal Growth.
    Crystallization Mechanism and Kinetics [in Russian], part 2, Izd. AN Arm. SSR, Erevan
    (1972), p. 81.
12. L. G. Lavrent'eva, V. D. Redkov, N. N. Bakin, and V. A. Ermolaev. In: Gallium Arsenide
    [in Russian], Vol. 2, Izd. Tomsk. Univ. (1969), p. 185.
13. L. G. Lavrent'eva, I. S. Zakharov, I. V. Ironin, and L. M. Krasil'nikova. Proceedings of
    the Fourth All-Union Conference on Crystal Growth. Crystallization Mechanism and
    Kinetics [in Russian], part 2, Izd. AN Arm. SSR, Erevan (1972), p. 70.
14. H. C. Abbink, R. M. Broudy, and G. P. McCarthy. J. Appl. Phys., 39(10):4673 (1968).
15. L. G. Lavrent'eva, M. D. Vilisova, I. V. Ivanin, L. M. Krasilnikova, Yu. M. Rumyantsev,
    and M. P. Yakubenya. Izv. VUZ, Fizika, No. 2 (1973).

16. W. Honigman. Crystal Growth and Form [Russian translation], IL, Moscow (1961).

17. L. G. Lavrent'eva, L. G. Nesteryuk, and V. M. Sennikova. Izv. VUZ, Fizika, No. 6, p. 111 (1972).

18. A. A. Chernov. In: Growth of Crystals, Consultants Bureau, New York (1962), p. 31.

19. L. G. Lavrent'eva, M. P. Yakubenya, O. M. Ivleva, and V. A. Moskovkin, In: Growth of Crystals, Vol. 9, Consultants Bureau, New York (1975), p. 249.

20. L. G. Lavrent'eva, Yu. G. Kataev, and V. A. Moskovkin. Izv. VUZ, Fizika, No. 11, p. 141 (1970).

21. A. E. Blakeslee. Trans. Metallurg. Soc. AIME, 245:577 (1969).

22. L. G. Lavrent'eva and I. S. Zakharov, Proc. Internat. Conf. Phys. and Chem. Semicond., Heterojunction and Layer Structures, Budapest 1:253 (1970).

SHAPE OF TWINS WITH THE DIAMOND STRUCTURE

N. G. Sokolova and M. D. Lyubalin

Interest has long attached [1-3] to the details of twinned crystals; recently, interest in such crystals has increased in connection with the production of planar twinned dendrites. The complicated shape is due to skeletal directional growth, and from it it is fairly difficult to judge the effect on the growth from the twin boundary. For this reason, there is still debate [4-6] on the roles of the reentrant and convex angles at the boundary.

Here we examine in more detail than in [7] the various cases of influence from the twin boundary on the growth forms of twins with the diamond structure, the analysis being from the geometrical viewpoint. Experimental evidence is used to show that the conditions determine whether the reentrant angles at the boundary do or do not extend.

Crystals with the diamond structure characteristically have twins of intergrowth type with two octahedra meeting on (111) planes via the spinel law; we consider first polyhedral growth, in which the twin boundary does not change the face growth rates of the individual crystals. Using this assumption, the twin boundary may be considered as a purely mechanical obstacle to expansion of each of the individuals. The limiting form for such growth will be a figure consisting of two half-octahedra with a regular hexagonal boundary (Fig. 1).

This twin has three types of faces: 1) ones parallel to the intergrowth boundary (pinacoid), 2) intersecting the boundary with reentrant angles (negative bipyramid), and 3) with convex angles (positive bipyramid). The reentrant angle may be completely lost if there is an increase in the relative growth rate of the negative dipyramid faces.

Other relationships between the growth rates of these types of faces can occur under varying conditions; the corresponding shapes of the twins are indicated in Table 1.

There may be more than two individuals in a twinned intergrowth; Fig. 2 shows the growth shape of three octahedral individuals twinned in pairs on parallel (111) planes. Although the growth rates are the same for all faces, the growing crystal does not remain of the same shape, since the middle individual cannot change in thickness. The limiting form in that case is an octahedron with a thin twinned layer. The form consisting of three individuals is notable in that the reentrant angles at the twin boundaries are not lost for any change in the relation between the growth rates. This clearly applies also to intergrowths with parallel twin boundaries where the number of individuals is greater than 3.

Here, as distinct from an intergrowth of two individuals, there are actually two sorts of faces: (1) parallel to the boundary (pinacoid) and (2) intersecting the boundary at reentrant and convex angles. No matter which faces accelerate or slow down in growth (at the reentrant or convex angles), their effects are geometrically equivalent. The shape of the intergrowth as a whole (the degree of flattening) and also the shape of the faces on the end individuals will be dependent on the relationships between the growth rates of the bipyramids and pinacoid.

161

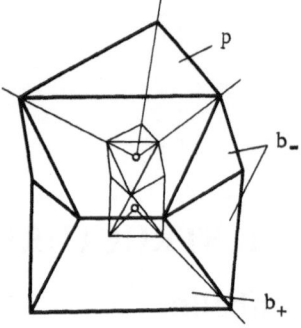

Fig. 1. Growth form of two octahedra combined in the spinel position. Equal growth rates for all faces. Three types of face on the twin: p, pinacoid, b_+ and b_-, bipyramid.

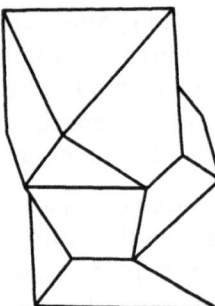

Fig. 2. Growth form of three octahedral crystals twinned in pairs on parallel (111) planes. Equal growth rates for all faces.

If we have a cyclic disposition of the individuals as in Fig. 3a, the twinned intergrowth has two types of faces: at the reentrant and convex angles. Increase in growth rate along the reentrant angles causes elongation of the intergrowth along the common [110] edge. The figure shows that reentrant angles of this shape do not become filled in.

Another possible cyclic twin is the fiveling of Fig. 3b; in this intergrowth, increase in the face growth rates around the reentrant angles can lead to complete loss of the latter. The faces at the reentrant angles in the fiveling are not equivalent; some adjoin one twin boundary, while others adjoin two.

Then the shape of a twinned crystal is determined partly by the relative growth rates of the different faces and also by the geometry of the intergrowth. In turn, the relation between the growth rates may vary with the growth conditions.

A relative increase in the growth rate in the direction of a reentrant angle should occur when there are only slight deviations from equilibrium in the crystal−medium system. If the deviations are large and the growth-limiting factor becomes transport of heat or material, the growth rate can increase in the direction of the convex edge. We have observed these two cases in the crystallization of silicon and germanium.

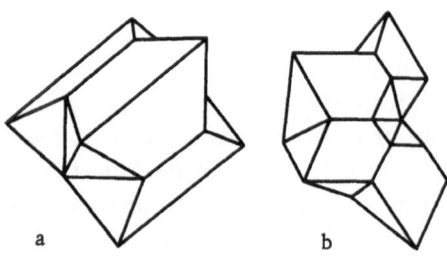

a   b

Fig. 3. Cyclic twins: a) a threeling, angles of 109°28' between twin planes; b) fiveling, 70°32' between twin planes.

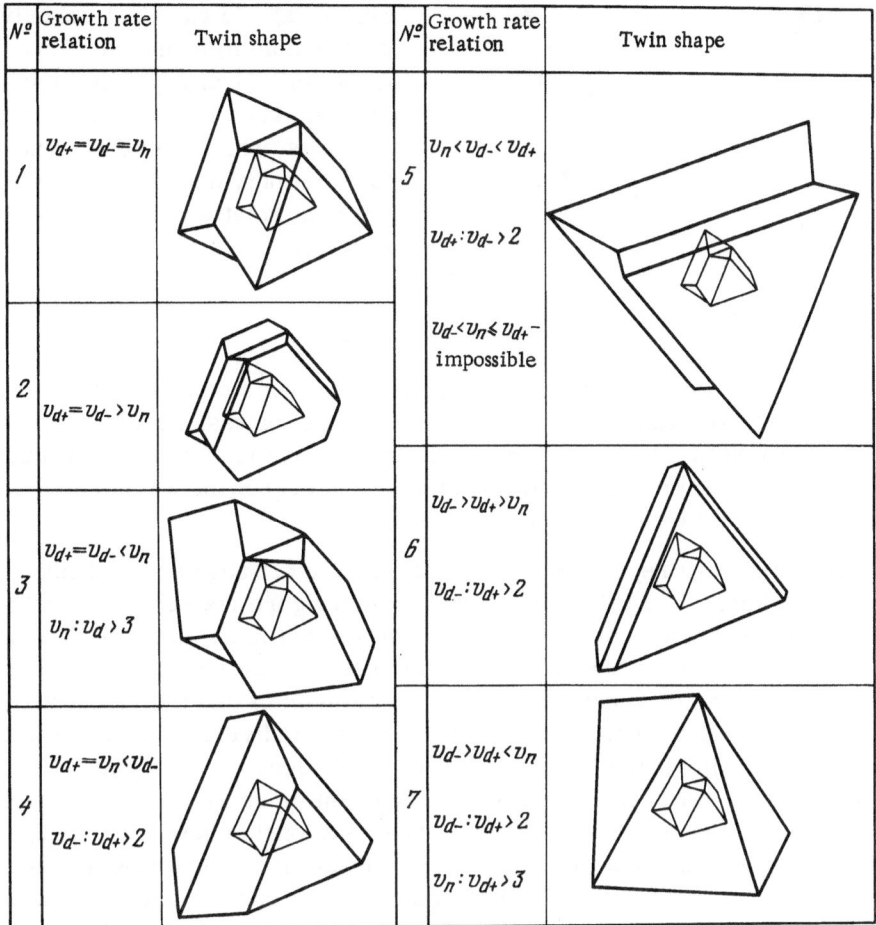

№	Growth rate relation	Twin shape	№	Growth rate relation	Twin shape
1	$v_{d+} = v_{d-} = v_n$		5	$v_n < v_{d-} < v_{d+}$ $v_{d+} : v_{d-} > 2$ $v_{d-} < v_n \leqslant v_{d+}$ — impossible	
2	$v_{d+} = v_{d-} > v_n$				
3	$v_{d+} = v_{d-} < v_n$ $v_n : v_d > 3$		6	$v_{d-} > v_{d+} > v_n$ $v_{d-} : v_{d+} > 2$	
4	$v_{d+} = v_n < v_{d-}$ $v_{d-} : v_{d+} > 2$		7	$v_{d-} > v_{d+} < v_n$ $v_{d-} : v_{d+} > 2$ $v_n : v_{d+} > 3$	

Fig. 4. Shapes of spinel twins with various relations between the face growth rates v.

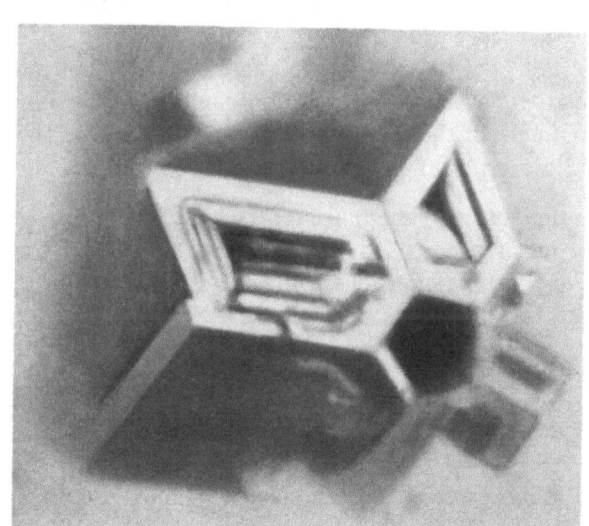

Fig. 5. Fiveling of silicon with unfilled reentrant angles.

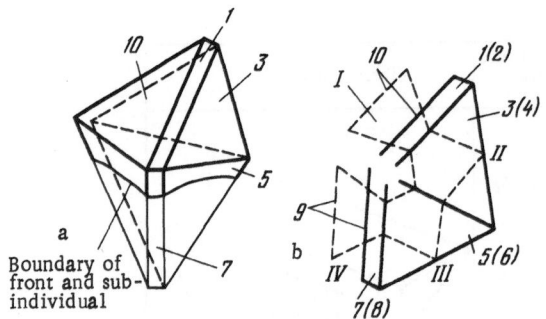

Fig. 6. a) Tetrahedral subindividual in germanium; b) mutual disposition of the individual crystals in an intergrowth. The arabic numerals denote the faces, while the roman ones denote the individuals; part b is symmetrical relative to the plane of the drawing.

Silicon crystals* were grown by the solution-in-melt method in quartz tubes by heating a mixture of Zn and Si [8]. Under identical conditions (in a single tube) we obtained euhedral crystals of various sizes and appearance; isometric crystals, plates, strips, needles, and types intermediate between these. The microscope and the goniometer revealed that all the crystals were facetted by {111} planes, while most of the intergrowths were spinel twins with various numbers and mutual dispositions of the individuals.

The single crystals were represented by small and almost regular octahedra, sometimes skeletal. The two individuals in a twin intergrowth form a figure via scheme 2 of Fig. 4. Large hexagonal plates are formed by superposition of several individuals, with accentuated growth parallel to the boundary. Elongated plates and needles are produced by accentuation of one of the ⟨110⟩ directions of a plate containing a minimum of one individual via the scheme of Fig. 3. We observed all cases with three, four, and more individuals involved in these schemes, which will result in elongation of the intergrowth along the common [110] axis and in turn will increase the extent of the edges of the reentrant or convex angles in each individual when these are parallel to the common direction. The relationship between the edge lengths of the reentrant and convex angles at the end face of a needle remained roughly equal, as in a simple intergrowth of two individuals. This enables us to say that the decisive influence on the form comes from the geometrical factor here, namely, the mutual disposition and number of individuals in an intergrowth with an unchanged relationship between the growth rates of the various types of faces.

Some of the crystals have a skeletal structure, which shows that they were produced mainly by faces composing a convex angle. Usually a depression replaces the reentrant angle. The most active sites for material deposition are the vertices of the twin boundary, from which the convex edges begin to expand, together with the faces intersecting at them, which confirms the growth scheme suggested in [5]. In general, the silicon crystals were produced under conditions such that the reentrant angles were not filled in. This is very clearly seen from the shape of the fiveling (Fig. 5). The intergrowth shows no filling of the reentrant angles adjoining a single twin boundary or two such.

An unusual tetrahedral small crystal of germanium (Figs. 6 and 7) was observed at the growth front of a large single crystal produced by Czochralski's method. A few seconds before the crystal was detached, the heater was switched off, which resulted in supercooling in the bulk of the melt. Examination under the microscope, with the goniometer, and by etching showed that the tetrahedral subindividual was a cyclic twin consisting of four individuals with good {111} bounding planes. Out of all the octahedral faces, those persisting were only the ones at the convex angles. The extreme individuals (I and IV in Fig. 6), each of which had

---

*We are indebted to A. A. Obukhov and V. N. Gurin for providing specimens for morphological examination.

Fig. 7. Subindividuals at the growth front
of a germanium crystals.

only one neighbor, preserved also faces parallel to the intergrowth plane (faces 9 and 10 in
Fig. 6). In these same individuals, the faces in the convex angles 1 (2) and 7 (8) were poorly
developed and took the form of narrow strips elongated along the convex edge. In individuals
II and III, which have two neighbors each, the analogous faces 3 (4) and 5 (6) show simultaneously two convex edges at angles of 60°, and they are therefore triangular in shape. This
case corresponds to schemes 4 and 6 for the relationship between the growth rates given in
the table (absence of reentrant angles), while the shape of the bases is due to the mutual disposition of the individuals. The germanium intergrowth was produced with moderate supercooling, as is clear from the planar faces not only of the tetrahedral subindividual but also of
the other subindividuals at the front (Fig. 7). Silicon intergrowths, on the other hand, were
formed when there was inadequate material supply to the growing faces (these were screened
by the zinc and the high viscosity of the melt). Therefore, growth either of reentrant angles or
convex ones at the twin boundary can occur under different conditions.

Literature Cited

1.   V. Goldschmidt. Z. Kristallogr., 43:582 (1907).
2.   G. Tertsch. Tracht der Kristallen, Berlin (1926).
3.   A. E. Fersman. Crystallography of Diamond [in Russian], Izd. AN SSSR, Moscow (1952).
4.   R. S. Wagner. Acta Metallurgica, 8:57 (1960).
5.   A. A. Bukhanova and D. A. Petrova. Izv. AN SSSR, Neorg. Mat., 4:1439 (1968).
6.   M. Ya. Dashevskii, A. N. Poterukhin, A. V. Zakharova, and L. I. Kaikova. In: Growth
      of Crystals, Vol. 9, Consultants Bureau, New York (1975), p. 206.
7.   M. D. Lyubalin and N. G. Sokolova. In: Crystal Growth Mechanism and Kinetics [in
      Russian], part 2, Izd. AN Arm. SSR, Erevan (1972), p. 160.
8.   A. P. Obukhov, V. N. Gurin, I. F. Kozlova, Z. P. Terent'eva, and T. I. Mazina. Izv. AN
      SSSR, Neorg. Mat., 4:527 (1968).

CRYSTALLIZATION MECHANISM OF GERMANIUM FROM SOLUTION IN GOLD

S. A. Grinberg

Formulation

Zone melting with a temperature gradient [2] can provide useful information on the mechanisms and kinetics of crystallization from solution in a melt, in particular the vapor—liquid—crystal process [1].

This method constitute movement of a liquid zone such as Ge—Au in a solid (Ge crystal) in response to a temperature gradient; the process consists of three stages: a) dissolution of the crystal at the high-temperature boundary of the zone in accordance with the phase diagram of the Ge—Au system, b) diffusion of the solute atoms to the low-temperature boundary, and c) crystallization at that boundary on account of supersaturation. The method was proposed by Pfann [2], and it is discussed theoretically in [3-5], and it also has been used to grow crystals and examine crystallization mechanisms [6-9]. It has been suggested [10, 11] that it should be applied not in the form of the ordinary (bulk) style but in the motion of drops of germanium dissolved in gold over the surface of a germanium single-crystal plate in response to a longitudinal temperature gradient; droplet movement on crystals had been reported in [2, 12-14], but no systematic studies had been made. The proposed method made it possible to compare the speeds of large numbers of drops of different sizes under identical conditions, and hence to draw conclusions on the kinetics for a wide range of moving forces. The method also enables one to determine the supersaturations at which growth or dissolution are possible processes [10]. Finally, the droplet style enables one to examine the recrystallization trace and draw conclusions on the mechanism from the micromorphology.

A study has been made [10] of the motion of droplets of Ge dissolved in Au over Ge on annealing in vacuum. Here we report a similar process in hydrogen, and some essentially new trends are observed.

The following aspects were examined: (a) which of the two major stages is the limiting one for the droplet migration speed (kinetic processes at the atomic level at the droplet boundaries or component diffusion in the solution); (b) if atomic kinetic processes are important, then which of the two boundaries (dissolution or crystallization) is the rate-limiting one; (c) the growth and dissolution mechanisms of Ge in molten Au under these conditions; and (d) the supersaturations for which growth and dissolution of Ge are possible.

Methods

Films of gold from ~50 to ~1000 Å thick were deposited on single-crystal plates of Ge of (111), (110), and (100) orientations by evaporation at ~10^{-5} mm Hg; the heating was performed either under vacuum at 10^{-5}-10^{-6} mm Hg (in a resistance furnace) or in a flow of purified hydrogen (in an induction furnace). Above the eutectic temperature (~360°C), the liquid film of Ge−Au alloy split up into many drops of sizes from about 1 to about 200 μm; if the longitudinal temperature gradient and the drop diameter were sufficiently great, the drops moved towards the hotter side. After annealing, the specimens were examined in an optical microscope with interference contrast.

Results and Discussion

Droplet Motion

Theoretical analysis of inclusion motion in crystals [3, 5] shows that

$$v \sim D \frac{dC}{dT} G$$

when the zone speed v is limited by bulk diffusion through the zone, where D is the diffusion coefficient of the solute or solvent in the liquid zone, dC/dT is the slope of the liquidus curve, and G is the temperature gradient in the drop.

Under these conditions, the zone migration rate is not dependent on the size; in so far as the diffusion or zone length l increases, so does the driving force (the concentration difference ΔC, which is proportional to the temperature difference over the length of zone $\Delta T = Gl$). If, however, the zone migration speed is restricted by atomic kinetic processes at the boundaries (kinetic state), then the migration will accelerate as the zone thickness increases.

We used heating temperatures of 650-850°C with gradients of 10-100 deg/cm and droplet diameters of 10-100 μm, the medium being vacuum or hydrogen; under these conditions, the migration rate increased with the size of the droplet in the direction of motion (Figs. 1 and 2). These results indicate that the atomic kinetic processes at the dissolution of crystallization front are the rate-limiting ones under these conditions.

Fig. 1. Droplet migration rate as a function of diameter. Annealing under vacuum, Ge (100) surface, mean annealing temperature T_{av} = 860°C, temperature gradient G = 45 deg/cm.

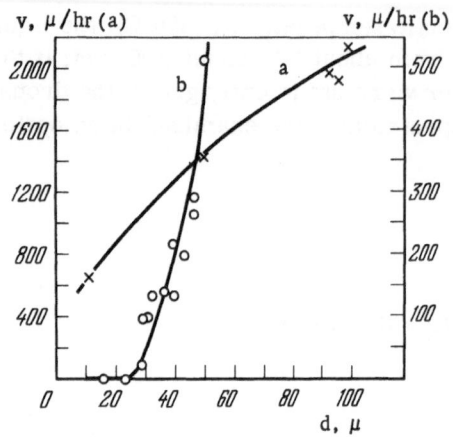

Fig. 2. Increase in droplet migration rate with diameter; annealing in hydrogen, $T_{av} \approx$ 820°C, (110) surface: a) droplets moving along [1$\bar{1}$0]; b) droplets moving perpendicular to [1$\bar{1}$0].

Droplet Shape

The shape of the droplets is substantially dependent on whether vacuum or hydrogen is used.

Under vacuum, droplets on various faces and moving in various directions were circular (Fig. 3); in hydrogen, the shape was dependent on the orientation of the substrate and on the direction of motion. For all the orientations, namely (111), (110), and (100), the drops had a shape corresponding to the symmetry of the substrate (Fig. 4).

This difference is due to reduction by the hydrogen; the crystal−solution boundary is purified, and this allows the anisotropy of the surface to make itself felt.

Anisotropy in the Direction of Motion

A further difference as between vacuum and hydrogen concerns the direction of droplet motion relative to the temperature gradient. Under vacuum, the drops always move along the gradient, while in hydrogen substrates of all orientations give anisotropy in the direction of motion.

The anisotropy in the migration direction was examined in hydrogen with a radial distribution of the temperature gradient on the substrate. This gradient was provided by placing a graphite cylinder at the center of a circular part, which was maintained by induction heating

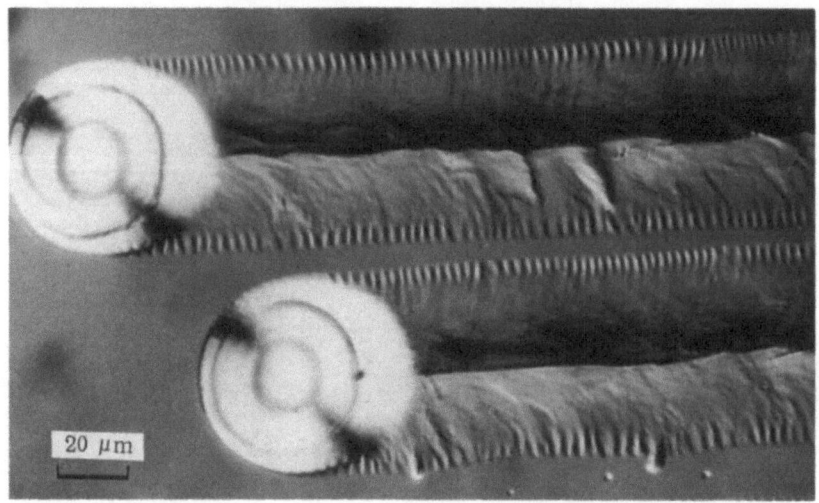

20 μm

Fig. 3. Shapes of drops moving on a surface under vacuum.

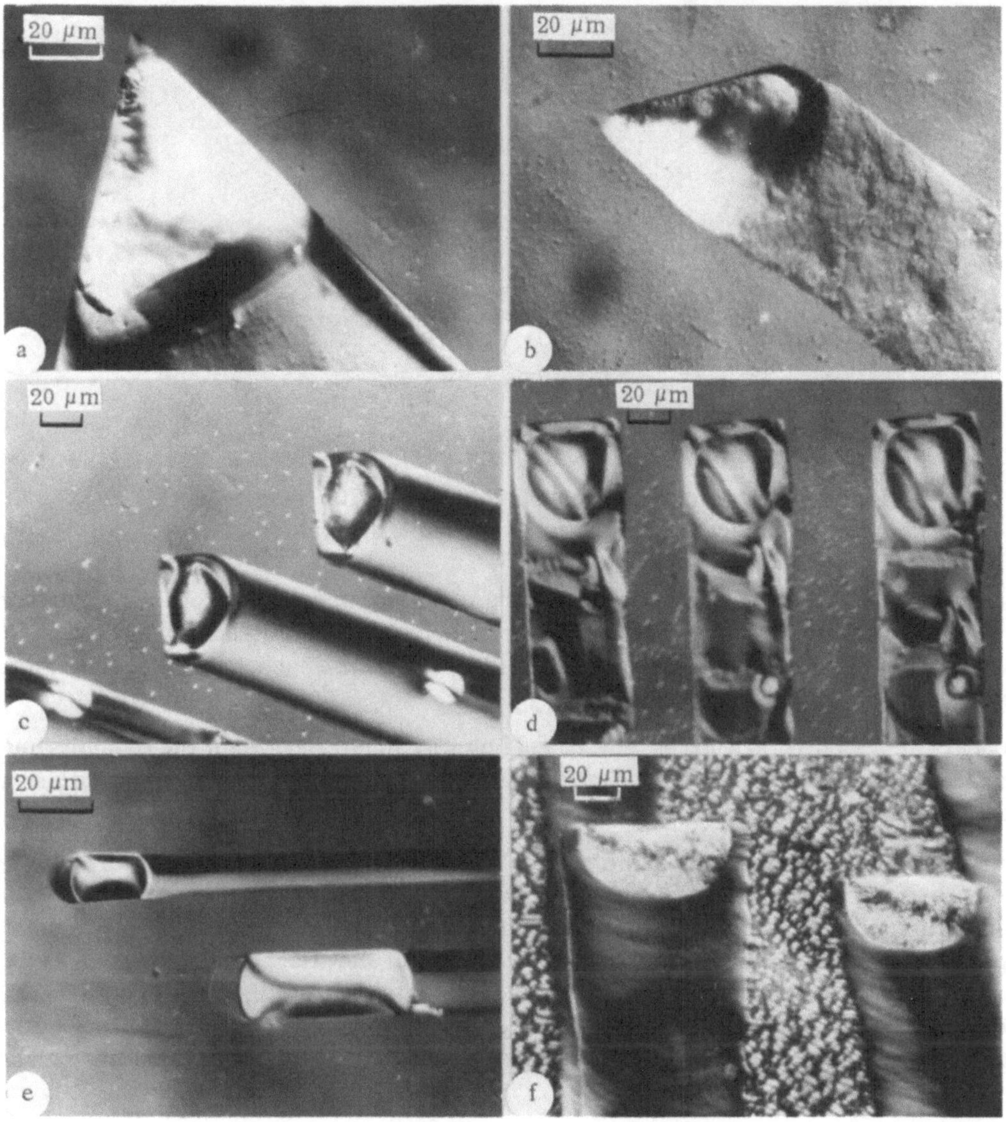

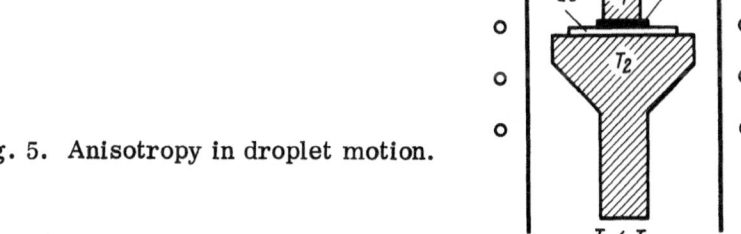

Fig. 4. Shape of droplets on annealing in hydrogen: a, b) on (111); c, d) on (100); e) on (110) moving along [1$\bar{1}$0]; f) on (110) moving perpendicular to [1$\bar{1}$0].

Fig. 5. Anisotropy in droplet motion.

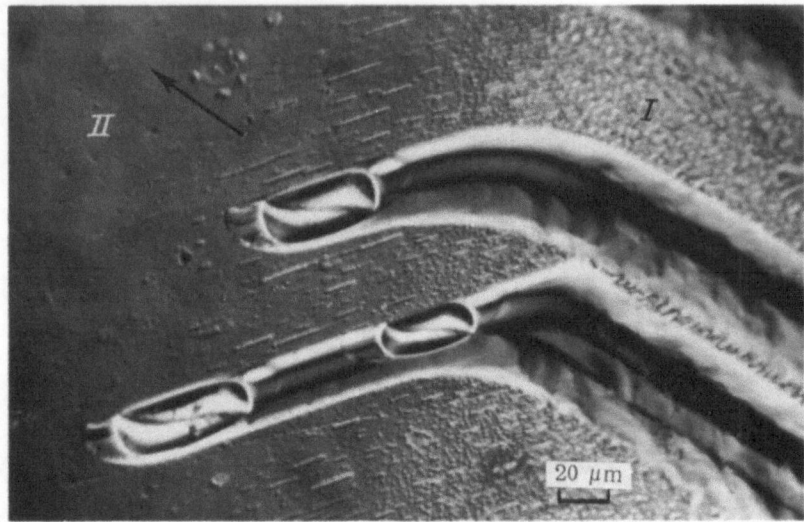

Fig. 6. Anisotropy of droplet motion on (110) in hydrogen. The
arrow indicates the direction of the temperature gradient.

at a temperature T_1 less than T_2 (the temperature of the lower graphite crucible), as shown in
Fig. 5. The heat lost through the upper cylinder produced a radial gradient in the specimen,
so one could study simultaneously the motion of droplets with respect to the center along all
directions.

Measurements on (110) showed that at sufficiently small gradients the droplets moved
along the $[1\bar{1}0]$ directions of the strong bonds even when the angle between the gradient and
this direction was 60-70° (Fig. 6).

This arises because there is some critical driving force necessary to overcome the
barrier on the close-packed dissolution of crystallization fronts. Consider a (110) surface,
which is the most anisotropic one in the diamond structure. If the projection of G on a direction perpendicular to $[1\bar{1}0]$, namely $G_\perp$ (Fig. 7) is so small that the supersaturation at the crystallization front in this direction is insufficient for crystallization, then the drop responds to
the other component parallel to $[1\bar{1}0]$, namely $G_\parallel$, and moves in the $[1\bar{1}0]$ direction, where there
is no barrier to crystallization; the same applies to the dissolution front in an analogous
manner.

The direction of motion is dependent not only on G and the angle between this and the
chains of strong bonds but also on the droplet size perpendicular to $[1\bar{1}0]$. Figure 8 shows
droplets of various sizes D_1, D_2, d_1, and d_2 in the first stage of annealing with the relatively

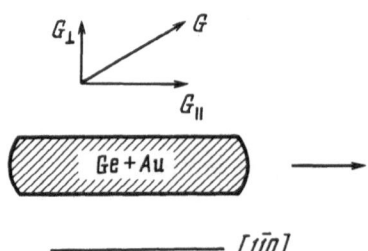

Fig. 7. Droplet motion on (110) in hydrogen.

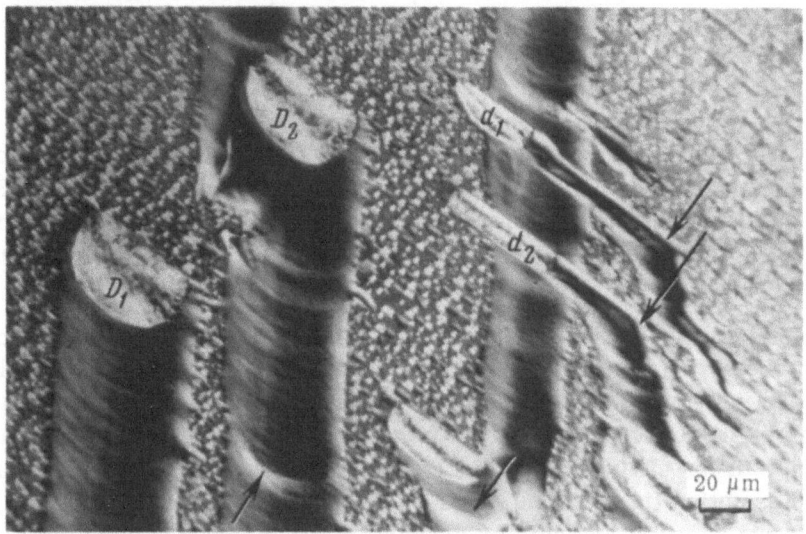

Fig. 8. Effects of droplet size along the temperature gradient in
the direction of motion for a (110) surface in hydrogen. The arrows
denote the start of the second stage of annealing with a lower temperature gradient.

high gradient $G_\perp \approx 100$ deg/cm, which move along the gradient; in the second stage (start indicated by arrows), the gradient was reduced to $G_\parallel \approx 50$ deg/cm, and the larger droplets D_1
and D_2 continued to move in the previous direction, while the smaller ones d_1 and d_2 turned in
the $[1\bar{1}0]$ direction. On (100) surfaces (Fig. 9) and on (111) surfaces, there is also anisotropy
in the direction of motion, but it is less pronounced than that on (100).

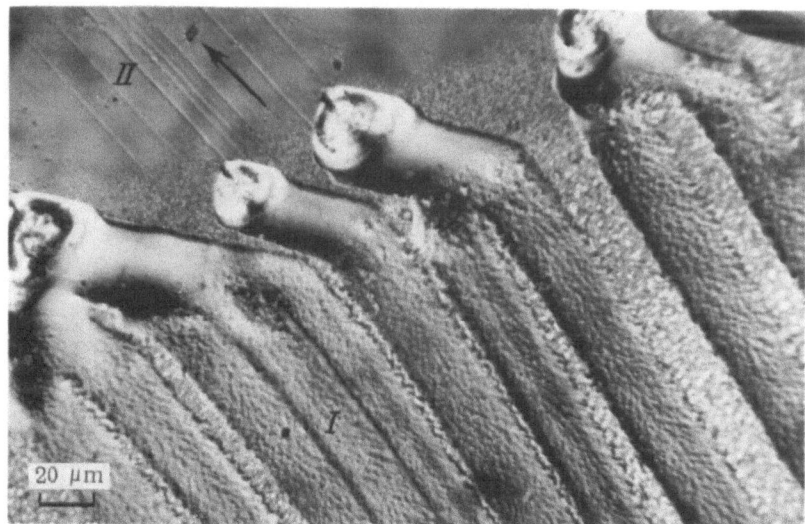

Fig. 9. Anisotropy of droplet motion on (100) in hydrogen. The
arrow indicates the direction of the temperature gradient.

The motion thus reveals competition between two factors: crystal anisotropy, which makes itself felt via a close-packed boundary, and the supersaturation or saturation deficit, which is dependent on the magnitude of the gradient and on its direction relative to [1$\bar{1}$0] as well as on the droplet size.

Limiting Boundary and Process Mechanism

When a droplet moves in any direction different from [1$\bar{1}$0] on a (110) substrate, the dissolution front is facetted, while the crystallization front remains rounded (Fig. 4). so dissolution is the rate-limiting stage. There is a critical droplet size in these directions, so one assumes that dissolution at a close-packed boundary occurs by two-dimensional nucleation in a hydrogen atmosphere.

Two-dimensional nucleation clearly relates to a facetted dissolution boundary, which consists [15] of one or more {111} planes. This is indicated by the reduction in the threshold for the moving force when a drop moves on parts I covered with very small (~1 μm) drops of the same Ge−Au alloy, the reduction being relative to the threshold for clean surfaces II (Figs. 6 and 9). The small drops clearly provide additional steps and thereby reduce the nucleation barrier at the dissolution front, so on part I the drops move along the gradient, while on the clean surface II they move at an angle on account of the dissolution barrier in the direction perpendicular to [1$\bar{1}$0].

A rounded dissolution front together with a growth front of this form (Fig. 4e) would appear to indicate that the process has a normal mechanism.

Smooth (111) faces and growth layers are visible in the recrystallization trace of a droplet that has moved over a (111) surface (Fig. 10); this indicates that the growth of {111} faces also occurs by two-dimensional nucleation.

Evaluation of the Threshold Driving Force

The thermal migration method used here enables one to estimate the critical driving force for recrystallization of germanium from solution in molten gold.

The front is rounded in hydrogen when the droplets move on a (110) surface along the [1$\bar{1}$0] direction of the strong bonds, while it is facetted in the transverse direction; on (100)

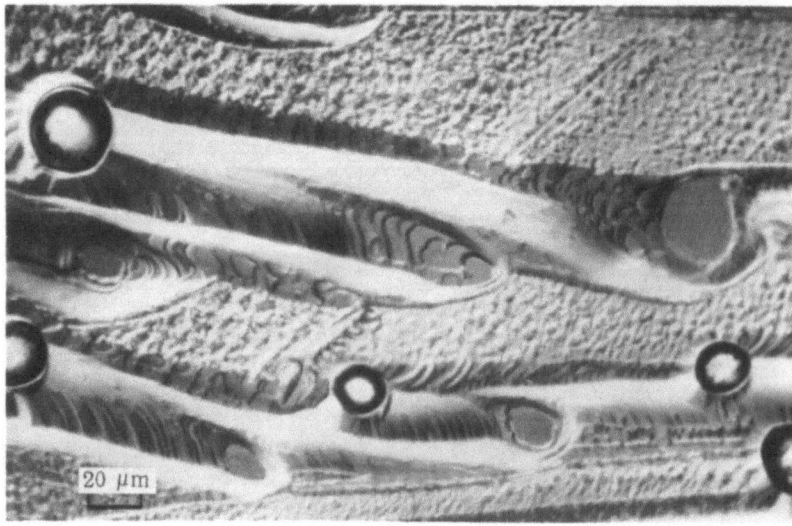

Fig. 10. Formation of smooth faces and growth layers in the recrystallization trace of a drop moving on (111).

and (111) substrates, the dissolution front is facetted for any direction of motion. Correspondingly, there was a critical size for droplet motion on (110) substrates in directions perpendicular to [1$\bar{1}$0], and for any direction on (100) and (111) substrates. No threshold was observed for [1$\bar{1}$0] directions on (110); even the smallest drops (diameters down to 2-3 μm) were seen to move in the optical microscope.

The critical size perpendicular to [110] was $d_{cr} \approx 25$ μm (Fig. 2) on a (110) surface with a mean temperature of $T_{av} \sim 820°C$ and $G \approx 60$ deg/cm; the threshold force (the relative saturation deficit at the dissolution front) was as follows for a droplet of critical diameter:

$$\frac{\Delta C}{C_e} = \frac{\frac{dC}{dT} \Delta T_{cr}}{C_e} = \frac{\frac{dC}{dT} \frac{dT}{dx} d_{cr}}{C_e},$$

where C_e is the germanium concentration in the solution corresponding to the equilibrium composition at that temperature, while dT/dx is the temperature gradient along the surface of the crystal (and, as is assumed here, within the drop, which has been confirmed [16] for this case). The temperature difference across a drop of critical diameter is $\Delta T_{cr} \approx 0.15°C$, while $\Delta C/C_e \approx 0.10\%$.

Conclusions

It has been found for droplets of germanium dissolved in molten gold on Ge crystals that movement occurs in a temperature gradient in a hydrogen atmosphere, and the following results have been obtained:

1. The speed is limited by atomic kinetic processes at the drop boundaries under the conditions used (droplet size of 10-100 μm, annealing temperature of 650-850°C, gradient of 10-100 deg/cm, orientations (111), (110), and (100) for the substrate.

2. When the droplets are formed under vacuum, they are circular and migrate isotropically; in hydrogen the droplets are anisotropic in shape and in direction of migration. This difference is due to reduction by the hydrogen acting on the germanium oxide, which can block the crystal–solution interface.

3. When a droplet moves in hydrogen in directions different from [1$\bar{1}$0] on (110) substrates, the dissolution front is facetted, which means that the dissolution stage is the rate-limiting one; there is a threshold force (saturation deficit) in these directions, and it is assumed that dissolution occurs by a two-dimensional nucleation mechanism on {111} close-packed planes. Rounded dissolution fronts and growth fronts are formed in motion along [1$\bar{1}$0] directions on (110) faces, and a normal mechanism is assumed.

I am indebted to E. I. Givargizov for a valuable discussion and to L. N. Obolenskii for assistance in the experiments.

Literature Cited

1. R. S. Wagner and W. C. Ellis. Appl. Phys. Letters, 4(5):89 (1964).
2. W. G. Pfann. Zone Melting, New York (1966).
3. W. A. Tiller. J. Appl. Phys., 34(9):2757 (1963); 36(1):261 (1965).
4. Ya. E. Geguzin and M. A. Krivoglaz. Motion of Macroscopic Inclusions in Solids [in Russian], Metallurgiya, Moscow (1971).
5. V. N. Lozovskii. Zone Melting with a Temperature Gradient [in Russian], Metallurgiya, Moscow (1972).
6. R. G. Seidensticker. J. Electrochem. Soc., 113(2):152 (1966).
7. Trudy Novocherkassk. Politekh. Inst., No. 170 (1967); No. 180 (1968); No. 208 (1970).

8. R. W. Hamaker and W. B. White. J. Appl. Phys., 39(3):1758 (1968); J. Electrochem. Soc., 116(4):478 (1969).
9. T. R. Anthony and H. E. Cline. J. Appl. Phys., 42(9):3380 (1971).
10. S. A. Grinberg and E. I. Givargizov. Kristallografiya, 18(2):380 (1973).
11. S. A. Grinberg. Abstracts for the Fourth All-Union Conference on Crystal Growth, Tsakhkadzov, 1972; Crystal Growth Mechanism and Kinetics [in Russian], part 2, Izd. AN Arm. SSR Erevan (1972), p. 89.
12. E. Biedermann. J. Electrochem. Soc., 114(2):207 (1967).
13. G. M. Gavrilov. Kristallografiya, 16(4):834 (1971).
14. R. N. Erlikh, N. V. Belov, and T. S. Kondrat'eva. Abstracts for the Fourth All-Union Conference on Crystal Growth, Tsakhkadzov, 1972; Crystal Growth Mechanism and Kinetics [in Russian], part 2, Izd. AN Arm. SSR Erevan (1972), p. 93.
15. J. W. Faust, A. Sagar, and H. F. John. J. Electrochem. Soc., 109(9):824 (1962).
16. T. R. Anthony and H. E. Cline. J. Appl. Phys., 43(5):2473 (1972).

SCANNING ELECTRON MICROSCOPE OBSERVATION OF DOMAIN FIELDS IN SbSI CRYSTALS GROWN FROM THE VAPOR STATE

L. A. Zadorozhnaya, V. G. Galstyan, and V. A. Lyakhovitskaya

Scanning electron microscopy has been applied to the domain structure of SbSI crystals; etching and decoration have been used previously [1] to reveal the domains, but these showed only the static pictures.

We used SbSI crystals made from the vapor with well-developed faces in the (hk0) vertical belt and the (101) rhombic prism.

The secondary electron flux from a point on the surface is modulated by the local features, including the chemical composition, the geometry, and the potential relief, which ultimately determine the secondary emission coefficient. Elastic reflection gives information only about the surface relief, while slow secondary electrons provide information not only about the topography but also about the local electric fields.

In SbSI, the bound charges in domains of opposite directions of spontaneous polarization vector lie in sections perpendicular to the c axis and can be observed on {101} faces, whose planes are inclined at 67°57' to the polar axis. The potential relief due to these charges can be visualized in the scanning microscope by secondary emission.

Figure 1 shows a photomicrograph of a (101) face of an SbSI single crystal in the ferro-electric state; the fields of the bound charges are seen as light spots of elliptical or more complex shape, the latter when several domains of one sign join up. There are point defects at the center of nearly every elliptical domain, which indicates that the domain structure is related to the defect structure of the crystal.

The potential relief obtained by this method is similar to the pattern revealed by selective etching on a (101) face (Fig 2).

Parts a–c of Fig. 3 show some stages in polarization reversal in response to an electron probe acting on domains on a (101) face; the photographs represent the same part of the matrix in the positive domain at the center of which lies a domain of negative sign (Fig. 3a). We determined the sign of the bound charge from the contrast formation mechanism for the microscope. The time to record a frame was 25 sec. Parts b and c of Fig. 3 show the states of the (101) face at 25 and 50 sec respectively. The sense of the spontaneous polarization vector of the negative domain is reversed after the first scan, i.e., the domain contrast in the matrix is altered (Fig. 3). In the subsequent scan, we get an additional (dark) region of oval shape (Fig. 3c), which indicates that the matrix has produced a further domain of positive sign on account of increase in the polarizing charge from the probe electrons.*

---

*A similar effect has been reported for TGS crystals, but only for a matrix of negative sign [2].

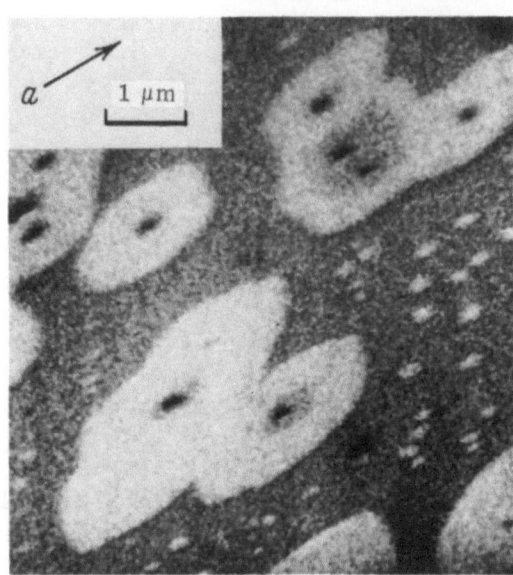

Fig. 1. Scanning micrograph in secondary emission showing the potential relief on a (101) face of SbSI.

We used secondary emission from a face in the vertical (hk0) belt in the ferroelectric state to record the sequence of light and dark bands of Fig. 4, which are oriented along the polar c axis throughout the length of the crystal. The bands are 2–8 μm wide. The conditions of observation have been stated previously [3].

To elucidate this banded contrast we examined part of a (100) face under various conditions; in the ferroelectric state, elastically reflected electrons revealed the surface relief (Fig. 5a); the secondary-emission image showed characteristic black-and-white bands oriented along the polar c axis (Fig. 5b).

When the crystal was heated above the transition point, the banded contrast in the secondary-emission state vanished completely, and the image was exactly the same as that obtained with elastic reflection (Fig. 5c).

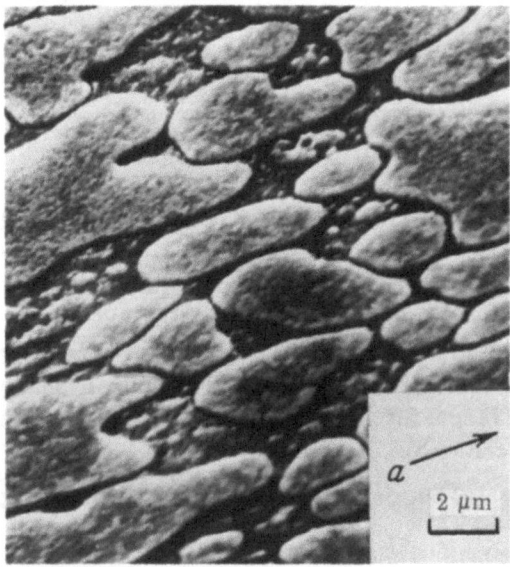

Fig. 2. Domain structure of a (101) face revealed by etching.

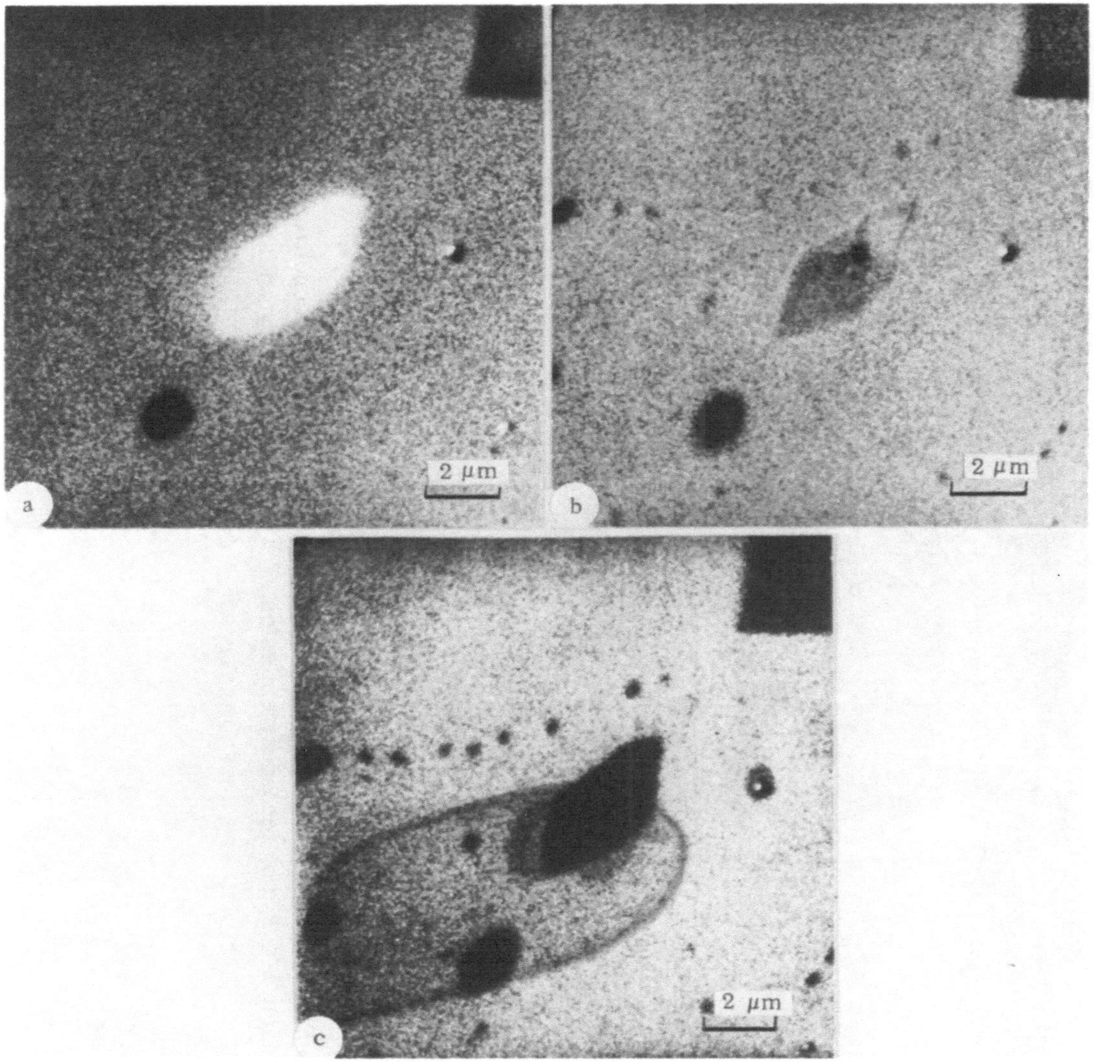

Fig. 4. Banded contrast on a (100) face in secondary emission in the scanning electron microscope, $T < T_c$.

Fig. 3. Polarization reversal of a domain of negative sign on (101) in response to electron probe (secondary emission mode in a scanning electron microscope):   a) initial state; b) after 25 sec; c) after 50 sec.

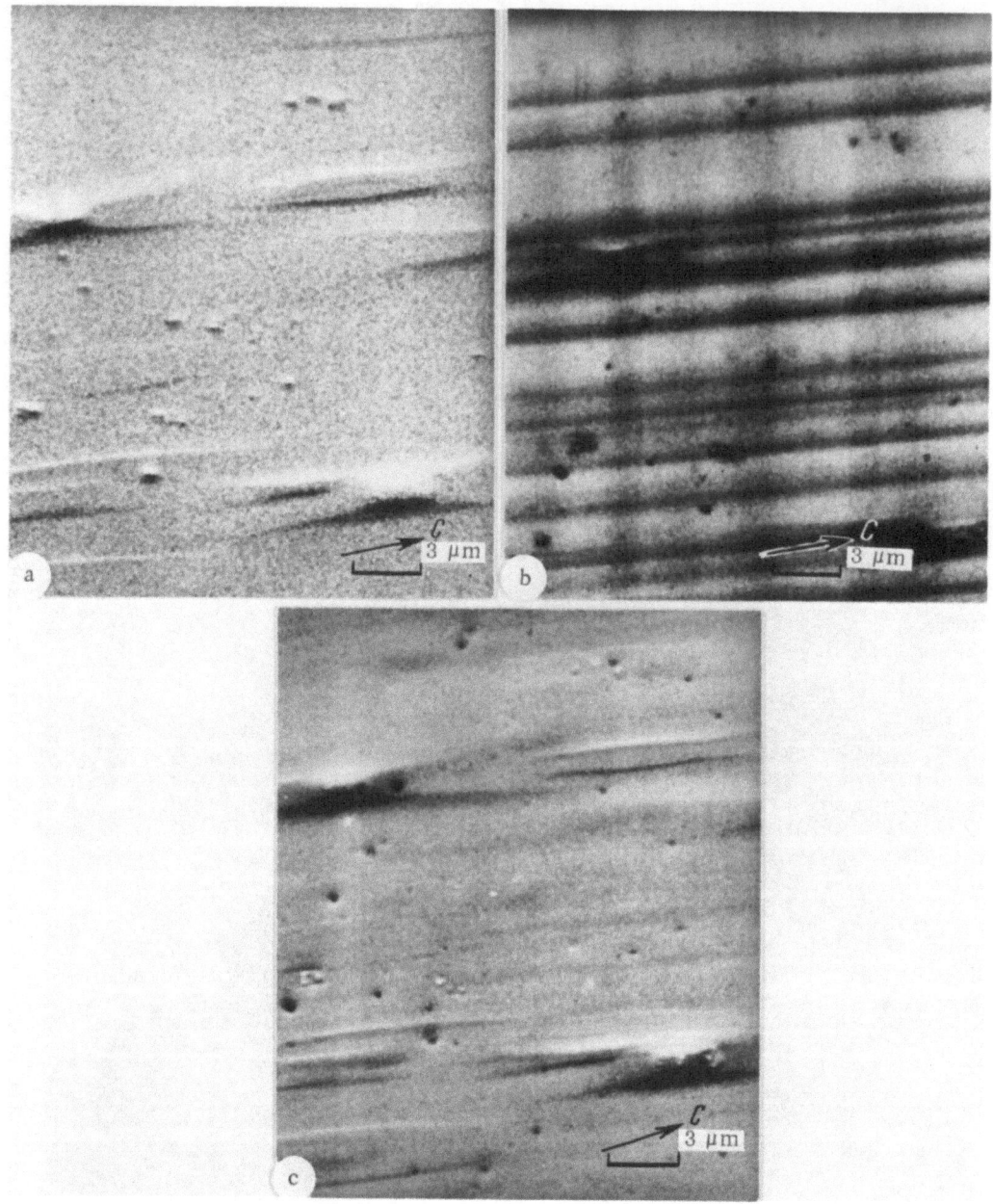

Fig. 5. Electron micrograph of part of a (100) face under various conditions: a) elastic reflection, $T < T_c$; b) secondary emission, $T < T_c$; c) secondary emission, $T > T_c$.

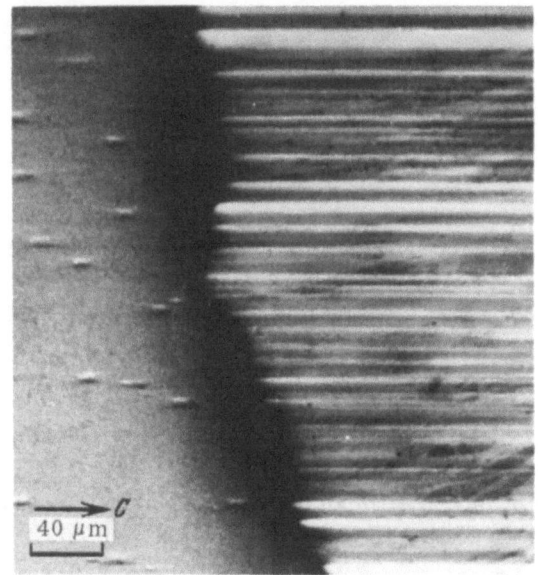

Fig. 6. Banded contrast absent from a part with deposited film in secondary emission, $T < T_c$.

One therefore assumes that the banded contrast is due to the fields of charges localized at the side faces of the SbSI crystal in the ferroelectric state.

To confirm this, we deposited on part of a (100) face a layer of gold 200-300 Å thick; secondary emission showed that no bands occurred on the area with the film (Fig. 6), the only contrast being that due to the surface topography. This means that under these conditions the part of the (100) face without the gold is seen via the normal component of the charge field, not the tangential component, since, if this were not so, the banded contrast would persist on the part with the conducting film.

Conclusions

1. The scanning electron microscope provides local examination with high resolution for the domain structure of SbSI crystals.

2. The fields of domains on (101) faces have been visualized, and polarization reversal has been produced.

3. The potential relief can be seen on the side faces of SbSI crystals, and the contrast production mechanism has been examined.

Literature Cited

1. M. O. Kliya and V. A. Lyakhovitskaya. Kristallografiya, 15(1):75 (1970).
2. R. Le Bihan and M. Maussion. Abstrs. Second Europ. Meet. on Ferroelectricity, Dijon (1971), 137.
3. V. G. Galstyan, M. O. Kliya, V. A. Lyakhovitskaya, and E. M. Fadyukov. Abstracts for the Eighth All-Union Conference on Electron Microscopy [in Russian], Vol. 1, Moscow (1971), p. 100.

MASS CRYSTAL GROWTH

L. I. Kvater and I. V. Frishberg

Vapor−solid transition with simultaneous nucleation and mass growth of crystals is based on high supersaturations in the vapor; the resulting crystals are usually of irregular form, and the imperfections frequently bear detailed information about the interaction between the medium and the growing material [1].

We have examined the mass growth of crystals for the systems naphthalene−air, iodine−air, magnesium−helium at supersaturations of 10^2-10^6; the essential system [2] was the same for all conditions: a closed volume of gas was kept saturated with the vapor (vapor pressure 10-100 mm Hg, total pressure 25-760 mm Hg), and a horizontal cooled crystallization surface was introduced (corresponding vapor pressure 10^{-3} to 10^{-5} mm Hg). The resulting high supersaturation initiated growth on the basal surface for a single set of crystals, each of which essentially reflects the complex origin of the whole set. Therefore, although we have split up the description into sections on the medium and the crystal, we have constantly borne in mind that the two have an indissoluble relationship.

Gas Medium

The behavior of the vapor could be examined because very small condensed droplets were formed (in our case, of naphthalene), which served as decoration; their paths were made visible via the Tyndall effect, and this gave a good illustration of the classic concept of the mobility of the atomsphere surrounding a crystal grown from the vapor [3].

Before the crystals began to form, the decorating droplets were seen to be in weak random motion; when the surface was introduced, all the droplets streamed toward it, and the gas flows associated with the onset of crystallization were visualized. These flows pass around the crystals and deposit solid material, then tending upward, and they entrain the naphthalene droplets and also leave indirect evidence of themselves in the facial sculpture of the crystals.

These flows resemble the concentration flows accompanying crystallization from solution [4, 5]; they also resemble the crystallization flows in phase transitions in melts [6], and also the convected flow arising in vapor−liquid or liquid−vapor transitions [7]. In addition, they move also on account of gravitational forces, which are additional to those from the phase transition.

Closer consideration shows that the phase transition itself, no matter whether it is vapor−solid, vapor−liquid, or liquid−solid, causes the medium to flow as a whole and supports the motion when the degree of supersaturation is relatively high.

If all these flows are of the same physical nature, we have a general explanation for this spontaneous motion: the lower-density layer of solution leaving the crystal is of lower density because the excess solute has been deposited and the latent heat has been released, so it rises upwards and is replaced by cooler and more concentrated solution [5].

Crystallization Kinetics

In mass growth, we recorded the growth rate for the whole crystalline assemblage; the molar rate S_μ may be represented for all states by the single empirical equation

$$10^3 \cdot S_\mu = \frac{A}{\gamma P} \exp{-\frac{\Delta H}{RT}}, \tag{1}$$

where A is a constant that is an unknown function of the geometrical parameters of the system, P (atm) is the total pressure in the system, γ (6.67 $\cdot$ 10^{-11} m^3/kg-sec) is the gravitational constant, H (kcal/mole) is the activation energy of the process and is close in magnitude to the heat of sublimation of the solid, R (1.987 cal/mole-deg) is the universal gas constant, and T (°K) is the evaporation temperature.

Formula (1) reflects the tendency of the crystallization rate to increase with the vapor concentration, and on the whole it takes the traditional form; however, it contains also a new element, namely the gravitational constant γ, which is a coefficient of proportionality in Newton's gravitational formula. The gravitational constant is considered in physics as a quantity independent of the medium separating the gravitating masses, and also independent of the physicochemical properties and motion. This constant γ is present, on our view, by virtue of the quantitative expression of the qualitative observations on the convected motion arising from the phase transition, i.e., on a process owing its origin to gravitation.

The similarity between the activation energy and the heat of sublimation agrees with microscopic observations on the crystals, which indicate that the vapor passes directly into the solid without intermediate stages.

The results of all the experiments lie on a single straight line in coordinates of γ P against P (for instance, Fig. 1 for naphthalene and magnesium), which enables one to predict the crystallization rates for any intermediate states.

Crystal Microgenesis

All the crystals grew normally on the basal surface (orthotropism in the growth), and they were seen superficially as thin platelets standing on edges. The shape of the crystals gradually changed from dendritic to simple hexagonal form as the total pressure in the system was reduced; this was accompanied by gradual combination of the flows in different directions into a single flow normal to the crystallization surface.

These details are felt not only in differential development of the faces on the crystal but also in the face sculpture; only the (001) faces were directly accessible to observation, but the accessories provided abundant information on the crystal genesis.

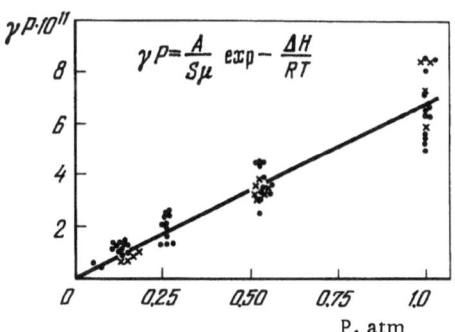

Fig. 1. Plot of γ P against P.

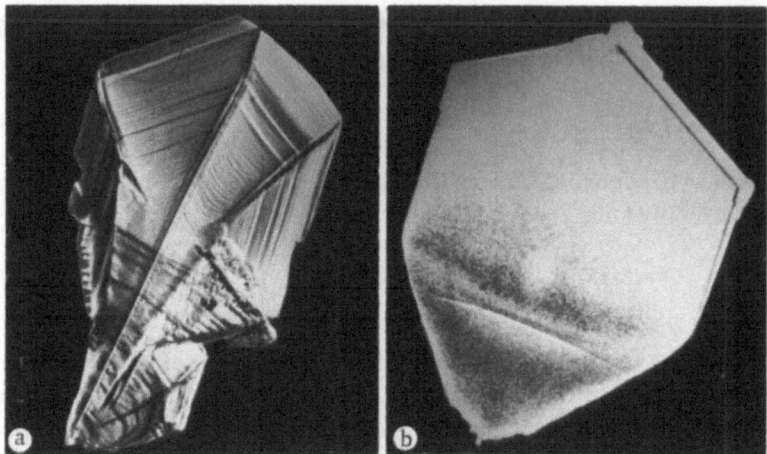

Fig. 2. Sculpturing of pinacoid faces for crystals of: a) naphthalene; b) magnesium.

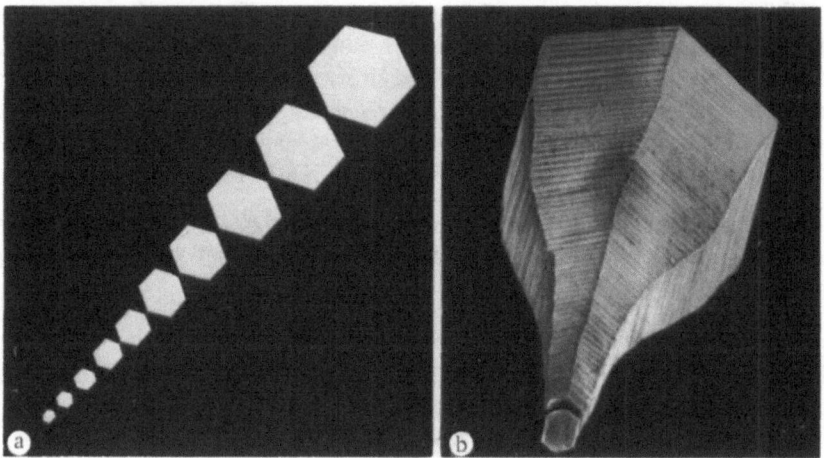

Fig. 3. Crystal overgrowth mechanism: a) sequentially deposited
layers; b) crystal composed of layers.

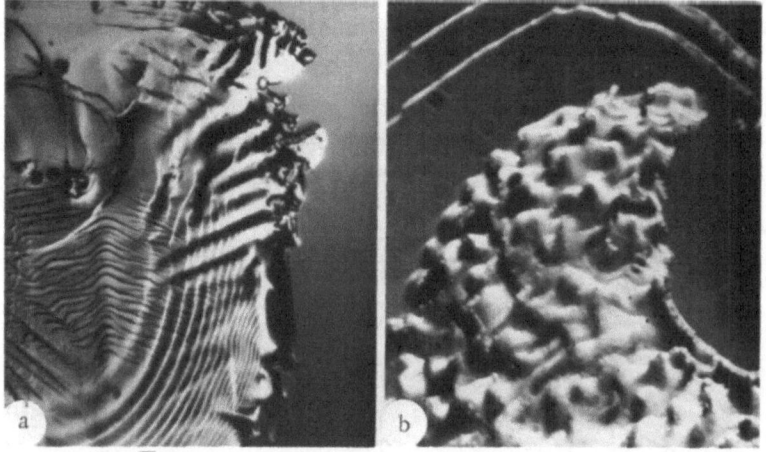

Fig. 4. Sculpturing of face parts for: a) naphthalene; b) magnesium.

The pinacoidal face was formed as a rule by vicinal planes intersecting at the edges (Fig. 2a); the edges took the form of plateaus of variable width or ridge structure, and they were always produced, without exception, only on one surface of the crystal, whether naphthalene or magnesium. The reverse side of a platelet can be either almost smooth (Fig. 2b) or covered with terraces.

This characteristic and at first sight inexplicable feature of the face sculpture enabled us to perform a reconstruction of the crystals via the history of formation.

The crystals grow sequentially by discrete deposition of thick layers of hexagonal outline (Fig. 3a), and each successive layer is somewhat larger and grows tangentially against the supply flow [8]. Figure 3b shows an example of a crystal composed of hexagonal platelets, which reproduces real crystals (Fig. 2, a and b), and these results simultaneously illustrate the viewpoint of [9] that vicinals are formed by planar layers parallel to the main face.

In this way we were able to reconstruct almost all the main types of crystals; it was found that the terraces are formed on both sides of a plate on account of sequential deposition of incomplete segmented layers.

The vicinal planes arise from crystallization or concentration flows [5, 10], and the regularly facetted (hexagonal) layers acquire secondary relief as the flows develop [11]. The waves incident from above can conceal the vicinal lines of the layers (Fig. 4a), or else may break away from the edges of the crystal and split up into a set of eddies (Fig 4b) [12].

All these observations throw light on the mass crystallization mechanism, and in their various aspects they reproduce the views of individual workers on the different stages in phase transitions.

The production of a high supersaturation (by introducing a cold surface) initiates simultaneous formation of a mass of crystals on that surface.

The conversion of the vapor to solid is accompanied by release of the latent heat, which reduces the density of the nearby gas, and thereby excites the circulation.

The material brought in by the flow is deposited on the crystals as discrete layers of polymolecular or polyatomic thickness; each successive layer is larger than the previous, which provides for upward growth of the crystal. The attachment of the layers occurs selectively on one side of the plate, and governs the thickening.

The material is deposited as such thick layers on account of a hyperconcentrated transition layer at the surface [13]. The resulting energy situation causes rapid ordering of the structure in the entire layer lying on the crystal [14].

The transition layer itself is formed on account of relatively stable molecular complexes in the vapor, which are of the type of colloidal particles [15]. The long-range orienting forces on the crystal surfaces draw in these particles and produce the hyperconcentrated layer.

This mechanism involves indissoluble interaction between the medium and the crystals; the mechanism appears to explain much of the variety of forms and the extremely high growth rates in mass crystal growth at high supersaturations.

Literature Cited

1. I. I. Shafranovskii. Lectures on Crystal Morphology [in Russian], Vysshaya Shkola, Moscow (1968).
2. L. I. Kvater, I. V. Frishberg, and A. S. Myakulinskii. Trudy Inst. Metallurgii, Ural. Fil. AN SSSR, 19:125 (1969).
3. H. Buckley. Crystal Growth [Russian translation], IL, Moscow (1954).

4.   T. E. Lovits. Tekh. Zh., 2(1):50 (1805); Selected Papers on Chemistry [in Russian], Izd. AN SSSR, Moscow (1955).

5.   G. W. Wulff. Selected Works on Crystal Physics and Crystallography [in Russian], Gostekhteorizdat, Moscow (1952).

6.   B. Chalmers. Theory of Solidification [Russian translation], Metallurgiya, Moscow (1968).

7.   L. A. Frank-Kamenetskii. Diffusion and Heat Transfer in Chemical Kinetics, Plenum Press, New York (1969).

8.   G. G. Lemmlein. Sector Structure in Crystals [in Russian], Izd. AN SSSR, Moscow (1948).

9.   N. N. Sheftal'. Trudy Inst. Krist. AN SSSR, 3:55 (1947); 3:71 (1947).

10.   Z. Weyberg. Z. Kristallogr., 36:40 (1902).

11.   W. Goldschmidt. Beitr. Kristallogr. und Mineral., 2:167 (1924).

12.   S. S. Kutateladze. Heat Transfer in Condensation and Boiling [in Russian], Mashgiz, Moscow (1952).

13.   Z. Gyulai and S. Bileck. Acta Phys. Hung., 1:199 (1952).

14.   E. S. Fedorov. Priroda, 1471 (December 1915).

15.   R. O. Grisdale. Theory and Practice of Crystal Growth [Russian translation], Metallurgiya, Moscow (1968).

TRENDS IN REAL CRYSTAL FORMATION AND SOME PRINCIPLES FOR SINGLE CRYSTAL GROWTH

N. N. Sheftal'

Crystallography deals with the world of crystalline materials, particularly with emphasis on practical uses; the specific methods are symmetry and geometrical representation, on the basis of the simplicity and singularity inherent in crystals. Crystallography also deals with crystal growth at the elementary and holistic levels. The latter terms means that a crystalline individual is considered as a single whole. The history of the crystal is examined, and the distinctive features are established for the morphology and interaction with the environment, together with the changes from point to point during growth. The study is directed to finding ways of growing almost perfect bulk, film, and whisker single crystals. The individual crystals may be either microscopic or macroscopic.

These two crystallographic approaches are necessary and supplement one another.

Here I present some results from crystal growth studies in the following aspects:

(a) The preferred growth direction of crystals and the observation of artificial epitaxy;

(b) Growth via complexes and crystallites;

(c) The relation of the equilibrium form to the growth form for almost perfect single crystals;

(d) Natural crystal selection for uniformity; and

(e) Long-range effects.

Preferred Growth Direction and

Artificial Epitaxy

There are two basic conceptions in crystal growth mechanisms; the first is the earlier and represents crystallization as the accumulation and attachment of groups of crystalline structure to the growing crystal, apart from single atoms. This effect can be detected directly under the microscope in highly supersaturated solutions, and one of the first observations was due to Fedorov [1]. The second conception is due to Kossel [2], Stranski [3], and Volmer [4], in which the basis of the growth process is two-dimensional nucleation directly at the surface of the growing crystal, with subsequent attachment of single atoms. The classical theory developed by Stranski and Kaishev deals mainly with the second conception. The process in that case can be referred to elementary atomic acts, and an analysis of the over-all effects

---

*Read at the meeting of the Scientific Council of the Institute of Crystallography, December 15, 1972.

185

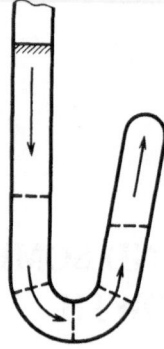

Fig. 1. Change in direction of rapid growth for a single
crystal growing in a bent tube.

of these is provided by thermodynamic and kinetic principles. The agreement between these
calculations and observation has been followed up by Kaishev and his students, especially by
electrical crystallization [5].

However, many effects, especially single-crystal growth, are governed by the formation
of three-dimensional nuclei, complexes, and microcrystals; this is particularly so when there
is a change in the preferred growth direction, and in the artificial epitaxy which I have particularly examined.

Gross and Möller [6] were the first to describe change in preferred growth direction;
if a bent glass tube is filled with molten material and seeded at one end, then a single crystal
starts to grow if the supercooling is not too great. The direction of the greatest growth rate
is that of the tube axis. The crystal turns at the bend, which causes the direction of maximum
growth rate to stay coincident with the tube direction (Fig. 1). It was considered that the rotation is provided by nuclei arising at the corner formed by the crystal and the tube wall; the
nuclei have the same form but with a slight change in orientation. As these nuclei appeared
one after another, they alter the orientation and adapt the direction of fast growth to the change
in direction of the tube.

Shubnikov [7] described these experiments and stated that there was no satisfactory explanation of the rotation, and the geometry of the crystal could not be explained.

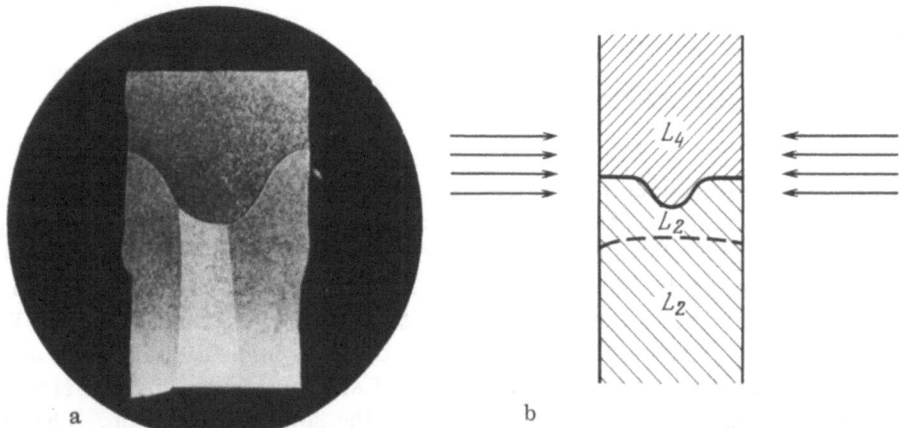

Fig. 2. Change from [110] to [100] in orientation of tungsten crystals during
electron-beam working in response to growth rate increase (etched length-
wise section): a) photograph due to M. V. Pikunov and V. V. Shishkov; b)
scheme.

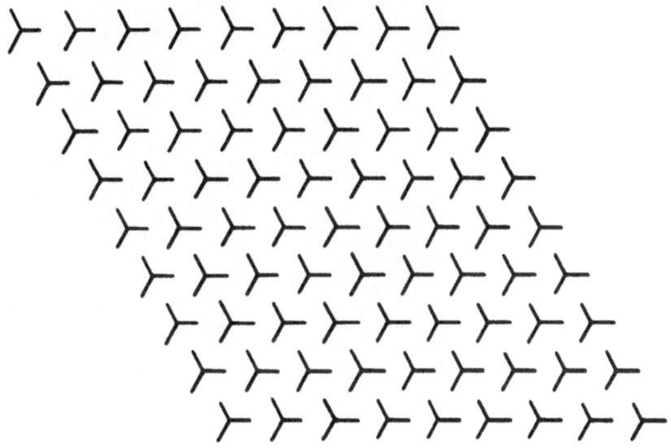

Fig. 3. Discrete system of figures on an
amorphous substrate.

We give an example of a change in preferential growth direction for single crystals from recent studies.

Pikunov [8] found that a single crystal of pure tungsten will grow on the [110] direction with a rate not more than 3 mm/min; if the zone speed is raised to 10 mm/min, the work continues in the former direction, since this cannot change, on account of the strictly vertical supply from above, but the single crystal changes in orientation over a short distance from a [110] (L_2 axis) to [001] (L_4 axis).

Figure 2 illustrates the orientation change in a longitudinal section; the new orientation arises at the center of the growing surface, where the supercooling is larger and the growth rate is higher, and within a minute or so it extends over the entire cross section without any geometrical selection. Nuclei of the new orientation grow laterally to fill the entire cross section.

It remains unclear where the nuclei of the new orientation arise, i.e., outside the crystal or at the surface, and whether they are two-dimensional or three-dimensional [9]. My experiments enable one to answer these questions via tests on artificial epitaxy [10].

If it is true that the actual nucleation involves not very small microcrystals, then a reentrant angle on an amorphous substrate should stimulate deposition, and perhaps also formation of such nuclei in a shape and orientation such that the faces forming an appropriate angle will be in contact with the sides.

The first series of such experiments was performed in 1963; a photochromic glass was fitted with a system of periodically disposed figures as three short strokes from one point at 120°; the sizes of the figures were some tens of microns (Fig. 3).

After irradiation and etching, the figures became convex, but they would appear to have been too large, and ordered crystallization did not occur.

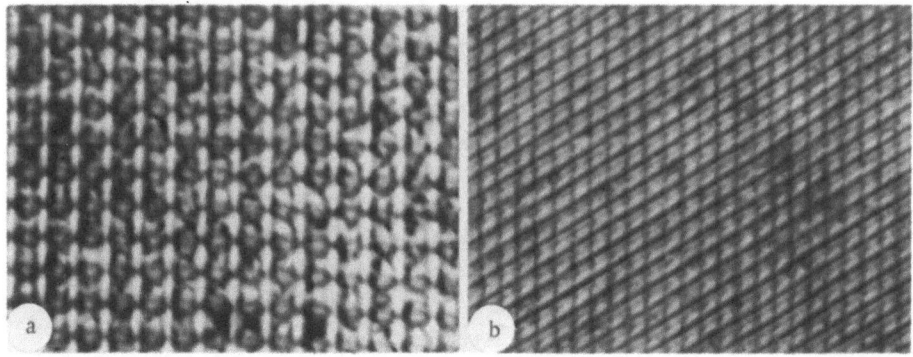

Fig. 4. Crossed diffraction gratings: a) square; b) hexagonal. Each system has 600 lines/mm.

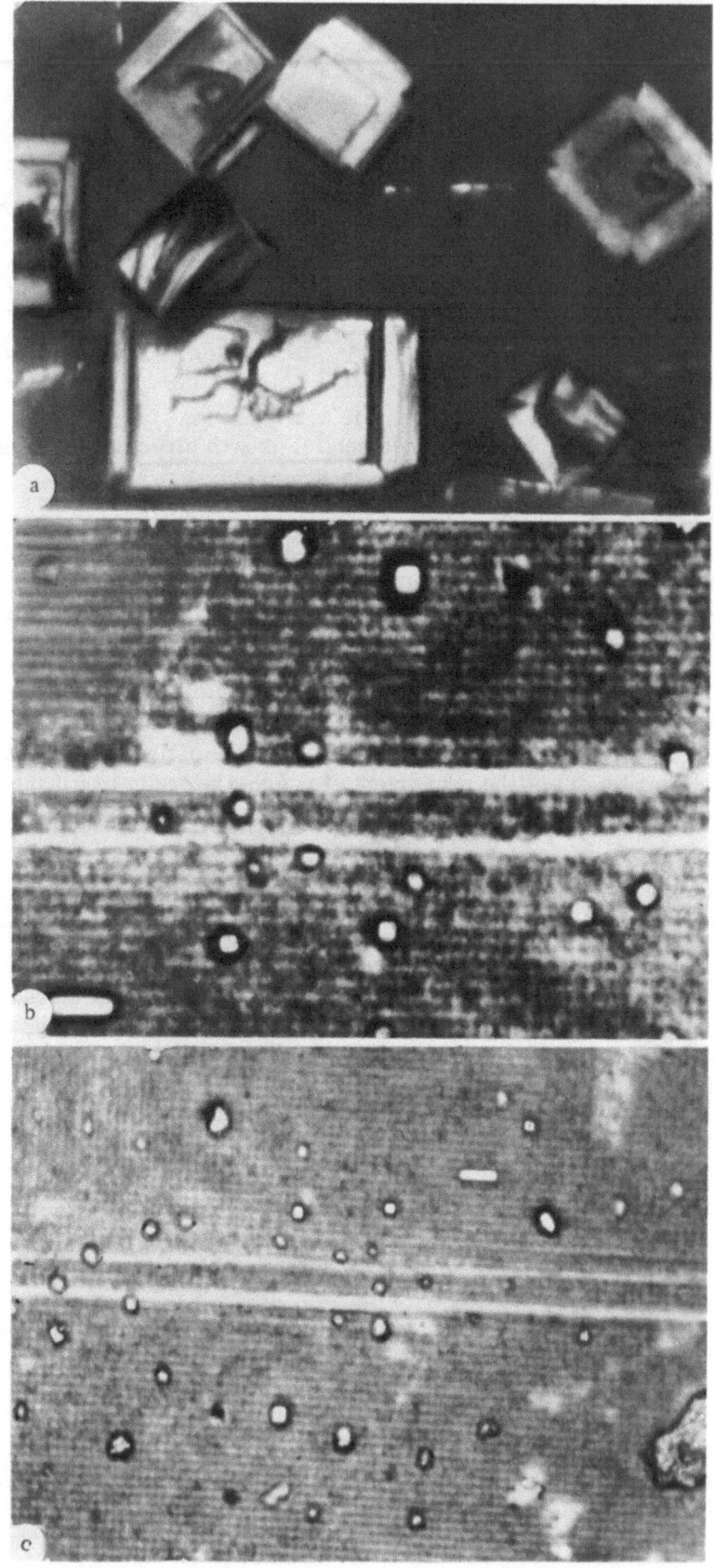

Fig. 5. Artificial epitaxy on: a–c) square

grid; d-f) hexagonal grid.

Fig. 6. Shape change in NH₄I crystals in response to substrate symmetry (hexagonal grid).

Fresh experiments were done in 1971 [10]. In that case, the substrate consisted of
optical diffraction gratings having 600 lines per mm, which had two such systems of lines, in
one case with the lines mutually perpendicular, and in another intersecting at 60° (Fig. 4).

These were used in a series of crystallizations using aqueous NH₄I solution; the orienting
action was weaker than that for mica but quite clear. The square pattern caused most of the
crystals to present the cube faces in a parallel fashion, with some crystals lying at 45°, which
would appear to arise because the formation of the second system of lines damaged the crest
of the previous angles, and some of the crystals were attached by the edges to the resulting
sides (Fig. 5).

In the case of the hexagonal grating, some of the crystals had their three-fold axes
perpendicular to the surface (Fig. 6).

This effect from the artificial periodic symmetrical relief showed that microcrystalline
nuclei play an important part in perferential crystal growth and in real crystallization generally. An important argument for such nuclei is that preliminary strong superheating of the melt
before casting reduces the preferential growth in any direction [11].

Artificial epitaxy confirms the important part played by crystallites, perphaps of tens
of microns in size, in the nucleation and growth of any crystal.

Such microcrystals appear to exist already in the medium and can be affected by factors
such as superheating, as well as by artificial surface structures, and these may be the principles that can be used in growing almost perfect single crystals.

Growth Mechanism Involving Complexes

and Microcrystals

Change in preferential growth direction and artificial epitaxy show that ordinary growth
involves complexes of particles forming parts of the crystalline structure, such as microcrystals, and these may be decisive; we now consider a more extensive range of effects that
confirm and detail this viewpoint.

Fig. 7. Growth of a quartz seed via small crystals: a) general view, ×3; b, c) grown crystals, ×8.

Figure 7 shows a quartz crystal grown in 1949 under conditions such that growth occurred via large crystallites attached to the seed crystal in an oriented fashion; this led to composite faces for the rhombohedra, which combined the parallel faces of small crystals.

The chloride method gives much more perfect single-crystal films of silicon and germanium than does vacuum condensation [12], which raised the question whether it is easier to obtain a perfect crystal from particles that can aggregate via random collisions than from the isolated atoms in a molecular beam.

Bulakh [13-15] made a very important advance in our ideas on complexes in crystal growth by reference to crystallization of cadmium sulfide.

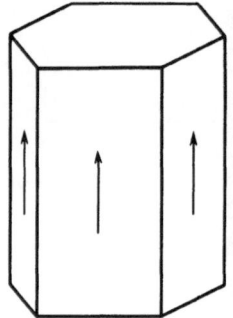

Fig. 8. Possible shape for an almost perfect crystal with the wurtzite structure (polar hexagonal prism with two monohedra).

A chamber for mass crystallization was supplied separately and in strictly controlled amounts with the vapors of cadmium and sulfur; it was found that in a given temperature range one obtained different shapes of crystals in accordance with the relation between the numbers of cadmium and sulfur atoms: prisms, pyramids, needles, plates, strips, or whiskers. It was assumed that the medium first of all gave rise to complexes of three kinds: CdS, Cd_2S, and Cd_3S. The relation between these was determined by the composition of the medium. The second stage in the preparation for crystallization was association of the complexes: $n(Cd_pS) \rightarrow (Cd_pS)_n$. At a certain stage of association, the features of a microcrystal began to appear, and the product could be considered as a nucleus. The marked deviation from stoichiometry in some of the initial complexes is partially compensated in the crystal by periodic overgrowth by stoichiometric complexes.

Simulation of crystal formation from associations of complexes corresponding to a given deviation from stoichiometry led to models for the morphologic forms that can give rise even to cavities, which were actually observed in the crystals.

The problems of growing single crystals presented chemists, long concerned with producing pure substances, with essentially new tasks; they had to produce crystals purified from dislocations, small-angle boundaries, slip lines, clumps of vacancies, twins, dissolved oxgen, and so on, while having a uniform impurity distribution. Bulakh's work showed that these properties of single crystals are largely determined by the microcrystallography of the medium.

The morphology of a single crystal is that of a giant macromolecule formed as the final stage of a chemical reaction, and the most sensitive indicator for the actual composition and micromorphology of the medium is provided by this, together with the character and mode of the chemical reaction.*

In these studies on the crystallization of CdS, the crystals were imperfect in shape, apart from the platelets; again, most of them were hollow, which also indicates internal imperfection and an inadequately ordered growth mechanism, which is governed in general by fluctuations in the component ratios, to which the crystallization is very sensitive.

A first general problem to be solved, therefore, in producing perfect single crystals is to define the conditions for obtaining bulk polyhedral crystals of maximal perfection, if possible in a form of limiting simplicity, which often goes with perfection. It can be supposed that one of the most likely forms for CdS crystals would be a polar hexagonal prism with two monohedra (Fig. 8). Such crystals will grow under conditions of high temperatures and with excess of cadmium, but the Cd : S ratio leading to perfect growth must be carefully determined by experiment.

Crystal growth long ago rapidly became important in chemistry; here there is a need to take into account more fully crystallographic principles, especially dynamic effects in the chemical reaction sphere.

---

*This involves complexing if there is no chemical reaction.

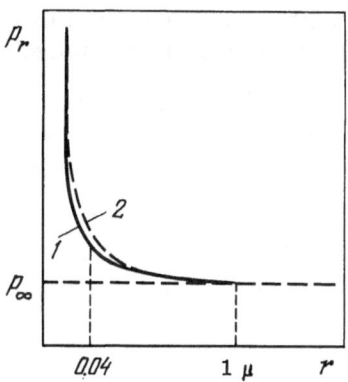

Fig. 9. Saturation vapor pressure as a function of r (radius of inscribed sphere) for crystals with: 1) equilibrium shape; 2) nonequilibrium shape.

A third of a century ago, Khodakov [16] in the USSR developed an original crystallographic theory of dominant chemical forms on the basis of calculations on the potential energy of spatial configurations, complexes, and molecules, taking into account ionic radii and charges. This deductive theory represented a transfer of ideas from chemical crystallography directly to chemical reactions in an aqueous medium, and the results agreed well with experiment. This theory appeared before it became vitally necessary; subsequent progress on crystal growth would appear to have involved its principles to a considerable extent, and the basic ideas were tested against new evidence, and it is now possible to extend this approach to the chemical study of medium crystallography not only for aqueous solutions but also for vapors and melts.

If one is to produce deliberately perfect single crystals, one must travel the road from stable complexes to aggregates and from these to the perfect single crystal.

The idea of growth from complexes enables us to use a principle for making perfect single crystals from experience in growing single crystals from the vapor, nonstoichiometric media with predominance of cations, and so on.

However, the principle needs detailed experimental testing in each particular case, since excess of cations, for instance, may lead to formation of hollow forms also.

The Growth Form of a Perfect Single Crystal*

It is important to relate the form to the perfection for freely growing crystals; from the very early days of studies on equilibrium forms, attempts have been made to relate these to the growth form, as in [18, 19], but these early attempts were unsuccessful.

The equilibrium form is determined by the minimum surface energy, and it can arise only at equilibrium, and the conditions of formation cannot define the growth form.

This is clear from Fig. 9; the solid and broken lines show the vapor pressures of crystals in relation to the radius of the inscribed sphere; the solid curve relates to a crystal of equilibrium form with faces having a specific surface energy σ_1, while the broken line relates to a crystal of the nonequilibrium form whose specific surface energy is σ_2, $\sigma_1 < \sigma_2$. The shapes of the curves show that the difference affects the vapor pressure only for crystals of dimensions of micron order.

The two curves come together for large crystal sizes, which shows that the crystal stability is indifferent to the surface energy, and that the vapor pressure is not dependent on the size and shape of the crystal.

_____

*See [17] for more details on the equilibrium form.

Honigman [20] considers that one also cannot say definitely in an isolated experiment whether one has certainly obtained the equilibrium form.

In spite of all these features, the equilibrium form has served as basis for a whole system of concepts and calculations that agree well with evidence on the actual shapes of the crystals and with views on the mechanisms and kinetics of nucleation and growth.

Kern [21] has pointed out that periodically doubts are expressed about the basic Gibbs-Wulff equations; these arise in relation to assertions that these equations are applicable only to extremely small crystals, and that the supersaturation should appear in relation to equilibrium.

We consider that the main reason for these doubts is the lack of clarity on the relation between the equilibrium form and the growth form.

Here we endeavor to show that a very close connection exists between the equilibrium form and only one of the growth forms, namely: the final growth form, which is the one to which nearly perfect crystals tend. This relationship is seen most clearly for large single crystals, for which it is found that the two forms are essentially identical, subject to a minor restriction, namely: if the crystal is nonpolar, it contains only the most important equilibrium faces, which are determined by interaction between the structural units and their nearest neighbors. The following expression characterizes an almost perfect single crystal having the minimum concentration of equilibrium defects:

$$\frac{\rho}{M} \Sigma v_i \mu_i = \min \ \text{for} \ \ V = v_1 + v_2 + \ldots + v_i = \text{const},$$

where v_i is the volume of growth pyramid i, μ_i is the chemical potential of this, ρ is density, and M is molecular weight.

To consider the possible theoretical growth form of such a crystal, we first have to elucidate whether there is any relation between the perfection and the shape; such a relationship was first observed in 1893 by Wulff [22], who pointed out that inhomogeneous crystals grow more rapidly.

A detailed study has been made [23, 24] of the relationship of homogeneity and shape in sucrose; faces whose growth pyramids produce defects increase in growth rate, which leads to a change in habit, the shape gradually becoming more isometrical on account of the increased growth rate of the principal faces. The overall shape therefore alters, and one gets minor faces, which lie at the edges when the main faces grow slowly (Fig. 10).

These changes occur in response to an increase of less than 0.1% or so in the supersaturation, and they are best seen on large single crystals, in which the perfection and shape are extremely sensitive to such changes, which affect the incorporation of the medium into the crystal, i.e., the formation of defects [23]. The essence of the bulk processes that occur in single crystals of reduced perfection is seen also in the face sculpturing, and in the formation of macroscopic and microscopic defects, together with spontaneous cracking on account of growth stresses. This spontaneous cracking occurs although the single crystals grow under strictly thermostatically controlled conditions [24].

It is therefore established that defects have a great effect on the face growth rates and shape, which leads us now to consider whether crystal perfection is related to a particular growth form.

Khaimov-Mal'kov [25-27] has resolved some debated questions about the nature and extent of the so-called crystallization pressure, which a growing crystal exerts on an obstacle;

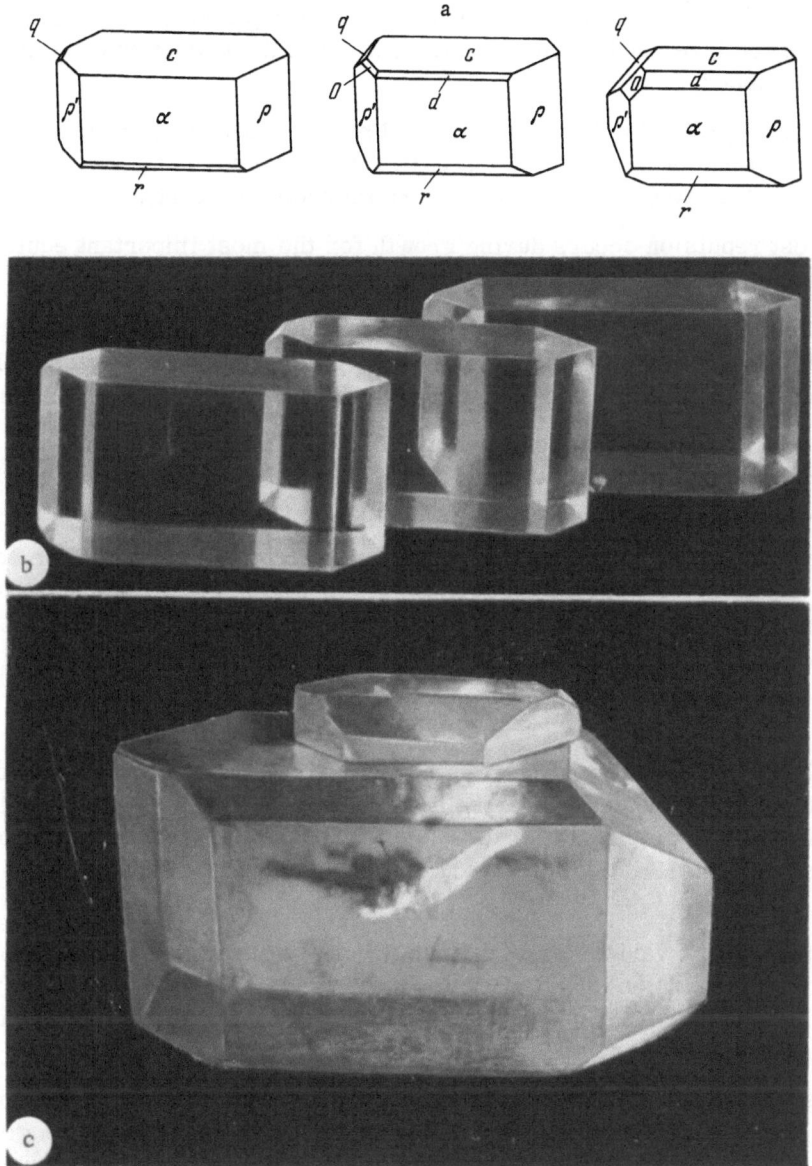

Fig. 10. Effects of homogeneity on the shape for sucrose crystals: a) schematic; b) homogeneous single crystal; c) inhomogeneous.

he has defined also the related self-purification processes, such as repulsion of foreign particles by the crystal, which would otherwise result in defects on trapping.

The crystallization pressure arises from the phase-transition energy, and not from the surface energy; it is proportional to the supersaturation. If the supersaturation coefficient $\gamma = 1.2$, the unilateral pressure on an obstacle from the entire face is about 20 kgf/cm^2 [25]. This limiting value for the crystallization pressure is not dependent on the indices of the face, since faces grow by a repetitive process, and the energy of the process is essentially that of the phase transition per particle, and this is what finally determines the repulsion.

The repulsion mechanism for small foreign particles is somewhat different; if there is a thin layer of the medium under the particle, repulsion or trapping is determined by the ratio of the growth rates for the parts of the face covered by the particle and free from it. The less the growth rate of the layer under the particle relative to the open parts, the more likely the particle is to be trapped.

The repulsion implied by this is determined by the capacity of the face for layerwise growth, i.e., by the ratio of the tangential growth rate to the normal growth rate, which controls the capacity of the layer to grow through the medium under the particle.

The greatest repulsion occurs during growth for the most important equilibrium faces, especially those with the greatest capacity for layerwide growth.

The same conclusion is reached by considering the minimum supercooling or supersaturation needed for face growth; the energy barrier to be overcome is largest in the growth of equilibrium faces, which require two-dimensional nucleation, so these grow more slowly.

Consequently, the crystallization pressure, but not the limiting value, will be the largest in the growth of the equilibrium faces, for which the supersaturation should be the highest directly ahead of the growth front under otherwise equal conditions.

But this means that a crystal to show the most perfect growth should be bounded by the most important equilibrium faces; consequently, the uniformity and the growth form of a perfect crystal are determined by two linked conditions:

$$\frac{\rho}{M} \Sigma v_i \mu_i + \Sigma F_i \sigma_i = \min \ \ \text{for} \ \ V = v_1 + v_2 + \ldots + v_i = \text{const},$$

The requirement for high perfection in a polyhedral crystal provides an additional condition governing the shape. At the boundaries between growth pyramids, one has incoherent links between layers with slightly different histories, so here, and even more so at the corners, one inevitably gets slight deviations in the lattice parameters and defects due to the lack of coherence.

This is confirmed by various observations [28] on the production of dislocations at the boundaries of growth pyramids in $NaBrO_3$ crystals (Fig. 11). It was shown that dislocations often occur in this way in other crystals, and that dislocations are particularly common at edges representing intersections of faces of different simple forms.

From this we conclude that the uniformity of a polyhedral crystal should be related to the limiting simple form produced by the equilibrium faces most important for a given crystallographic structure. It is more likely that one will obtain simple forms with few faces.*

If a crystal is of medium or low symmetry, the shape of a perfect crystal involves the normal habit; one expects that perfect crystals in such systems will show greater deviations from isometric shape than will inhomogeneous ones.

The shape of a homogeneous single crystal is unusually sensitive to imperfections; sucrose crystals show first of all a habit change, and then there appears an indubitable signal of future cracking arising from the accumulation of growth stresses, which produce a small (010) face on one of the vertices [24]. Then other faces develop (Fig. 12).

However, these phenomena are appreciable only on large single crystals; also, visual perfection is not always related to shape simplicity. For instance, visually homogeneous single crystals of Rochelle salt grown from aqueous solution are covered by the faces of the two rhombic prisms, together with the basal pinacoid and minor faces from the first and second pinacoids, although the limiting form has only one prism and pinacoid (Fig. 13). As the

─────────

* Belov [29] has drawn a similar conclusion from arguments from chemical crystallography.

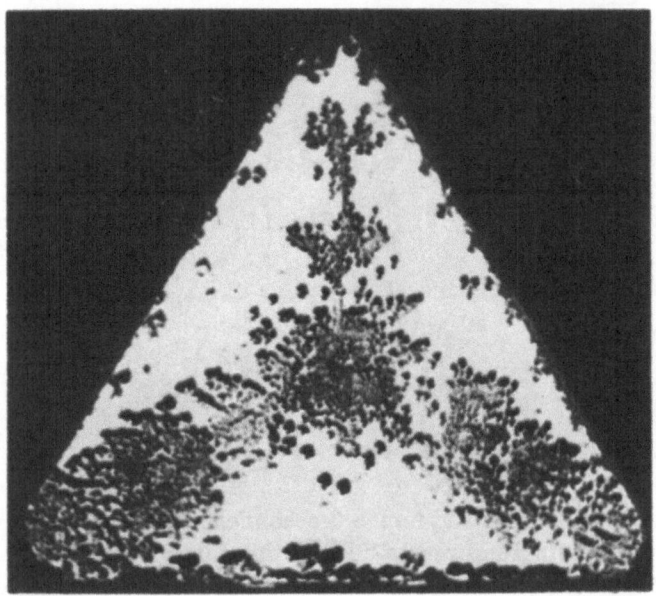

Fig. 11. Preferential formation of dislocations at
the vertices and edges of a NaBrO$_3$ crystal (after
Treivus, Petrov, and Kamentsev).

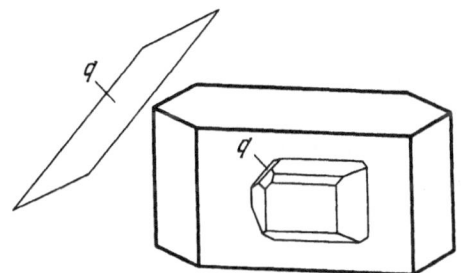

Fig. 12. Occurrence of face q(0$\bar{1}$0) on a sucrose single crystal indicating future cracking.

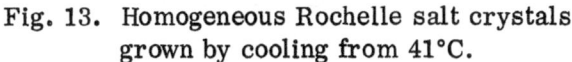

Fig. 13. Homogeneous Rochelle salt crystals
grown by cooling from 41°C.

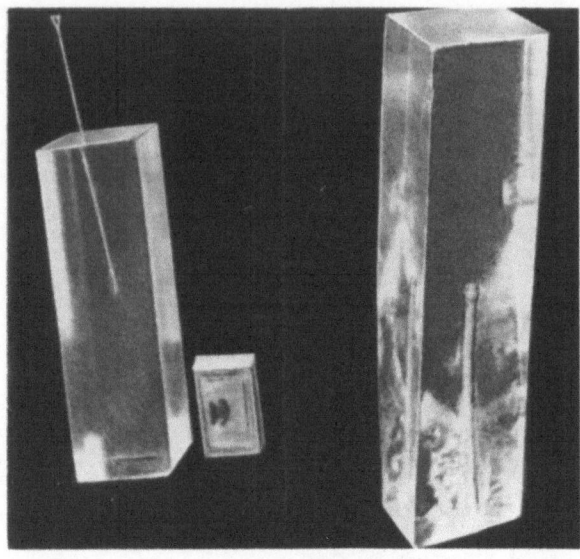

Fig. 14. Homogeneous Rochelle salt crystals
grown by cooling from 52.5°C.

initial growth temperature is raised, and hence the concentration of the solvent impurity is
diminished, the shapes of the single crystals become simpler; if the growth is from a solution
containing only about 10% water (initial growth temperature about 52.5°C), one finds that not
only are some major structural defects eliminated (very thin channels filled with mother
liquor), but also that the crystals acquire a limiting simple form consisting of the {110} rhombic prism and the basal pinacoid, being highly elongated on the c axis [30] (Fig. 14). Crystals
of potash alum also show a tendency for the shape to simplify as the growth temperature is
raised [30].

There are therefore reasons for supposing that the limiting simplicity is linked to visual
perfection and indicates a high degree of structural uniformity, while visual perfection found
in conjunction with a complex shape would indicate a poor degree of perfection.

We also conclude that perfect crystals are not formed very close to equilibrium, and so
that elevated supersaturations can sometimes facilitate the production of perfect crystals in
the limiting simple form. Deviation from uniformity is caused not by the supersaturation but
by the supersaturation differences at the surface [31]. However, a tendency to shape simplicity

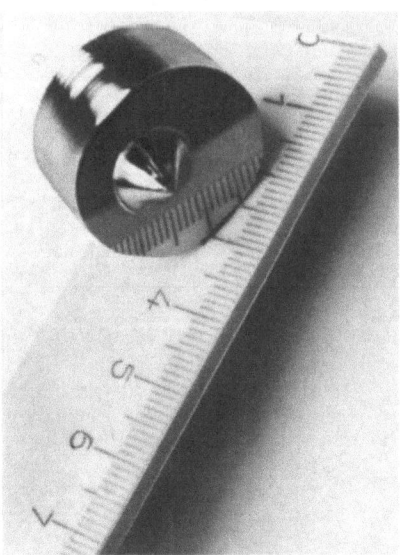

Fig. 15. Germanium single crystal pulled from
the melt with a (111) face at the growth front
covering the entire end. The cone was produced
by a drop of melt remaining attached to the crystal on removal.

is no more than a trend, which is far from always realized in view of the multiplicity of factors that influence the crystal shape.

Important evidence for understanding the trends in growth of equilibrium faces comes from the production of semiconductor single crystals from melts and of synthetic quartz from hydrothermal solutions; in the first case, the condition is the production of the equilibrium octahedron face at the end of the silicon or germanium single crystal pulled from a super-cooled melt, this face sometimes completely covering the growth front (Fig. 15).

A distinctive feature of an equilibrium face is the maximal uniformity of the related growth pyramid; electrophysical parameters measured across the cross section differ sub-stantially from those of other parts of the crystal (the face effect), while over the face itself they are very constant [32] (Fig. 16). The crystals also had low or zero dislocation densities in such faces, since each structural unit lies at the intersection of not less than two chains of strong bonds in the face.

The atoms of substitutional impurities enter into an equilibrium face at quite high con-centrations, but they are more uniformly distributed here than they are on entry through any other surfaces.

For instance, a silicon crystal will take up large amounts of Sb, As, and P (donor impuri-ties) via an equilibrium (111) face [33], the excess valences of these serving to strengthen the bonds in the crystal.

Acceptor dopes such as B and Al do not so readily enter the growth pyramid of the equilibrium face, and not in such high concentrations, and it would seem that they do not act as substitutional atoms, at least for silicon crystals.

Similar properties are found in the principal equilibrium faces of single crystals of synthetic quartz (faces of the large rhombohedron), where the Al and Na structural impurities result in a smoky coloration on x-irradiation, these entering the growth pyramids uniformly and in the highest concentration. Conversely, the nonstructural gross impurity of colloidally dispersed sodium silicate (heavy phase) is almost absent, since it is repelled by the very high crystallization pressure. This repulsion does not occur on interaction with a structural im-purity, since the latter is used to some extent for regular construction of the basic lattice; Al directly replaces Si in quartz, while Na lies nearby and compensates the charge [34].

We now consider how single crystals behave when the supersaturation or supercooling is increased, or else the dope concentration is raised; semiconductor crystals first of all show an increase in the thickness of the layers of dope, which is analogous to the zoned structure in polyhedral crystals (Fig. 17). Hummocky formations appear at the growth front, and the num-ber of these steadily increases, with onset of a cellular substructure [35]. When the growth is on [100], the cells are seen at the growth front as fragments extended laterally and projecting from the growth front (Fig. 18). The boundaries between these are represented by depressions,

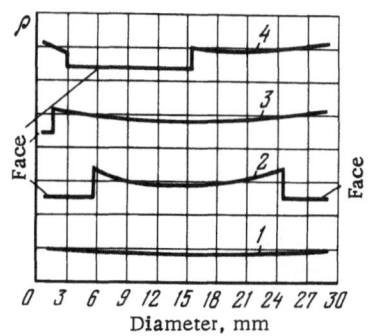

Fig. 16. Specific resistance ρ of a silicon crystal across the cross section for facetted and curvi-linear parts of the end face; 1-4 are the numbers of the diagrams.

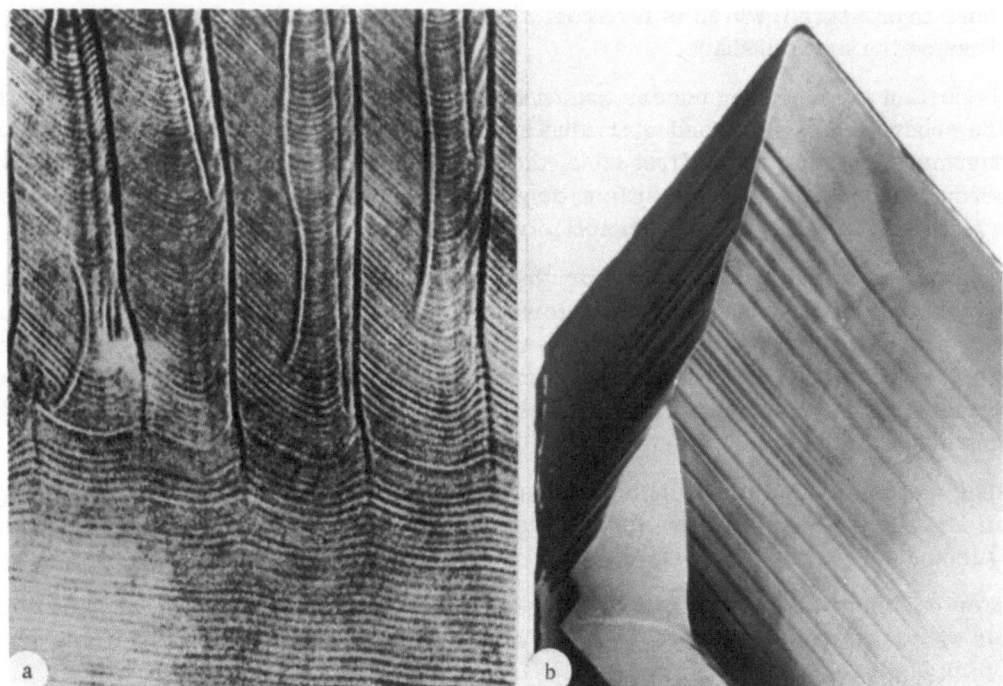

Fig. 17. Zoned structure in crystals: a) gallium arsenide (×300, after S. P. Grishina); b) quartz.

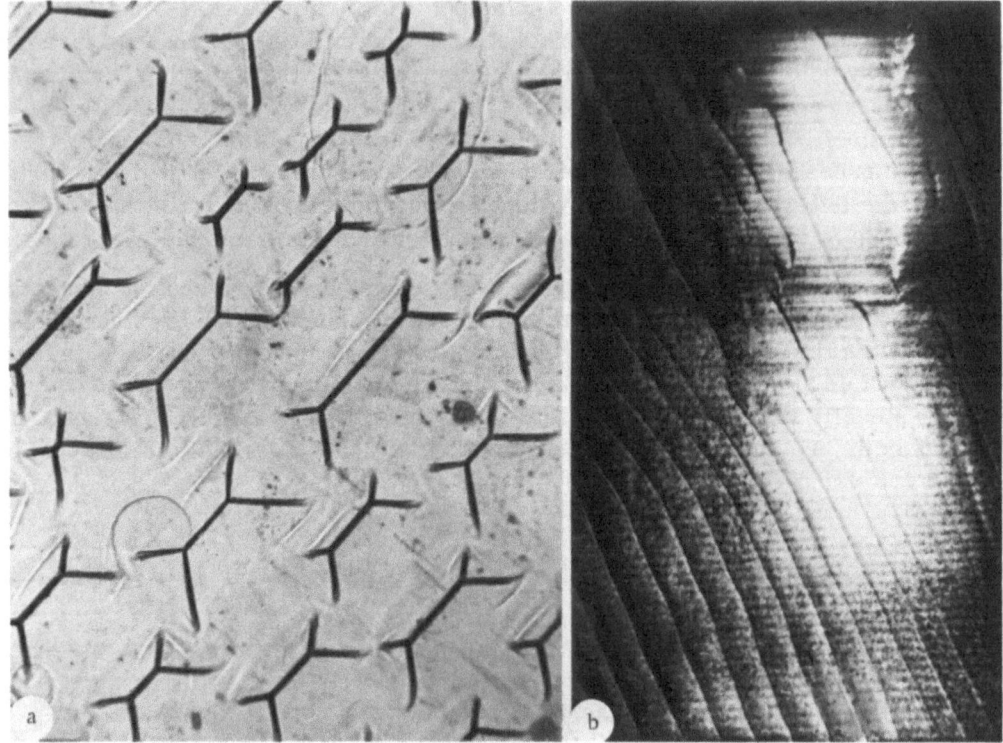

Fig. 18. Cellular structure in semiconductor crystals: a) fragmentation at the growth front for silicon crystals, ×300; b) cellular structure on the side surface of a gallium arsenide crystal; cell boundaries within the crystal visible in Fig. 17, ×100 (after S. P. Grishina).

which appear also on the side surface of the crystal (Fig. 18b). These contain the dope trapped by the crystal. The depth of the boundaries between the cells increases with the dope concentration and the growth rate. The developed cellular structure penetrates throughout the crystal and results in a long columnar structure along [110].

Analogous effects occur for quartz; the surface of the pinacoid faces has a macrocellular structure (Fig. 19), which is rather similar to the microcellular structure found in the semiconductors and also in metals, and which responds similarly to impurity concentration increase. However, this face differs from the equilibrium ones in always having a cellular structure [36].

The cells on the pinacoid face are independent growth regions (Fig. 19b); they take the form of cones, whose vertices lie at the seed, while the bases are on the growing surface. The impurity accumulates between the cones (Fig. 19c), and on etching one finds the deepest recesses in the gaps between the cones. If the intergrowth of the cones is inadequate, the links between the cones become so weak that plates cut on the pinacoid may sometimes show some of the cones dropping out.

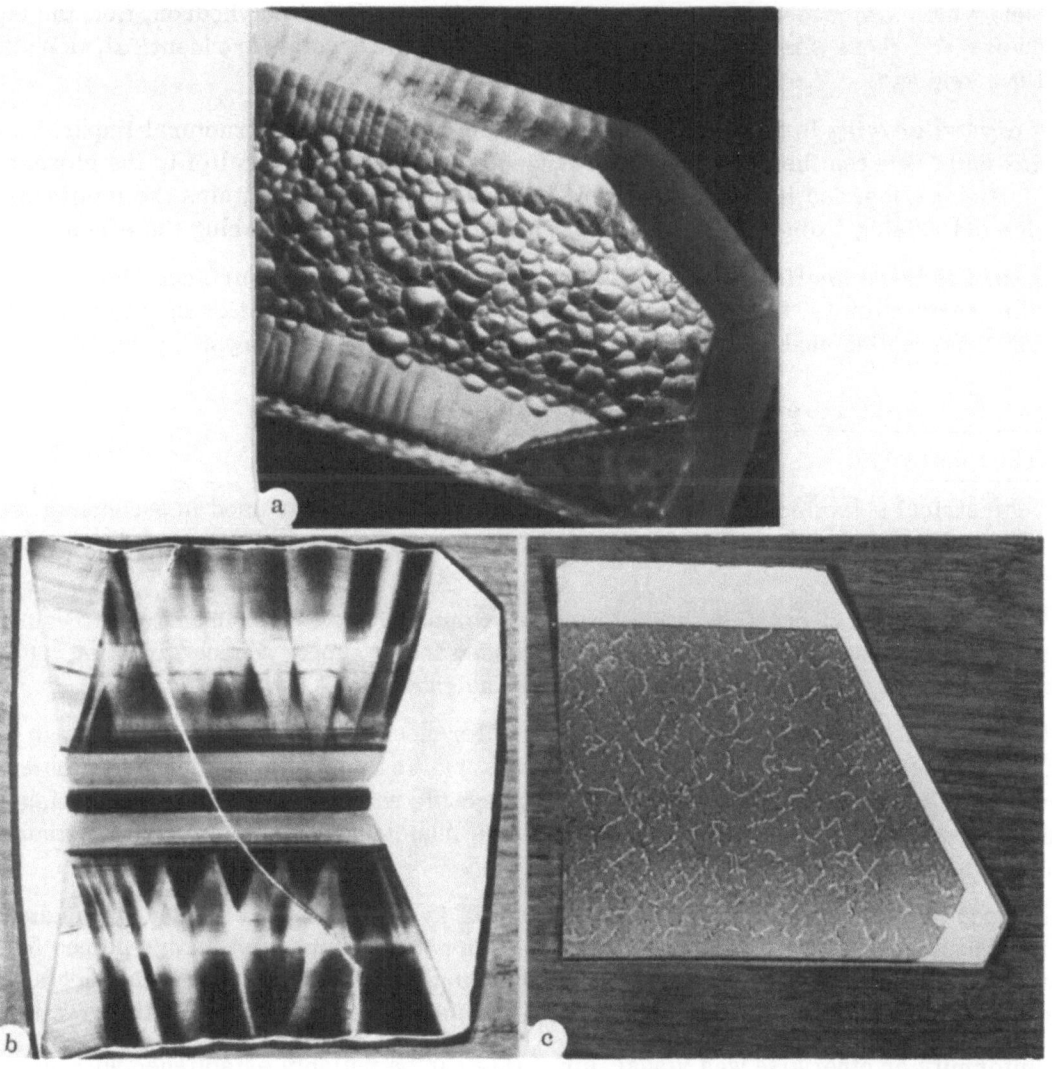

Fig. 19. Cellular structure in quartz crystals: a) on surface; b) in spatial section revealing growth cone; c) impurity uptake on cone faces (after P. P. Butorin).

Single crystals of semiconductors and synthetic quartz therefore enable one to observe a tendency for independent growth of individual areas; however, equilibrium faces have maximum tangential interaction of the particles and the most marked tendency for layerwise growth so the surfaces show a pronounced tendency to remain intact and to provide perfection in the single crystals, maintaining unity between the potentially autonomous regions. This would appear to be one of the principal functions of equilibrium faces.

We therefore have to consider whether we are restricted to producing perfect crystals with slowly growing faces.

We can see that this is not so from the reduction in the trapping of structural impurity Al in the growth pyramid of the large rhombohedron in quartz when the growth rate is increased; similarly, increase in the temperature gradient, which is equivalent to increased supersaturation [37, 38], displaces the region where the substructures occur in semiconductors to much higher dope concentrations.

Then a nearly perfect freely growing crystal tends to adopt the shape of the equilibrium crystal, which consists of a minimum of faces having the greatest capacity for layerwise growth and which are necessary and sufficient to produce a closed polyhedron, i.e., the equilibrium shapes and those shapes found on nearly perfect single crystals are identical, as Wulff [39] pointed out long ago.

A crystal growing in this form repels most effectively any nonstructural impurities and maintains unity between the potentially independent growth regions parallel to the closest-packed lattice series, and the small number of edges means that it contains the minimum number of defects arising from incoherent contact between the layers reaching the edges.

If care is taken to eliminate supersaturation differences at the surfaces high supercooling can, within certain limits, facilitate perfection in a crystal; a combination of these conditions with melt superheating enables one to improve the scope for perfect crystal production.

Natural Selection between Crystals Differing

in Uniformity

Geometrical selection is very common in nature and is widely used in technology with crystals growing on a single substrate and under other conditions giving a variety of orientations.

When such a set of crystals grows together, competition for growth material results in survival of only those crystals that are most extended in the supply source direction. The others are swallowed up by their more successful neighbors.

Geometrical selection arises in part from differences in orientation but is not the only way of natural selection in a community of crystals; particular interest attaches to the struggle for development conditions, i.e., for material for growth, which occurs between homogeneous and inhomogeneous crystals developing together (the inhomogeneous ones contain mother liquor, melt, or gas).

A study has been made [40] of this situation for two crystals, one homogeneous and the other inhomogeneous.* It was shown that if two such potash alum crystals are placed together in a saturated solution, whose temperature is periodically raised and lowered in such a way as to cause a cyclic dissolution and growth, then the inhomogeneous crystal will grow at the

---

*The uniformity or otherwise was visual, since this can be reliably established without a special morphological study; a uniform crystal is transparent whereas a nonuniform one is cloudy.

expense of the other one, which occurs when the inhomogeneous crystal (size 5 mm) is several times smaller than the other. For instance, a rotating crystallizer was used in a series of ten runs, each time with a pair of such crystals. Each run lasted about $1^1/_2$ months for a solution saturated at 40°C, and in all cases the inhomogeneous crystal consumed the other one.

It would appear that a sufficiently prolonged run would result in the uniform crystal dissolving completely as a consequence of growth of the other one.

It has long been known [22] that nonuniform crystals grow much more rapidly than uniform ones, but the above conclusion about the preferential development of a nonuniform crystal deserves particular attention, which we now turn to.

If the material supplied to a crystal is uniform under dynamic conditions, the main factor that can influence the competition between such crystals is the supersaturation; uniformity is lost in response to a very slight increase beyond certain permissible limits. One therefore supposes that it will be precisely the nonuniform crystals that will survive if crystals are grown after a period of dissolution under conditions such that only nonuniform growth is possible; if this were not so, it would appear that uniform crystals might have some advantage.

In the study of [40], the deviations from the saturation temperature were ±3°; the temperature shifts occurred every 20 min, so the temperature in the dissolution period was 43°C, while that in the growth period was 37°C. A 3° difference in our tests was too great to produce a homogeneous crystal; only inhomogeneous crystals grew under such conditions. One therefore supposes that this large range of temperature variation facilitated preferential development of nonuniform crystals.

I carried out seven tests on joint growth and dissolution of such pairs of crystals using small temperature fluctuations.*

The volume of saturated solution was 200 cm³, the saturation temperature being close to 35°C (35.1°, 34.9°, 35.0°). The solution contained eight crystals, each of which was about 1 cm across. These were attached to a holder of disc shape placed at the bottom of the crystallizer; the homogeneous and other crystals were taken in equal numbers, with approximately equal sizes and equal total volumes. These were distributed uniformly on the disc. There was a lucite stirrer (60 rpm) above the crystals. Usually, a run lasted 12 days, and the temperature was varied in the different runs by at most ±0.8°, and at least by ±0.4°. The times spent at the dissolution and growth temperatures were 2 hr, or in certain instances 1 hr. Our conclusion from all these experiments is as follows.

Nonuniform crystals have an advantage as regards growth rate even under conditions favorable for homogeneous growth; however, they develop by new growth as a more uniform layer, which should diminish the growth rate and bring it close to that for the homogeneous crystals. In this process, although the conditions were favorable for homogeneous growth, the homogeneous crystals could dissolve completely, the material passing entirely to the inhomogeneous crystals. However, it was possible also to encounter a situation where the surface of an inhomogeneous crystal in places became almost homogeneous in this sequence.

This is the basic conclusion from this experimental study, which confirms the results of [40], and would appear to indicate that one cannot provide conditions under which a homogeneous crystal will grow while consuming the material of inhomogeneous ones.

However, one cannot accept such a conclusion without reservations, since the formation of a perfect crystal is governed by the Kossel principle, namely the energy advantage of re-

---

*The experimental part of the work was performed jointly with I. A. Shpil'ko and G. F. Dobrzhanskii.

gular sequential growth [3], which expresses as regards crystal growth the general principle of minimum free energy. Kossel's principle is realized over a wide range of growth rates when there is reasonably uniform supply of material to the growing surface.

Therefore, the advantage of inhomogeneous crystals has not finally been elucidated.

Let us imagine, for example, that the amount of initial saturated solution has been increased; the coupling between the crystals will then become weaker, and in the limits they will grow and dissolve independently of one another, retaining their sizes. On the other hand, reduction in the solution volume and direct contact between the crystals will greatly increase the coupling between them, which will occur particularly via the common adsorption layers. Such processes are involved in the aging of polycrystalline materials and in the production of single crystals by recrystallization.

Tamman [41] described such processes and wrote: "A metal after very extensive deformation shows a secondary crystallization, since the fragments of crystallites in a piece deformed in the cold constitute extremely small new crystallites; consequently, the start of secondary crystallization amounts not to the growth of large fragments at the expense of small ones but to the new formation of crystallites, which are initially extremely small." This is namely why unstressed homogeneous crystals grow at the expense of large stressed but defective crystallites and become much larger than the latter.

It is therefore possible for homogeneous crystals to grow on the basis of inhomogeneous ones under conditions favorable to the growth of homogeneous crystals when there is direct interaction between crystal individuals via the adsorption layer, especially when there is a scope for defect elimination, as from elevated temperature.

On the other hand, all this additional evidence does not dispose of the fact that inhomogeneous crystals grow at the expense of homogeneous ones under conditions of free growth at ordinary temperatures; new interesting evidence on this has come from systematic studies by Bazhal et al. [42-46] on recrystallization.

It had previously been established [47] that a body of small crystals in a saturated solution would recrystallize in response to systematic oscillation of the temperature around the saturation point; the larger crystals grew at the expense of the smaller ones. The recrystallization gave rise to crystals up to some centimeters in size. There was no recrystallization without the temperature oscillation, which indicates that the effect occurred not from differences in crystal solubility but from differences in dissolution and growth rates. In fact, individual mass experiments on growth of crystals of various sizes [40] would leave no doubt that small crystals dissolve more rapidly than large ones and grow more slowly. This difference increases with the size difference. Increase in the amount of liquid by a factor 2-5 does not increase the crystallization rate, which would appear to show [46] that the growth and dissolution rates are related to the size of the crystal but not to the interaction between crystals, i.e., these rates are functions of crystal size.

There are also two stages in recrystallization: a first, fast one, which corresponds to crystals a fraction of a millimeter in size, and a second, slow one, which had previously been observed.

It has also been shown [44] that the effects of size on recrystallization rate are the more marked the greater the amplitude of the temperature variation, i.e., the more extensive the dissolution and growth.

The effect has been studied on crystals of various substances: ammonium alum, potassium nitrate, potash alum, etc.

There is no satisfactory explanation [45] for recrystallization of large crystals.

Recrystallization equations can be derived from the Thomson–Gibbs equation $\ln C_r/C_\infty = 2\sigma v/rRT$, and these explain the recrystallization for sizes not more than 1 μm, while some consider that they apply only to sizes smaller by 1-2 orders of magnitude. The theory thus does not give an explanation of the firmly established recrystallization of large crystals, and one therefore needs to start from the indisputable fact that inhomogeneous crystals have advantages over homogeneous ones during growth under oscillatory recrystallization conditions.

The results of [40] go with ours to indicate not only that there are differences in rates of growth and dissolution for homogeneous and inhomogeneous crystals, which favor the latter, but also that there may be differences in solubility; if the saturation temperature is somewhat lower for an inhomogeneous crystal, then the supersaturation will be larger, so the growth will be more rapid and the dissolution will be slower.

An apparatus and method have been developed [48] for precision saturation temperature measurement using the observation of concentration currents in growth and dissolution; a red color is produced during crystallization, while a green one is produced during dissolution. The sharpness of the effect enables one to determine the saturation state to ±0.02°C.

At our request, this worker made a comparison of the solubilities of homogeneous and inhomogeneous potash alum crystals; he found that the two types had the same saturation point when placed together in a single solution, while when the supersaturation was gradually increased, the crystallization effect was more pronounced for the defective crystal and even for the defective face of an otherwise uniform crystal, and the resulting layer was thicker and was more brightly colored, which was indicative of the higher growth rate. Similarly, it was found that the dissolution of a defective crystal was slower.

Then the advantage of the inhomogeneous crystals in our experiments was due not to differences in solubility but to differences in growth and dissolution rates.

Then when two crystals compete in a single vessel, there are two factors that act in opposite senses: the size and the nonuniformity, the latter increasing the growth rate and giving the advantage to the defective crystal even if this is smaller than the other one, while the size gives the advantage to the larger crystal. This means that the uniform one should grow at the expense of the nonuniform provided that it has a sufficient excess in size when an oscillatory recrystallization state is employed, i.e., free growth can give conditions for survival of a homogeneous crystal, not merely an inhomogeneous one.

This conclusion should be tested by experiment, although it is a consequence of reliable experimental results obtained individually.

Finally, we may also make a supposition that is supported by evidence on single crystal growth, especially from the vapor state, namely that the capacity for growth and the general stability of inhomogeneous crystals tend to fall as the size is reduced beyond a certain limit; in other words, we suppose that while the defects of a sufficiently large crystal improve the capacity for growth and also the stability against dissolution, the converse applies for a very small crystal. If oscillatory recrystallization conditions are used, a small crystal may be consumed via the growth of a more uniform neighbor. Evidence for this comes from the standard method used in growing crystals from the vapor state [49], where the temperature oscillations around the saturation point suppress not only the smaller crystals but also the more inhomogeneous ones, with survival of one or two of the most homogeneous crystals, which steadily grow.

We may add also that this applies not only to entire crystals but also to the individual defective parts within a crystal; if such parts have not obtained a considerable size and thus have not become firmly established, they can be eliminated by suitable recrystallization.

We have thus provided the basis for another principle for making perfect single crystals, namely temperature variation around the saturation point, which should produce more perfect single crystals.

One form of realization of this principle is to grow single crystals from a boiling solution [50].

Under these conditions, no thermostatic control is required, since the boiling point is constant, though there are temperature fluctuations around the saturation point in the surface layers, where evaporation occurs, and near the heater, where there is superheating. One material is provided by the evaporation, which has to be controlled. Of course, the growth rate under these conditions is substantially higher than that under ordinary ones.

Apart from such temperature cycling, one can also use concentration or other chemical cycling in the medium, which must be such as to produce a systematic sequence of dissolution and growth cycles, and one can imagine using other fluctuating conditions for the same purpose.

Long-Range Effects

These effects are involved in growth via complexes and crystallites, and also have implications for the higher growth rates and lower dissolution rates of large crystals relative to small ones.

The theory provides no mechanism by which a coupling between particles in the crystal could act out to macroscopic distances; Volmer [51] has proposed such a mechanism for a crystalline surface in terms of a transition layer between the medium and the crystal involving an adsorption layer.

Studies on long-range effects were begun in the nineteenth century, and there has recently been increased interest in them [52]. Distler [53] has made a considerable contribution to this, and he has devised methods of decorating clean crystal surfaces and ones coated with thin films. Decoration has provided much detail on the nonuniformities of crystal surfaces. It has

Fig. 20. Preferential growth of potassium nitrate crystals on the edges and corners of an Iceland spar crystal, ×3.

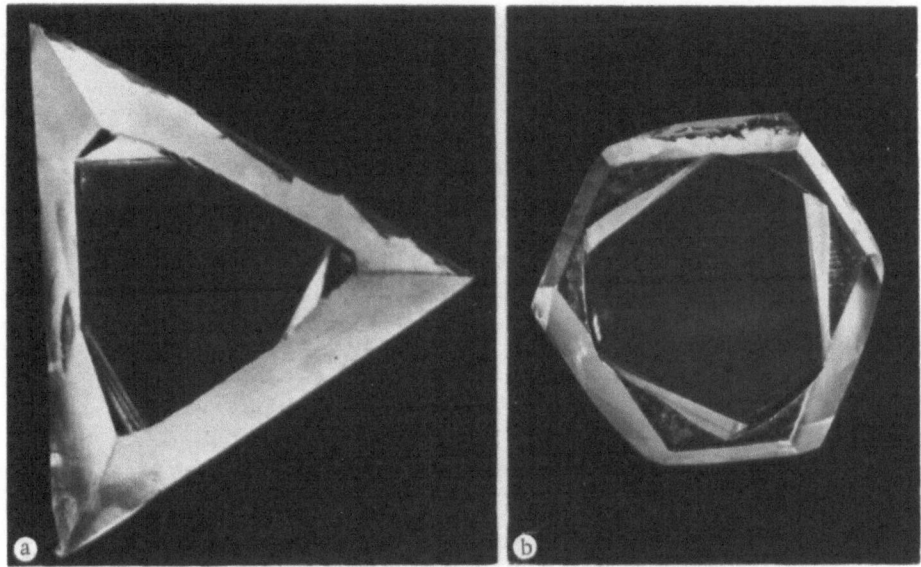

Fig. 21. Enclosure of holes (a and b) in quartz plates by the most rapidly growing faces (from V. T. Ushakovskii, K. F. Kashkurov, and A. V. Simonov).

been found that clumps of point defects play a major part, these being active centers in crystallization. A perfect crystal lacking such centers is probably ill adapted to growth.

On the other hand, the role of the lattice remains unclear. The crystal appears divided into two components: active centers and a passive matrix or lattice.

This division appears to be not fully justified, and we consider it a little further here.

In these experiments on long-range action, i.e., action of crystal forces at a distance, not all parts of the crystal appear to be equivalent.

In Gadolin's early experiments, an alum crystal was coated with lacquer and immersed in a saturated solution, and it grew only at the corners and edges.* A comparable instance from nature is that crystals of rhodochrosite may bear epitaxial crystals of dolomite only at the corners, while in other instances the overgrowth will occur at the corners and edges [55].

A. P. Dobrovol'skaya and I have performed experiments on the growth of potassium nitrate crystals on calcite; Figure 20 shows preferential overgrowth of the nitrate on the edges and corners, and the edges of the acute two-faced angle show appreciably more marked overgrowth than do edges of larger angle.

The results all show that the overgrowth is most pronounced where there are most unsatisfied bonds, whether short-range or long-range, since the number of unsatisfied bonds at the surface of a crystal is greatest at the corners, especially at sharp corners, though there are many also at sharp edges.

Some interesting experiments in this respect are ones on hydrothermal growth of quartz plates with holes (Fig. 21) [56].

These holes accumulate material much more rapidly than the rest of the crystal; they are filled in by the most rapidly growing faces. The cause of this rapid growth is apparently the same, namely the high concentration of unsatisfied bonds.

From this we have to recognize that the active centers of lattice defects are simply centers where collective electrical forces from the lattice arise from the unsatisfied bonds,

*Vadilo [54] has repeated these experiments.

Fig. 22. Growth of pentaerythritol crystals floating on
a melt: a) first stage; b) second stage (after A. V. Shubnikov
and V. F. Parvov).

which are broken by defects and which tend to form again. This viewpoint agrees with the fact
that active centers are readily introduced into an existing crystal by x-rays.

Further research is needed on the mechanism and nature of long-range actions as effects
not only of submicroscopic order but possibly also macroscopic order; but particularly one
needs evidence of such clarity as Distler has presented for active centers.

To illustrate the long-range action problem, we give two photographs (Fig. 22) of mutually repelling floating crystallites of pentaerythritol, which is taken from [57]. The photographs
clearly show that large crystals repel one another more strongly than do small ones. Our
task is to explain whether there is an analogous and more common situation involving mutual
attraction between crystals during growth, the attraction between larger crystals being
stronger.

In conclusion I wish to thank P. P. Butorin, S. P. Grishina, V. F. Parvov, M. V. Pikunov, E. V. Treivus, V. V. Shishkov, and V. T. Ushakovskii for providing photographs, and also A. P. Dobrovol'skaya, G. F. Dobrzhanskii, and I. A. Shpil'ko for assistance in the experiments.

Literature Cited

1. E. S. Fedorov, Priroda, 1471 (1915).
2. W. Kossel. Nachr. Ges. Wiss., Göttingen, Math.-phys. Kl., 135 (1927).
3. I. N. Stranski. Z. phys. Chem., 136A:259 (1928).
4. M. Volmer. Kinetik der Phasenbildung, Leipzig (1939).
5. E. Budewski, W. Bostanoff, T. Vitanoff, A. Kotzewa, and R. Kaischev. Phys. Status Solidi, 13:577 (1966).
6. R. Gross and H. Möller. Z. Phys., 19:375 (1923).
7. A. V. Shubnikov. How Crystals Grow [in Russian], Izd. AN SSSR, Moscow (1935).
8. M. V. Pikunov, V. V. Shishkov, Yu. A. Kostyukhin, Yu. V. Batanov, and S. V. Zhidovina. In: Single Crystals of Refractory and Rarer Metals [in Russian], Nauka, Moscow (1969), p. 7.
9. M. V. Pikunov, V. V. Shishkov, A. I. Desipri, V. M. Koltygin, and V. G. Lyuttsau, In: Growth and Defects of Metal Crystals [in Russian], Naukova Dumka, Kiev (1972), p. 257.
10. N. N. Sheftal' and A. N. Buzynin. Vestnik MGU, ser. geol., No. 3, 102 (1972).
11. D. Walton and B. Chalmers. Trans. AIME, 215:447 (1959).
12. N. N. Sheftal', N. P. Kokorish, and N. V. Krasilov. Izv. AN SSSR, ser. fiz., 21(1):147 (1957).
13. B. M. Bulakh. J. Crystal Growth, 5:243 (1969).
14. B. M. Bulakh. Thesis: Vapor-Phase Crystallization of Cadmium Sulfide, Semiconductors Institute, AN Ukr. SSR (1970).
15. B. M. Bulakh. This volume, p. 88.
16. Yu. Khodakov. Elements of Electrostatic Chemistry [in Russian], ONTI, Moscow (1934).
17. N. N. Sheftal'. Kristall und Technik, 8(1-3):149 (1973).
18. P. Curie. J. Phys., 3(3):393 (1894).
19. J. W. Wulff. Selected Papers on Crystal Physics and Crystallography [in Russian], Gostekhizdat, Moscow (1952).
20. B. Honigman. Crystal Growth and Form [Russian Translation], IL, Moscow (1961).
21. R. Kern. In: Growth of Crystals, Vol. 8, Consultants Bureau, New York (1969), p. 3.
22. L. Wulff. Z. Kristallogr., 22:473 (1893).
23. N. N. Sheftal'. Dokl. AN SSSR, 31:33 (1941).
24. N. N. Sheftal'. In: Growth of Crystals, Vol. 1, Consultants Bureau, New York (1959), p. 5.
25. V. Ya. Khaimov-Mal'kov. In: Growth of Crystals, Vol. 2, Consultants Bureau, New York (1959), p. 3.
26. V. Ya. Khaimov-Mal'kov. In: Growth of Crystals, Vol. 2, Consultants Bureau, New York (1959), p. 4.
27. V. Ya. Khaimov-Mal'kov. In: Growth of Crystals, Vol. 2, Consultants Bureau, New York (1959), p. 14.
28. E. B. Treivus, T. G. Petrov, and A. P. Kanentsev. Kristallografiya, 10(3):380 (1965).
29. N. V. Belov. Outlines of Structural Mineralogy [in Russian], Izd. L'vov. Univ. (1972).
30. N. N. Sheftal', I. A. Shpil'ko, and G. F. Dobrzhanskii. In: Growth of Crystals, Vol. 9, Consultants Bureau, New York (1975), p. 320.
31. P. S. Vadilo. Zh. Éksp. Teor. Fiz., 8:10 (1938).
32. M. S. Mirgalovskaya, M. R. Raukhman, and I. A. Strel'nikova. In: Growth of Crystals, Vol. 9, Consultants Bureau, New York (1975), p. 161.
33. M. G. Mil'vidskii and A. V. Berkova. Fiz. Tverd. Tela, 5(2):513 (1963).

34. L. I. Tsinober, V. E. Khodzhi, A. A. Gordienko, and M. I. Samoilovich. In: Growth of Crystals, Vol. 6A, Consultants Bureau, New York (1968), p. 25.

35. V. G. Fomin, M. G. Mil'vidskii, S. P. Grishina, N. S. Belitskaya, and M. I. Gurevich. Kristallografiya, 9(2):219 (1964).

36. G. V. Kleshchev, A. N. Bryzgalov, K. F. Kashkurov, A. V. Simonov, G. M. Safronov, and P. P. Butorin. Zh. Neorg. Mat., 4(3):362 (1968).

37. N. N. Sheftal'. In: Growth of Crystals, Vol. 3, Consultants Bureau, New York (1962), p. 3.

38. I. V. Stepanov, M. A. Vasil'ev, and N. N. Sheftal'. In: Growth of Crystals, Vol. 3, Consultants Bureau, New York (1962), p. 174.

39. J. W. Wulff. Priroda, 1091 (September 1915).

40. Yu. O. Tsunin. Zap. Vses. Min. Obshch., 94(4):459 (1965).

41. G. Tamman. Metallography [Russian translation], ONTI, Moscow (1931).

42. I. G. Bazhal and O. D. Kurilenko. Zh. Prikl. Khim., 40:25, 79 (1967).

43. I. G. Bazhal. Kristallografiya, 14:1106 (1969).

44. I. G. Bazhal, E. P. Dzyubenko, O. D. Kurilenko, and V. F. Chernenko, Ukr. Khim. Zh., 37(4):395 (1969).

45. I. G. Bazhal, O. D. Kurilenko, and E. N. Palash. Teor. Osn. Khim. Tekhnol., 3(6):931 (1969).

46. I. G. Bazhal, E. P. Dzyubin, O. D. Kurilenko, and V. F. Chernenko. Ukr. Khim. Zh., 37(5):508 (1971).

47. N. V. Gordeeva and A. V. Shubnikov. Kristallografiya, 12:186 (1967).

48. A. N. Kovalevskii. In: Growth of Crystals, Vol. 1, Consultants Bureau, New York (1959), p. 264.

49. H. Schafer. Chemical Transport Reactions [Russian translation], Mir, Moscow (1964).

50. K. Nassau. J. Cryst. Growth, 15(3):171 (1972).

51. M. Volmer. Trans. Faraday Soc., 28:359 (1932).

52. P. S. Vadilo. Abstracts for the Third Conference on Crystal Growth [in Russian], Izd. AN SSSR, Moscow (1963), p. 9.

53. G. I. Distler. In: Growth of Crystals, Vol. 9, Consultants Bureau, New York (1975), p. 230.

54. P. S. Vadilo. Crystallization of Metals [in Russian], Metallurgiya, Moscow (1964).

55. I. I. Shafranovskii. Lectures on Crystal Morphology [in Russian], Vysshaya Shkola, Moscow (1968).

56. V. T. Ushakovskii, K. F. Kashkurov, and A. V. Simonov. Kristallografiya, 13(3):559 (1968).

57. A. V. Shubnikov and V. F. Parvov. Crystal Nucleation and Growth [in Russian], Nauka, Moscow (1969).

EFFECTS OF GROWTH CONDITIONS ON
SINGLE CRYSTAL SHAPE

V. F. Bevz, M. I. Osovskii, and E. S. Fal'kevich

Some features have been described [1] for the shape of dislocation-free silicon single crystals that appear no matter what the mode of growth.

Here we present some new results on the shape of dislocation-free and other silicon single crystals grown under various thermal conditions, and an explanation is offered for the differences. The crystals were grown by Czochralski's method in the directions [111], [100], [110], [112], [115].

We found that parts of screw orientation on the side of such a crystal grown on [111] had on the explicit faces* straight lines when the frontal (111) face lay at the edge of the crystallization front (a frontal face is perpendicular to the [111] direction).

The asymmetry of the cross section of such a crystal is related to inclination of the crystallographic axis to the pulling direction, not to asymmetry in the thermal distribution, which conflicts with the assumption made in [1]. The degree of asymmetry is dependent on the shape, area, and point of emergence of the (111) face on the crystallization front.

Figure 1 shows the detachment surface of a dislocation-free single crystal grown with the [111] axis deviating from the growth axis by 6°15'. The edge described in [1] is clearly seen, which corresponds to the point of emergence of the (111) face on the crystallization surface.

If the frontal (111) face lies at the edge of the crystallization front, a dislocation-free single crystal has not only a larger area for such a face [1] but also a larger radius (Fig. 1) in the direction of disposition of this face.

The shape of the screw pattern in the region of the explicit faces is less affected in a dislocation-free crystal than in other crystals, which shows that the morphology is very sensitive to temperature fluctuations during growth in a crystal with dislocations.

All the features of the shape were observed on the single crystals no matter what the type and extent of the doping, but the degree of occurrence was dependent on these two factors.

Before we expand these shape features for dislocation-free crystals, we consider the change in shape for a frontal (111) face not emerging at the periphery of the crystallization front, and also the shape change in the boundary of the channel, which is the internal part of the single crystal whose crystallization front coincides with a (111) face. This enables us to

---

* Explicit faces are the pseudofaces in the terminology of [2], which are formed by {111} planes at angles less than 90° to the crystallographic growth direction.

Fig. 1. Detachment surface of a dislocation-free single crystal grown from a seed with 6°15' deviation from [111]. A region of emergence of a (111) face on the crystallization front is seen.

simplify the analysis by eliminating from consideration processes involving the capillary effect.

For our further discussions we adopt the following symbols: O_t the thermal axis of the system, O_p the pulling and rotation axis, O_c the crystallographic axis, and T_1 the temperature corresponding to the supercooling needed to produce a two-dimensional nucleus in the growth of a dislocation-free crystal; T_2 is the same for the growth of a single crystal with dislocations, and T_3 is the temperature corresponding to the supercooling needed for tangential layer growth. Here $T_3 > T_2 > T_1$.

In considering the conditions of crystallization we make the following assumption:

1. When a silicon single crystal grows from the melt along a [111] direction, it grows in the normal direction by two-dimensional nucleation, while in the tangential direction it grows via layer mechanism [3]. Dislocations are active centers that facilitate layer nucleation [4, 5], so a larger supercooling is needed to produce nuclei when dislocation-free crystals are being grown.

2. The temperature distribution is symmetrical and constant, while the isothermal surfaces (T_1; T_2; T_3) are conical with an angle 2α at the vertex.* The vertex of the cone is turned toward the melt.

3. The axial temperature gradient is constant between the T_1 and T_3 isotherms.

There are then several characteristic cases of single crystal growth.

I. The O_p, O_t, and O_c axes coincide. If a crystal with dislocations is being grown, a new layer arises at the point of intersection between the (111) face and the T_2 isotherm at point A (Fig. 2). The layers on the (111) face will meet the T_3 isotherm eventually. Then the

---

* In fact, the isothermal surfaces are parabolic; a conical form is assumed to simplify the analysis, and it does not essentially affect the results.

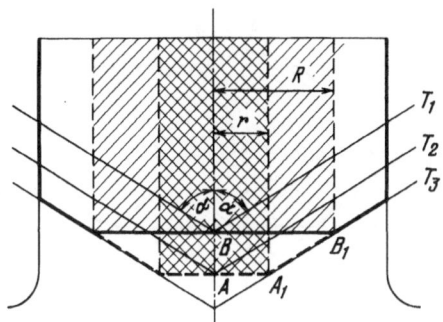

Fig. 2. Analysis of the shape of a channel in a single crystal when the O_t, O_p, and O_c axes coincide (channel region hatched).

radius of the (111) face will be AA_1. Similarly, in a dislocation-free crystal, a new layer will arise at point B on the T_1 isotherm, while the radius R of the (111) face will be equal to the length BB_1. The shape of the (111) face at the crystallization front will be symmetrical (a circle) with a constant radius, which does not vary along the length of the crystal. Figure 1 shows that $BB_1 > AA_1$, so the radius and area of the (111) face at the crystallization front will be larger in a crystal free from dislocations. The following is the difference between the radii of the (111) faces for dislocation-free and other crystals:

$$\Delta r = R - r = \frac{T_2 - T_1}{\operatorname{grad} T_r},\tag{1}$$

$$R = \frac{T_3 - T_1}{\operatorname{grad} T_r},\tag{2}$$

$$r = \frac{T_3 - T_2}{\operatorname{grad} T_r}.\tag{3}$$

It follows from (1) that the difference in the radii of the (111) faces at the front is the larger the lower the radial temperature gradient grad T_r. Therefore, the production of dislocations during growth of a previously dislocation-free crystal should be accompanied by reduction in the diameter of the (111) face at the crystallization front. The boundaries of the channel should not have a screw pattern in either type of crystal.

II. The O_p and O_t axes coincide, while the O_c axis is inclined at an angle β. In this case, the frontal (111) face will be inclined to O_p at an angle β, so the shape of the (111) face will be unsymmetrical relative to the pulling axis. The shape of the (111) face will remain constant during the growth of a crystal, so the boundary of the channel will take a constant form. The degree of asymmetry ΔR for the cross section of the channel may be expressed as the difference between the maximum and minimum radii relative to the pulling axis:

$$\Delta R = 2R \frac{\tan \alpha \cdot \tan \beta}{1 - \tan^2 \alpha \cdot \tan^2 \beta},\tag{4}$$

where R is the radius of the frontal (111) face, which is defined by (2) for the dislocation-free case, and by (3) for other cases.

It follows from (4) that the asymmetry will increase with β and as the radial temperature gradient decreases, while ΔR will be larger for the dislocation-free crystal. One can use (4) when $\alpha > \beta$ and $\alpha + \beta < 90°$.

III. The O_t and O_c axes coincide, while O_p is displaced in a parallel direction by d. This case has been considered in detail in [6], where it was shown that there will be a screw pattern of depth 2d and step V/ω on the side of the channel, where V is the pulling rate and ω is the angular velocity of rotation.

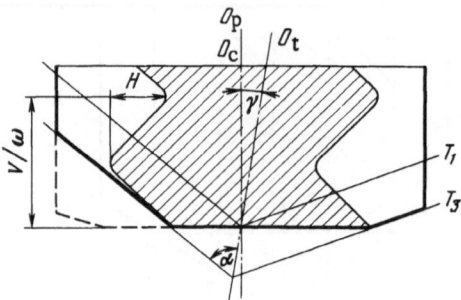

Fig. 3. Analysis of the shape of a channel in a dislocation-free single crystal when the O_p and O_c axes coincide, while the O_t axis is inclined to these at an angle γ.

However, the result 2d found in [6] for the depth is the maximum possible; the depth may be reduced if V/ω is varied. Then growth with the pulling axis displaced leads to a screw formation on the side surface, which is the same for dislocation-free and other crystals.

IV. The O_p and O_c axis coincide, while O_t is inclined at an angle γ (Fig. 3). As in the previous case, we get a screw structure with a pitch V/ω. The maximum depth H on the side surface may be defined from

$$H = \frac{\Delta T}{\text{grad } T_0} \cdot \cos \gamma \, [\tan(\alpha + \gamma) - 2 \tan \gamma - \tan(\alpha - \gamma)], \tag{5}$$

where ΔT is $(T_3 - T_1)$ for a dislocation-free crystal, as against $(T_3 - T_2)$ for one with dislocations; it follows from (6) that the depth of this on the side surface will be greater in the dislocation-free case, and that the depth difference will increase with γ.

Single Crystal Shapes under Real

Growth Conditions

To examine the shapes of real single crystals, we assume that all features of the shape characteristic of the channel boundary will appear when the frontal (111) face develops at the periphery of the crystallization front. Under real conditions, the pulling axis, the crystallographic axis, and the thermal axis do not coincide, while the isothermal surfaces have a complex parabolic form. Also, the capillary effect and the changes in the temperature distribution have definite effects on the actual shape. However, the above relationships for the channel boundary usually still apply. If the mutual disposition of the isotherms and axes is known, one can determine the shape of the growing crystal. Figure 4 is for growth on [111] without allowance for formation of the explicit faces, and it gives an analysis of the effects of the radial temperature gradient and inclination of the growth axis to [111] as regards the shape of single crystals when the isotherms near the growth front are convex or concave.

In cases 1–5 and 7–9 the finished crystals, whether with or without dislocations, will be symmetrical; in cases 1, 7, 8, and 9, the dislocation-free crystals will have a larger diameter and a larger area for the frontal (111) face than will those grown in the same thermal systems with dislocations. The differences in the diameters and areas of the (111) faces will increase in the sequence 9–8–(7, 1); in cases 2–5, the dislocation-free crystals will differ from the others only in having a larger area of the (111) face at the crystallization front. In cases 6, 10, 11, and 12, dislocation-free crystals will have a greater asymmetry in cross section. The differences in diameter and asymmetry will increase with β. We get a greater depth for the screw structure in dislocation-free crystals in cases 1, 7, 8, 9, 10, and 11 throughout the side surface, while in cases 6 and 12 this will occur only in part. In cases 1, 7, 8, 9, 10, and 11, the striations in the region of the explicit faces will take the form of straight lines; in cases 2–5, they will be curved lines, while in cases 6 and 12 striations as straight lines can occur only on one or two of the explicit faces.

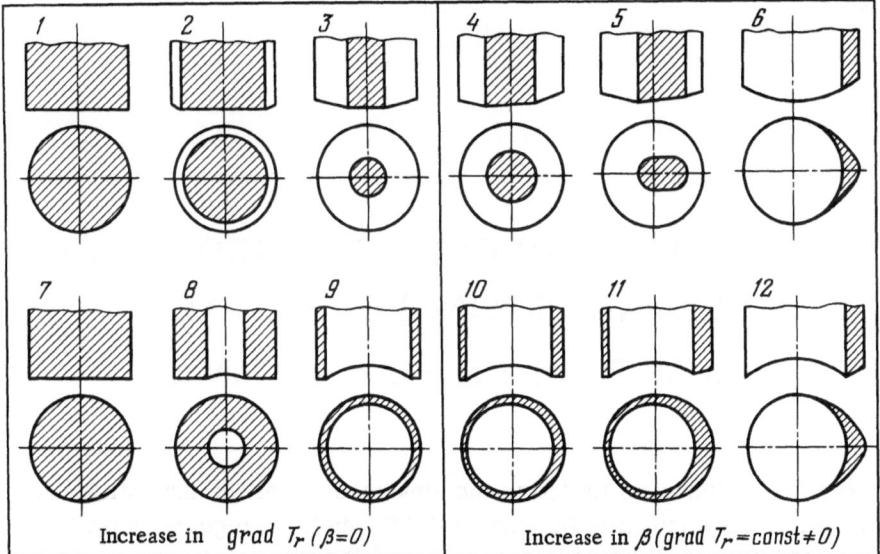

Fig. 4. Effects of a radial temperature gradient and of the angle β
between [111] and the pulling axis on the shape of the crystallization
front and the cross section of the crystal. Hatching shows the part
of the crystallization front that coincides with the (111) face. In
cases 1 and 7, grad T_r was close to zero; in 4 and 10, β was close
to zero. Isotherm shape: 1-6) convex to melt; 7-12) concave.

An examination of a large number of single crystals grown under various conditions has
confirmed the above analysis.

Therefore, the differences in the shape of dislocation-free single crystals are due to the
greater supercooling during growth.

It is of interest to compare these results with growth data for germanium single crystals.* In that case, dislocation-free single crystals also show greater asymmetry in the cross
section and broader explicit faces. A characteristic morphologic difference of dislocation-
free germanium single crystals is that the side surface shows clear rhombododecahedron
faces, which in isolated cases have been observed also on dislocation-free silicon single crystals.

Literature Cited

1.   É. S. Fal'kevich, N. I. Bletskan, K. N. Neimark, and M. I. Osovskii. In: Growth of
      Crystals, Vol. 9, Consultants Bureau, New York (1975), p. 214.
2.   M. D. Lyubalin. In: Growth of Crystals, Vol. 9, Consultants Bureau, New York (1975),
      p. 137.
3.   I. V. Salli and É. S. Fal'kevich. Production of Semiconductor Silicon [in Russian],
      Metallurgiya, Moscow (1970).
4.   G. I. Distler. In: Growth of Crystals, Vol. 8, Consultants Bureau, New York (1969),
      p. 91.
5.   L. S. Palatnik and I. I. Papirov, Oriented Crystallization [in Russian], Metallurgiya,
      Moscow (1964).
6.   K. Morizane, A. F. Witt, and H. G. Gatos, J. Electrochem. Soc., 114:738 (1967).

*This part of the work was performed jointly with Yu. M. Smirnov.

MELT SUPERCOOLING IN THE GROWTH OF
GERMANIUM SINGLE CRYSTALS

Yu. M. Smirnov and E. S. Fal'kevich

The object of this study was to define the supercooling at the crystallization front in growing perfect germanium crystals by Czochralski's method, together with theoretical calculation of the expected supercooling and comparison of the two.

The errors of measurement were minimized by growing the crystal at the bottom of the crucible (Fig. 1a). The direction of heat transfer was from the bottom of the crucible through the support. The crystal growth was monitored with a quartz rod attached to a manipulator. The growth rate was only 10-20 mm/hr in the axial direction, and the heat provided by the heater almost exactly balanced the heat loss. The temperature was measured with a platinum/platinum—rhodium thermocouple in a quartz jacket, this also being attached to a manipulator for placing at any point in the melt. The emf was recorded with an R-306 potentiometer. Check measurements were made with the same thermocouple coated with a composition based on zirconium dioxide. The thermocouple was calibrated from the melting points of high-purity tin, lead, NaCl, and copper. The overall error of measurement did not exceed ±1°C.

To define the crystallization temperature, the measurements were made at the solid—liquid interface under a layer of melt, with the thermocouple allowed to grow into the crystal. We measured the axial and radial temperature gradients in the melt. To determine the melt temperature near the crystallization front, we used a similar system as in Fig. 1b; the temperature was determined at 1-3 mm from the front under the edge of the crystal.

The values for the crystallization temperature with a crystal growing under a layer of melt varied from 936 to 938°C; bearing in mind the error of experiment, the crystallization temperature was taken as 937°C. We determined the axial temperature gradients in the crystal and the melt, which were fairly similar and were 200-300 deg/m. The radial gradients attained 600 deg/m in the surface layer at a depth of 5-10 mm in the melt. The supercooling was not more than 1°C when growing a single crystal with dislocations by Czochralski's method. When we grew dislocation-free single crystals, we found the supercooling was 2.5 ± 1°C. In this system of measurment, we were not able to determine separately the supercooling in the normal direction and that in the radial direction. The value given is the over-all supercooling. It is also quite likely that the cooling may vary along the crystallization front, which for a dislocation-free single crystal is a {111} face.

One can estimate theoretically the supercooling for this case via the concepts of [1], where it was shown that silicon single crystals grown in this way from the melt under industrial conditions show growth in the normal direction via two-dimensional nucleation, while the tangential growth involved a layer mechanism. Growth was therefore [1] assumed to occur as

216

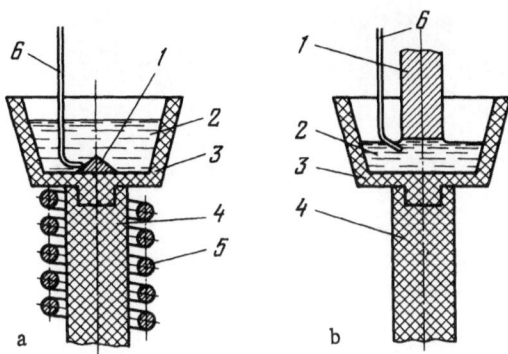

Fig. 1. Apparatus for crystal growing: a) on bottom of crucible; b) by Czochralski's method; 1) crystal; 2) melt; 3) crucible; 4) support; 5) cooler; 6) thermocouple.

follows. When a two-dimensional nucleus is formed on a face, one gets two radii of curvature for the interface: in the normal direction of face growth and in the tangential direction of expansion for the nucleus. The first is hardly different from infinite when a single nucleus arrives, so the growth rate in the normal direction remains high. The tangential growth is comparatively slow, since the high curvature of the boundary corresponds to a low growth rate. As the two-dimensional nucleus expands tangentially, other nuclei are formed on the surface, and this continues until the radius of curvature has been reduced sufficiently in the normal direction to reduce the growth rate. At this point, the radius of curvature in the tangential direction increases, and the growth in that direction runs ahead of the normal growth. The process restarts when the layer has spread entirely over the face.

It would seem that single crystals of germanium growing from the melt have a similar mechanism; a good confirmation of this is provided by Fig. 2, which shows the detached surface of heavily doped germanium (antimony concentration $2 \cdot 10^{18}$ cm^{-3}). The height of the terraces (90–100 μm) indicates that the observed pattern is due to growth processes.

The break on decantation occurs always in the liquid phase, and the depth of the layer of melt crystallizing after the break is 10–20 μm in this method, so the terraces seen in Fig. 2 would indicate the crystallization mechanism, especially since they are frequently seen also in the growth of dislocation-free single crystals.

The considerable size of the terraces arises from interference with the lateral motion of the steps from deposits of a second phase; these constitute barriers that retard the steps, and therefore favor grouping of steps [2, 3].

Fig. 2. Detachment surface (×2) of a germanium crystal heavily doped with antimony.

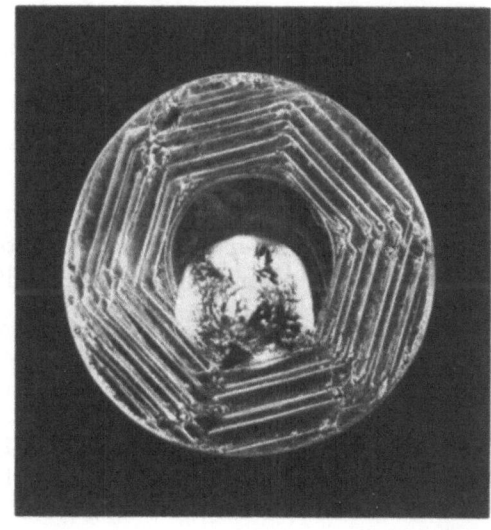

Fig. 3. A large germanium single crystal.

Figure 2 shows that the growth begins at the center, with tangential expansion out to the diameter of the single crystal.

We use the expression of [1] for the layer growth rate V_l, inserting the values for germanium,* to get

$$V_l = 0{,}162 \cdot \Delta T^2 \text{ cm/sec,}$$

where ΔT is the supercooling.

Existing equipment for growing germanium single crystals by Czochralski's method enable one to produce diameters of 25-30 mm with growth rates of 2 mm/min; from these values one gets the tangential growth rate as

$$V = 0.050 \text{ cm/sec.}$$

Substitution of this into the above expression gives $\Delta T = 0.6°C$, which is close to the observed value. This supercooling characterizes the value in the tangential direction; the supercooling in the normal direction may differ substantially.

It has been pointed out [5] that it is necessary to find technical means of realizing large supercoolings in order to obtain especially homogeneous and structurally perfect single crys-

---

* The expression contains the surface tension at the solid−liquid interface, and no measured value for this quantity appears to be available. The value 0.35 J/m² was therefore assumed, which was calculated from the formula of [4].

tals. Our experience with growing dislocation-free single crystals of germanium and silicon confirms this; a growth pyramid of {111} face is essentially involved in the dislocation-free growth of a germanium crystal along [111] with a nearly zero radial gradient along the crystallization front and considerable supercooling at that front (2.5°C). The crystallization front was a {111} face, and the distribution of the dope was exceptionally uniform [6]. The uniformity of the crystal may be characterized via the mean coefficient of variation for the specific resistance. When gallium or arsenic was used as the dope under the above conditions, the coefficient of variation was only 1-2%, or 2-3% with antimony. When ordinary commercial germanium single crystals are grown with a crystallization front of complex curvature, the usual value of this parameter is 6-10%, which provides an experimental confirmation of the view that a planar isotherm parallel to a definite face provides a way of controlling impurity rejection and doping [7].

On the above basis, we undertook to produce a single nucleus in a germanium melt at high supercooling [8]; this was found to be possible with a supercooling at the center of the melt of 10-15°C. The outward growth from this central nucleus gave a large single crystal, though this has pronounced sector and zone structures (Fig. 3).

Literature Cited

1.   I. S. Salli and É. S. Fal'kevich. Production of Semiconductor Silicon [in Russian], Metallurgiya, Moscow (1970).
2.   Single-Crystal Growth [collection of Russian translations] (N. N. Sheftal', ed.), IL, Moscow (1963).
3.   R. G. Rodes. Imperfections and Active Centers in Semiconductors [in Russian], Metallurgiya, Moscow (1968).
4.   I. V. Salli. The Physical Principles of Alloy Structure Production [in Russian], Metallurgizdat, Moscow (1963).
5.   N. N. Sheftal'. In: Growth of Crystals, Vol. 3, Consultants Bureau, New York (1962), p. 3.
6.   Yu. M. Smirnov. Izv. AN SSSR, ser. fiz., 36:561 (1972).
7.   N. N. Sheftal'. In: Growth of Crystals, Vol. 5A, Consultants Bureau, New York (1968), p. 25.
8.   F. Stöber, Z. Kristallogr., 61:299 (1925).

EFFECTS OF MEDIUM INCORPORATION IN THE EVOLUTION OF FINAL GROWTH FORMS IN CRYSTALS*

N. N. Sheftal' and E. G. Kolomyts

It has been pointed out [1, 2] that the final growth form of an almost perfect single crystal is very simple; in many cases it is limitingly simple, i.e., consists of a minimum number of faces characteristic of the substance sufficient to form a closed polyhedron (Fig. 1). The shape becomes more complicated if the perfection deteriorates (Fig. 2) [3, 4]. It has been shown that this is due to the deorganizing incorporation of solvent or gas [2].

The effects can be followed for Rochelle salt; crystals grown by cooling over the range from 52.5°C to 51.5°C (i.e., almost from a melt) have a very simple shape and high uniformity [5]. Very often, the shape consists solely of two groups of faces: the $\{110\}$ rhombic prism and the $\{001\}$ pinacoid (Fig. 3). At the same time, the amount of trapped solvent is minimal, which is clear from the absence of nonuniformity of shower type, which appears a certain time after growth as a system of very fine parallel straight channels in several directions, which are filled with mother liquor. These are visible from their cloudy appearance [6]. Crystals grown at 45°C or lower always show this rain effect, while the shape is statistically the more complex the lower the start of growth. Crystals grown at 45°C mostly have only 210 additional faces (Fig. 4); to these one adds for crystals grown from 41°C the first $\{110\}$ pinacoid while the second $\{010\}$ pinacoid is added at lower initial growth temperatures. The solution then contains more solvent, and the crystal traps more of it.

Various substances show simplicity in the final growth forms of perfect single crystals and loss of this simplicity when the uniformity deteriorates, such as epsomite, urotropin, calcite, fluorite, EDTA, etc, [2, 3].

In practice, it is impossible to follow on single crystals the shape change when much solvent is incorporated; the the large sizes make the crystals more stable against such entry, but the effect can be observed on mass recrystallization on small crystallites under conditions unfavorable to the growth of perfect crystals.

One of us has obtained such evidence [7] from a systematic study of the growth of individual snow crystals in dry snow cover.

The composition of the snow consists initially of a layer of randomly piled primary crystals or fragments; these have entered a medium very different from that in which they were formed, and they have broken up in stages (Fig. 5). Some rounding has occurred, which is accompanied by splitting-off and evaporation of the dendrite branches, with evaporation of

*First published in Acta Phys. Acad. Sci. Hung., 33(3-4):335-351 (1973).

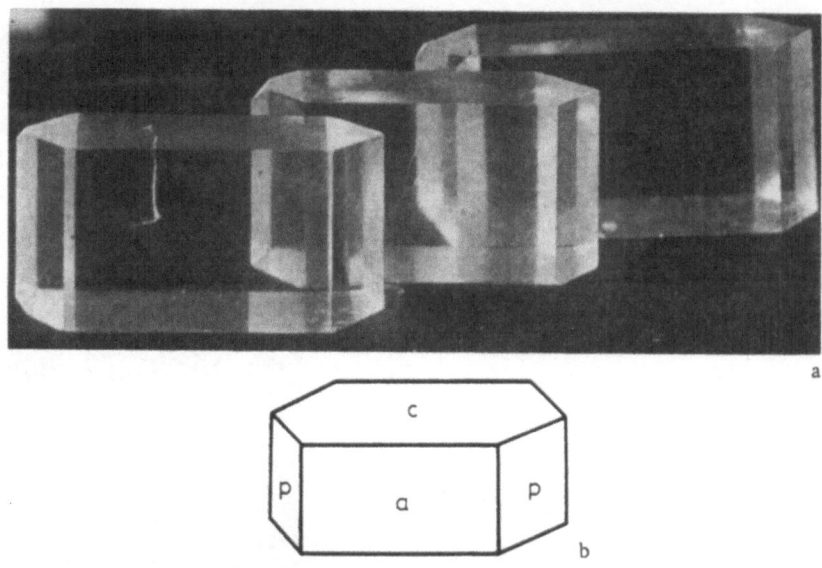

Fig. 1. Limiting final forms of homogeneous sucrose single crystals. The middle crystal has a small face at the top left, which is warning of imminent flaws: a) actual size; b) scheme.

Fig. 2. Complex growth lines of inhomogeneous sucrose single crystals: a) actual size of crystal with a subindividual; b) splitting crystal in the final inhomogeneous stage; c) scheme for shape of an inhomogeneous crystal.

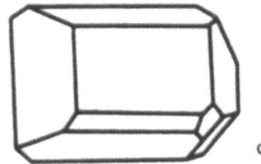

Fig. 3. Highly homogeneous Rochelle salt crystal (weight 971 g) formed mainly by {110} and {001} faces.

the smaller individuals and growth of the larger ones at their expense. The latter are transformed first into irregular polyhedral crystals of grains, whose mean diameters range from 1-2 mm to 0.1 mm or less.

Then the crystals, having reached 0.5 to 2 mm in size, become planar or columnar prismatic crystals, with {0001} and {10$\bar{1}$0} faces (Fig. 6). Crystals simplest in shape and perfect in structure can be compared with the limiting forms of homogeneous single crystals.

Defects appear on the faces as the snow crystals grow further, and these are invariably accompanied by complication of the crystal form. The initial signs indicating such defects are as for single crystals [2] (Fig. 1) in the form of various pyramids on the edges between the basal and prism faces, with the appearance much as shown in Fig. 7. Somewhat later, the faces of the crystal develop progressive caverns, which eventually develop into large unclosed cavities. The faces themselves at the same time become rougher and are covered in step

Fig. 4. Rochelle salt crystal grown at 45°C, with lower homogeneity and more complex shape (weight 3 kg): faces (110), (210), {100}, and {001}.

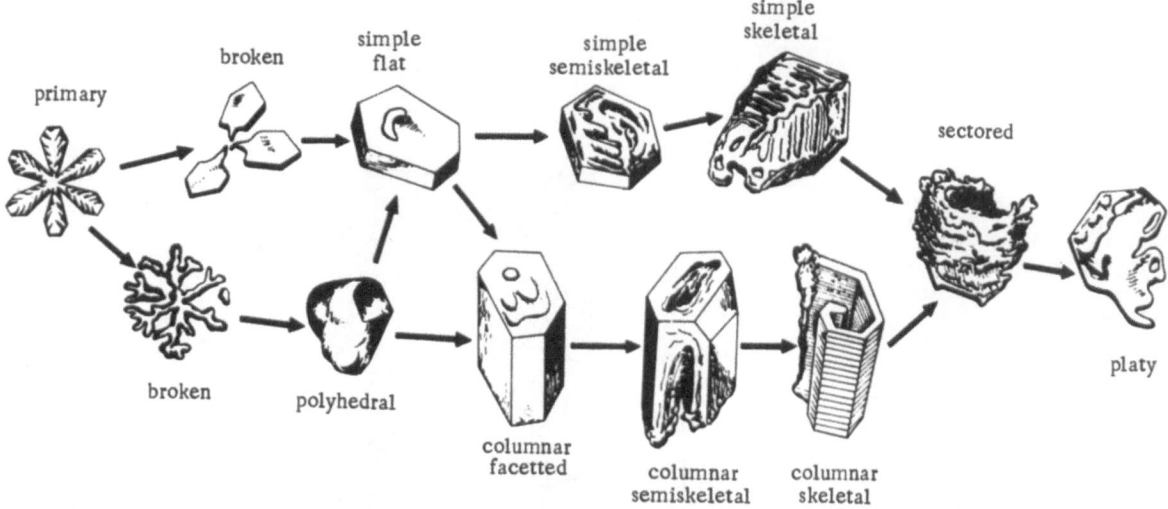

Fig. 5. Growth forms of snow crystals in dry snow. Left: breakage of primary snowflake. The growth forms of secondary crystals are shown: planar form (top) and columnar form (bottom). The evolution is shown in more detail in subsequent photographs.

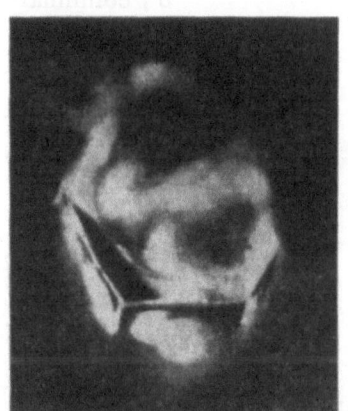

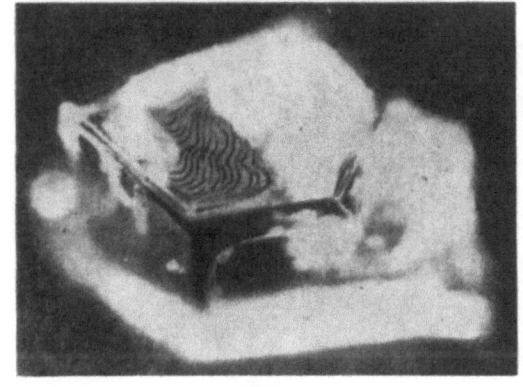

Fig. 6. Stage of facetted perfect growth of snow crystals (photographs ×2.7): a) low prism with smooth faces; b, b') columnar prism with macrosteps from layer growth on the basal plane.

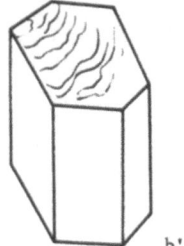

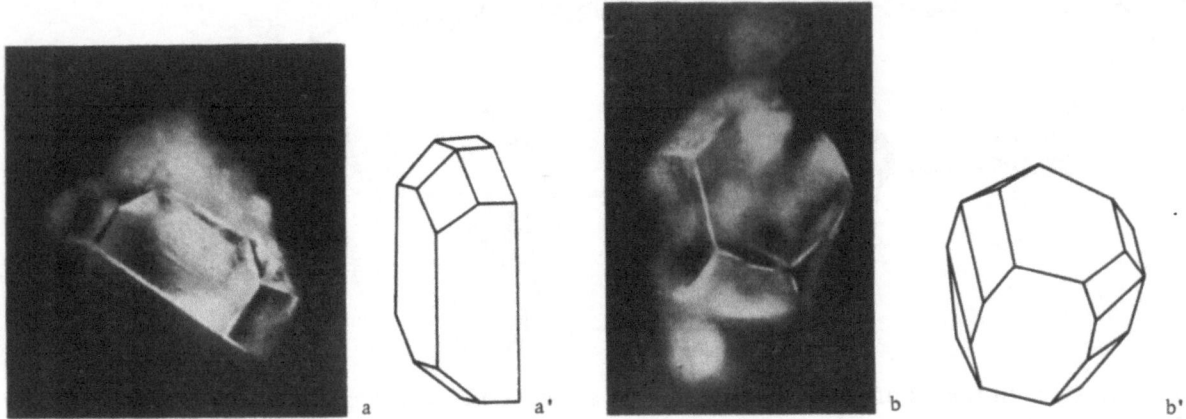

Fig. 7. Occurrence of minor pyramidal faces: a, a') shortened prism with pyramidal faces; b, b') columnar crystal with pyramidal faces.

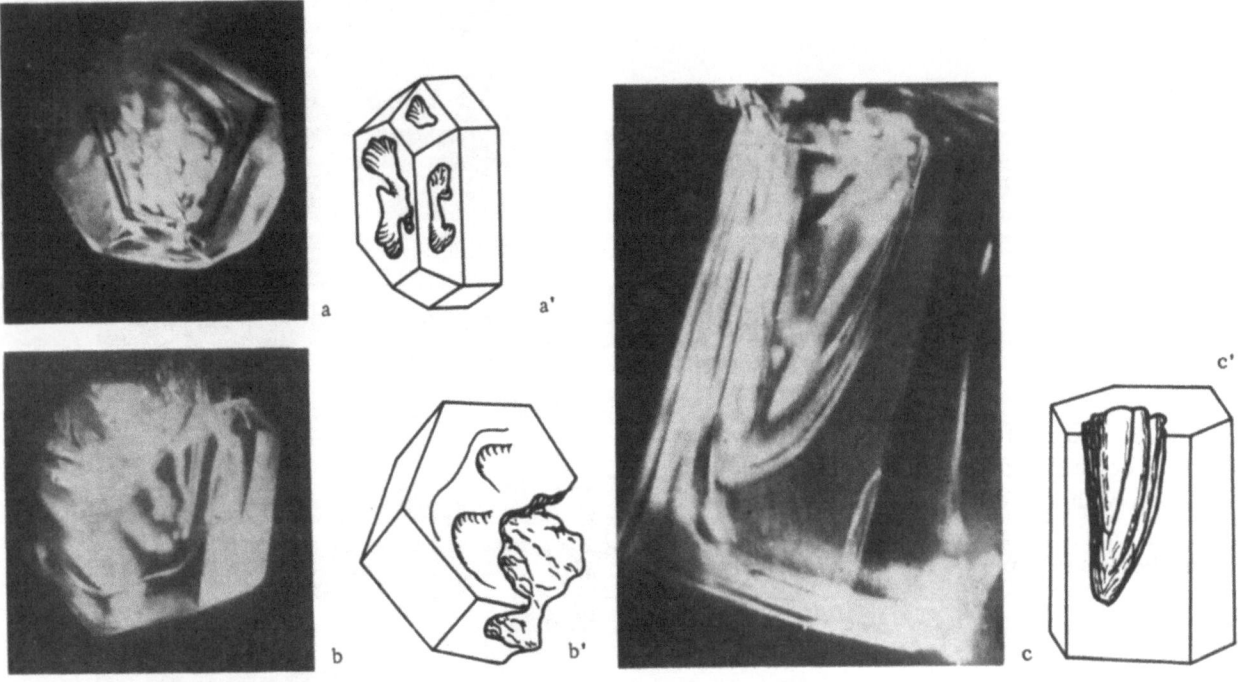

Fig. 8. Semiskeletal growth of snow crystals (photographs ×27): a, a') low prisms with small depressions in faces; b, b') semiskeletal prism half eliminated by cavities; c, c') columnar prism with faces fully developed at the sides.

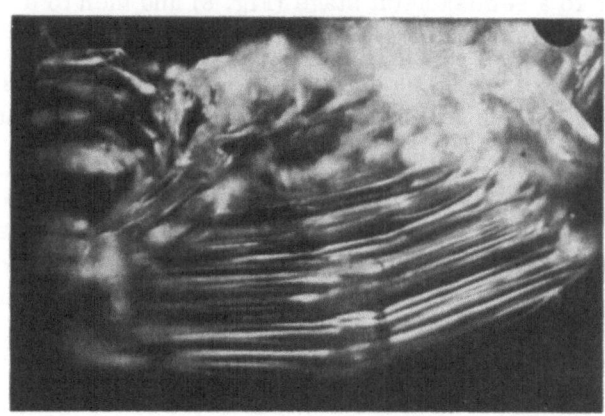
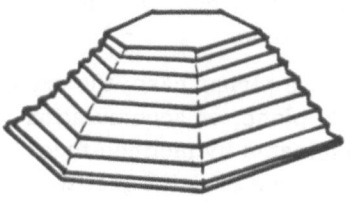

Fig. 9a. Stage of columnar skeletal growth of a snow crystal as a skeletal pyramid (photograph ×27).

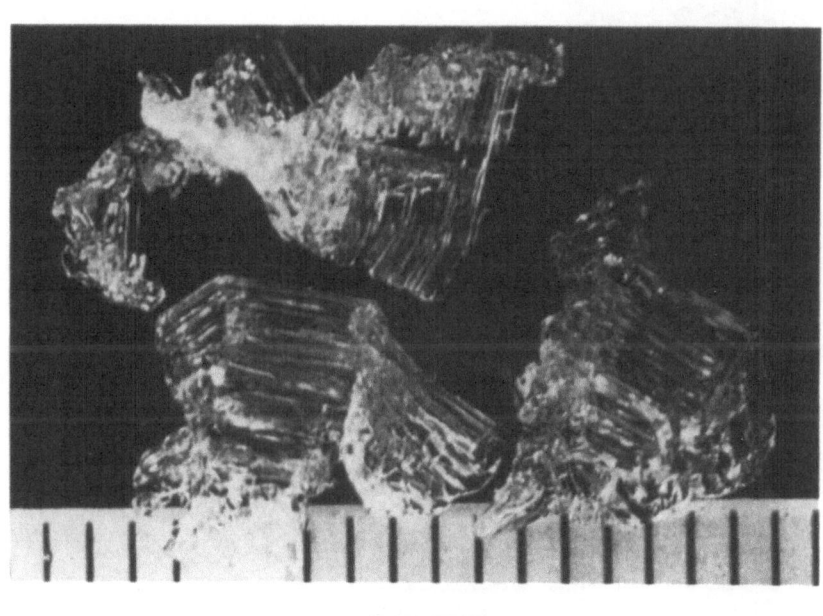

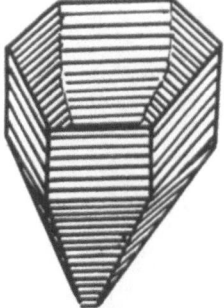

Fig. 9b. Stage of columnar skeletal growth of a snow crystal as a cup shape (photograph ×27).

microrelief, which subsequently goes over to a semiskeletal stage (Fig. 8) and then to a skeletal one (Fig. 9). In this period the mean diameter attains 4.0–5.5 mm.

The scheme of Fig. 5 shows that skeletal growth can occur in two different ways, giving either planar or columnar forms; the first occurs in the range roughly from −1 to −10° and the second from −15 to −25°.

In the planar case, the skeletal crystals are formed by overgrowth of the low steps on the basal plane, or accumulation of these at oblique angles to the principal axis. The prisms are terminated by inclined faces, while the pyramids become very unsymmetrical. An extreme case of planar skeletal growth is a form consisting only of two stepped prismatic faces linked by a very obtuse angle. The faces have striations parallel to the edges between the basal and prism planes (Fig. 10). Planar skeletal forms differ from columnar ones in having preferential growth on (0001), even if the basal faces are absent.

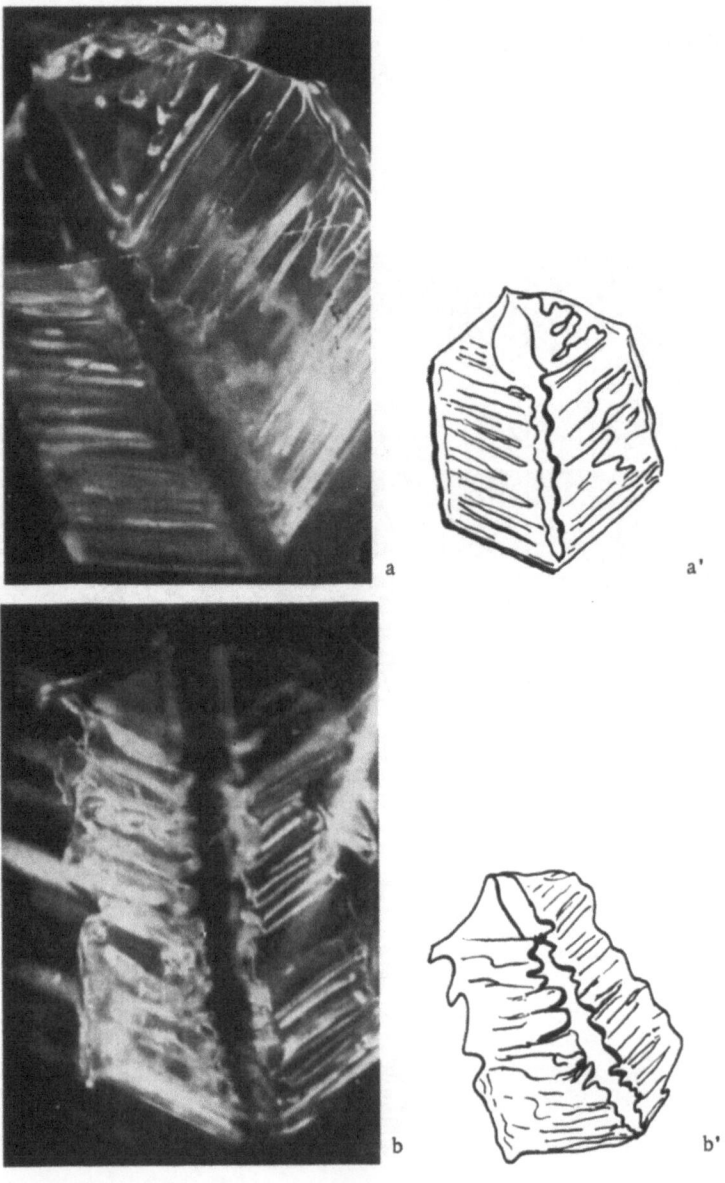

Fig. 10. Skeletal forms from two stepped prism faces (photographs ×27).

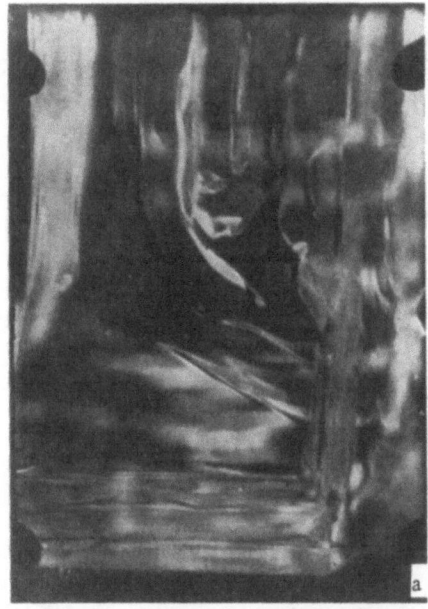

Fig. 11. Columnar skeletal prism with a large cavity (photograph ×27).

In the columnar case, the expansion of the cavities in the prismatic crystals leads to empty framework structures of prismatic forms with intact edges and thin platy faces of (10$\bar{1}$0) and (11$\bar{2}$1) types, which often remain smooth. From the facetted form of Figure 11 one gets predominantly pyramidal skeletal crystals such as shown in Fig. 9; progressive evaporation of several faces leads to the form shown in Fig. 12.

A second qualitative change in the evolution of snow crystals arises from the aging; with lapse of time, the regular growth of a skeletal crystal eases and the characteristic steps are lost, and a hooked form as in Fig. 13a may develop. The angles between the axes and the pyramid faces tend to increase, and the forms become more flattened. Simultaneously, the edges between the basal and prism faces develop thin platy projections perpendicular to the principal axis. Subsequently, the skeletal form breaks up into individual sectored parts, which take the form of irregular thin plates with curved edges, numerous small cracks, and through holes (Fig. 14).

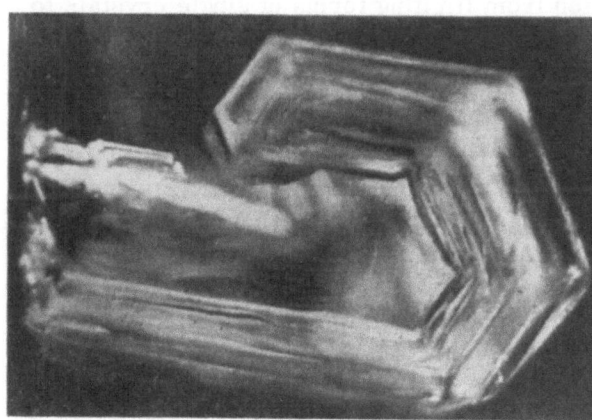

Fig. 12. Hooked prism (see Fig. 5, columnar
skeletal stage).

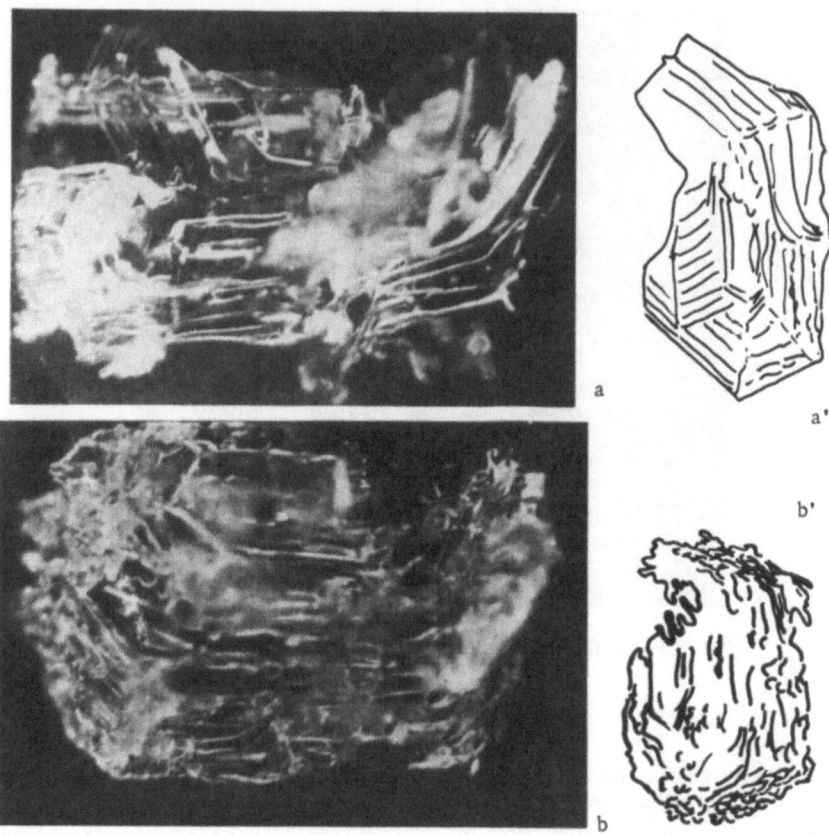

Fig. 13. Sectored and platy aging stages of snow (photographs ×27):
a, a') bent skeletal crystal and scheme; b, b') distorted skeletal cup
with platy outgrowths in the basal planes and scheme.

The crystals have by this time completely lost their macroscopic ordering; the mean diameter has fallen to 1.5-2.0 mm or less.

The entire transformation occurs purely by sublimation in the absence of melting. The sequence of recrystallization stages is not associated with any definite changes in the external medium, and this process of spontaneous development in snow cover is one for an association of crystals. The principal factor resulting in the change in shape is associated with internal processes in the snow cover.

One can consider the morphological evolution from limiting forms of single crystals to break-up of small skeletal crystals as an occurrence of Curie's symmetry superposition principle [8]. In the present case, this is the superposition of the symmetry of the medium and of the crystal developing from it, which goes through a sequence of stages. This can be elucidated as follows.

Curie's principle is that the symmetry of the generating medium is effectively superimposed on the symmetry of the body formed in that medium; the resulting shape of the body retains only those elements of its own symmetry that coincide with the symmetry elements of the medium imposed on it.* However, to apply this principle to this evolution, we have to bear in mind that the generating medium in which each individual crystal grows is quite unsymmetrical; consequently, the resulting crystal should not have any symmetry elements, since in

_____
* Formulated by I. I. Shafranovskii [9, page 53].

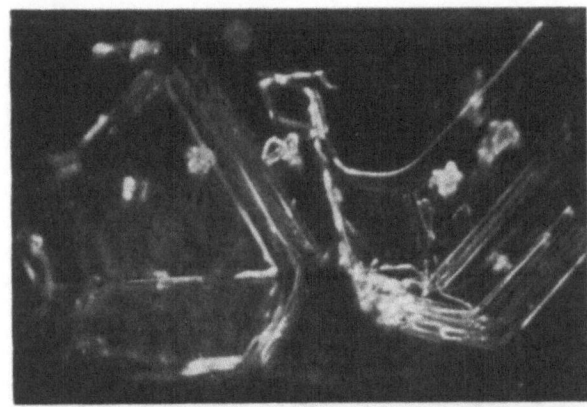

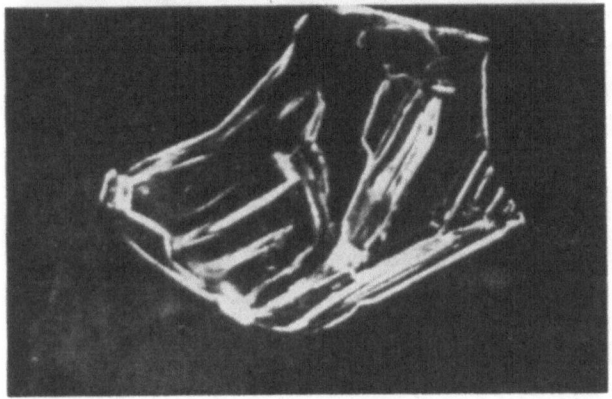

a

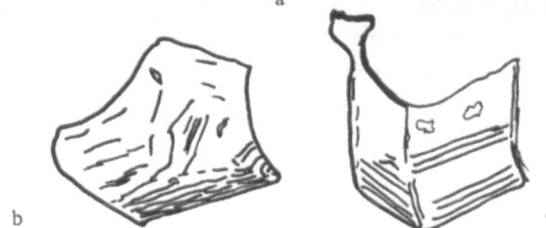

b   c

Fig. 14. Decomposed skeletal form: shape-less plates in the final stage of evolution of a snow crystal.

that case the medium has no symmetry elements, and there cannot be any in common. How-ever, there is only an approximate relationship between the symmetry of the internal structure and that of the macroscopic form of the crystal; in fact, the ideal symmetry is completely suppressed, for example, in a symmetrical figure consisting of several equal parts equally disposed [10], which occurs if one places on one of the parts a small point or spot absent from the other parts. The structural defects in the real crystal of course are not equal in the other-wise equal parts of the structure. We also have unequal surface defects in the macroscopical-ly symmetrical parts of a crystal. Therefore, these defects completely eliminate the symme-try, strictly speaking. However, we can neglect such defects, i.e., we always bear in mind the approximate or real symmetry of the crystal. Then the concept of approximate symmetry enables us to consider a crystal with defects only as less symmetrical, and all the resulting defects in the symmetry represent the struggle between the unsymmetrical medium and the symmetrical crystal.

If we consider the process as a whole from this viewpoint, we can say that initial homo-geneous crystals represent an example of a highly organized individual, which appears in many cases as shape simplicity, where the basic surface elements (faces) are present in minimal numbers, these being necessary and sufficient to produce a closed polyhedron. Subsequently, elements of inequality appear as new simple forms attached to the original one. The extent of incorporation of the medium into the crystal also increases. Even later, as is clear from the small crystals, one gets semiskeletal forms and then skeletal ones. Larger and larger inclusions are produced in the crystals, and eventually one gets closed periodic inclusions, and the crystal largely ceases to exist. The final products are unsymmetrical bodies with many internal defects, inclusions, and other features reflecting the asymmetry and statistically random incorporation of the parent medium, with random placing of joined crystals and irregular interfaces between them, which can arise from temperature and humidity gradients in the medium, amongst other causes (Fig. 6). These final formations retain no traces even of the approximate macrosymmetry, which has been completely overcome by the asymmetry of the external disordered medium.

Then the shape evolution of the final macroscopic forms of crystals is dependent on the incorporation of the internal medium and occurs via a sequence of stages governed by Curie's symmetry superposition principle. The asymmetry of the medium step by step overcomes the ordering of the macroscopic forms of the homogeneous crystals, as considered from the viewpoint of the real or approximate symmetry.

Literature Cited

1. I. N. Stranski. Discuss. Farad. Soc., Vol. 13 (1949).
2. N. N. Sheftal'. Growth of Crystals. Vol. 1, Consultants Bureau, New York (1959), p. 5.
3. N. N. Sheftal'. Growth of Crystals. Vol. 3, Consultants Bureau, New York (1962), p. 3.
4. I. V. Gavrilova. Growth of Crystals. Vol. 5B, Consultants Bureau, New York (1968), p. 9.
5. N. V. Alyavdin, N. N. Sheftal', and Z. I. Frolova. Kristallografiya 1:193 (1957).
6. A. V. Shubinkov. How Crystals Grow [in Russian], Izd. AN SSSR (1935).
7. E. G. Kolomyts. Sibirskiy geograficheskiy sbornik, No. 8, Novosibirsk (1972).
8. P. Curie. Selected works [Russian translation], (1963).
9. I. I. Shafranovskii. Symmetry in Nature [in Russian], Nedra (1968).
10. N. N. Sheftal'. Growth of Crystals, Vol. 4, Consultants Bureau, New York (1966), p. 183.

ELECTROCRYSTALLIZATION AND THE ELECTROLYTIC DEPOSITION MECHANISM FOR SILVER

E. Budevskii, V. Bostanov, and T. Vitanov

When a crystal has become covered with faces and has taken up a transitional or steady-state growth form, the subsequent growth rate is determined by the normal growth rate of the existing faces; if there are no other additional conditions that complicate the growth such as diffusion or adsorption, these faces will be the most close-packed, i.e., they belong to the equilibrium form. The theory envisages two basic growth mechanisms for such faces, one of which is the two-dimensional mechanism, in which the growth occurs by layerwise fluctuation production of each fresh layer. The second is the spiral growth mechanism, which occurs when there are screw dislocations, and this is the main mechanism found in nature.

A second aspect of crystal growth kinetics involves the mechanism and propagation speed of step growth, these steps arising from two-dimensional nuclei or from screw dislocations. Here again there are two limiting cases: a) the particles pass directly from the supersaturated mother phase directly to the growth sites, i.e., direct incorporation, and b) the particles pass initially to an intermediate adsorbed state on the surface, and then migrate to the growth points and are incorporated into the lattice (surface diffusion).

A study of the electrical crystallization of silver has shown that both mechanisms certainly occur; this has also enabled us to check the laws implied by the general theory, and to determine the parameters characterizing the process. In addition, it has provided evidence on some more detailed aspects of the transition, and it has provided means of determining which of the two mechanisms predominates in a given case.

Metal electrocrystallization is very convenient for quantitative study of the process; the growth rate v_N in moles/cm^2-sec is measured by means of the current density i in A/cm^2:

$$v_N = i/zF, \tag{1}$$

where F is the Faraday constant and z is the ion valence.

The supersaturation is σ = c/c$_0$, being a measure of the deviation of the concentration c from the equilibrium value c$_0$, and this is seen in electrocrystallization as the deviation of the potential from the equilibrium value, i.e., the overvoltage η:

$$\eta = \frac{RT}{zF} \ln \frac{c}{c_0}. \tag{2}$$

---

* Bulgarian Academy of Sciences.

The electrochemical method is convenient in that the current density and overvoltage (i.e., the growth rate and supersaturation) can be measured and controlled with high accuracy.

Growth of Dislocation-Free Faces

Introduction to a Theory

Crystal growth theory indicates that faces free from dislocations on the equilibrium form grow via two-dimensional nuclei, which are minute single-layer sets of atoms on the faces, whose growth probability is greater than the dissolution probability at a given supersaturation. The overvoltage controls the size of these nuclei, and consequently the number n_k of atoms in a nucleus, as well as the energy of formation A_k, the relationship being an inverse one [1, 2]:

$$n_k = \frac{A_k}{ze_0\eta},$$ (3)

$$A_k = \frac{4f\varepsilon^2 \cdot 10^{-14}}{ze_0\eta} \ \text{J/mole,}$$ (4)

where e_0 is the electronic charge, ε in ergs/cm^{-1} is the specific edge energy of a step, and f in cm^2 is the area taken up by one atom on the surface of a nucleus.

Nucleation is a fluctuation process, whose rate determines to a considerable extent the growth of dislocation-free faces; the following equation [3] defines the nucleation rate as a function of supersaturation:

$$I = k_1 e^{-\frac{Ak}{kT}} = k_1 e^{-\frac{k_2}{\eta}}, \ \text{sec}^{-1} \cdot \text{cm}^{-2},$$ (5)

where k_1 is a coefficient considered in more detail in the theory [4-6], while the ratio of the constant k_2 to η is proportional to the energy of formation.

The nucleation rate is negligible at low supersaturations, and a rapid increase begins only above a certain critical supersaturation; this is a characteristic feature of two-dimensional nuclei, namely a threshold in the growth rate.

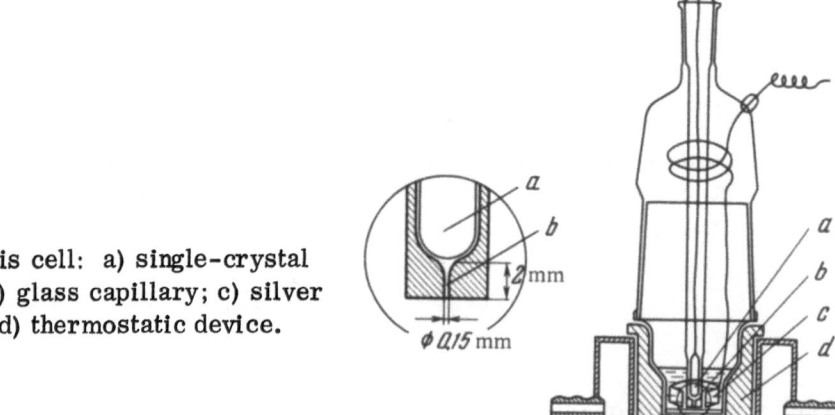

Fig. 1. Electrolysis cell: a) single-crystal silver electrode; b) glass capillary; c) silver cylindrical anode; d) thermostatic device.

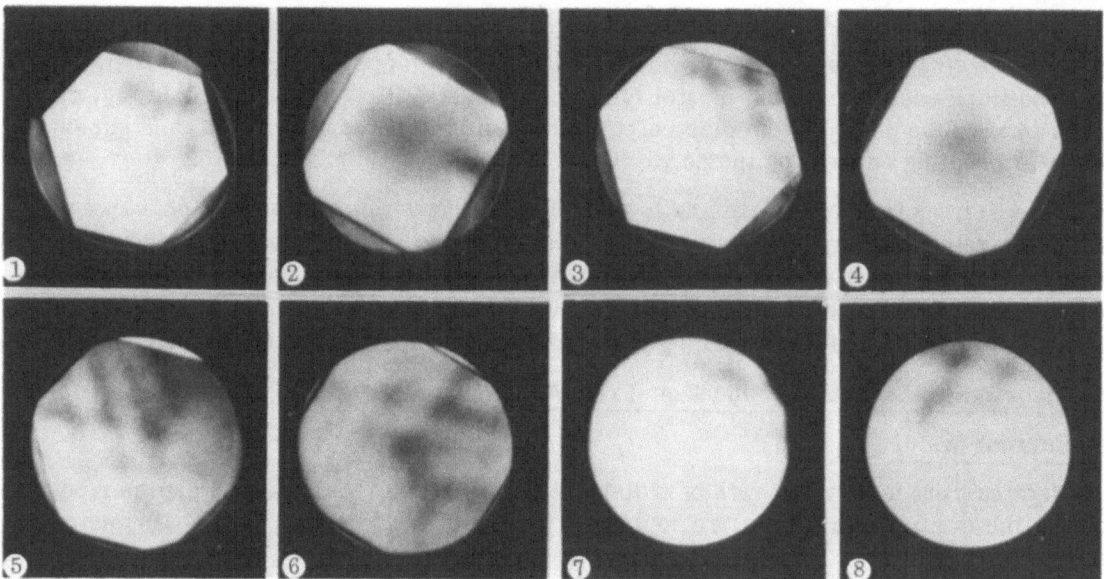

Fig. 2. Frontal (111) and (100) faces on a silver crystal growing to fill a capillary.

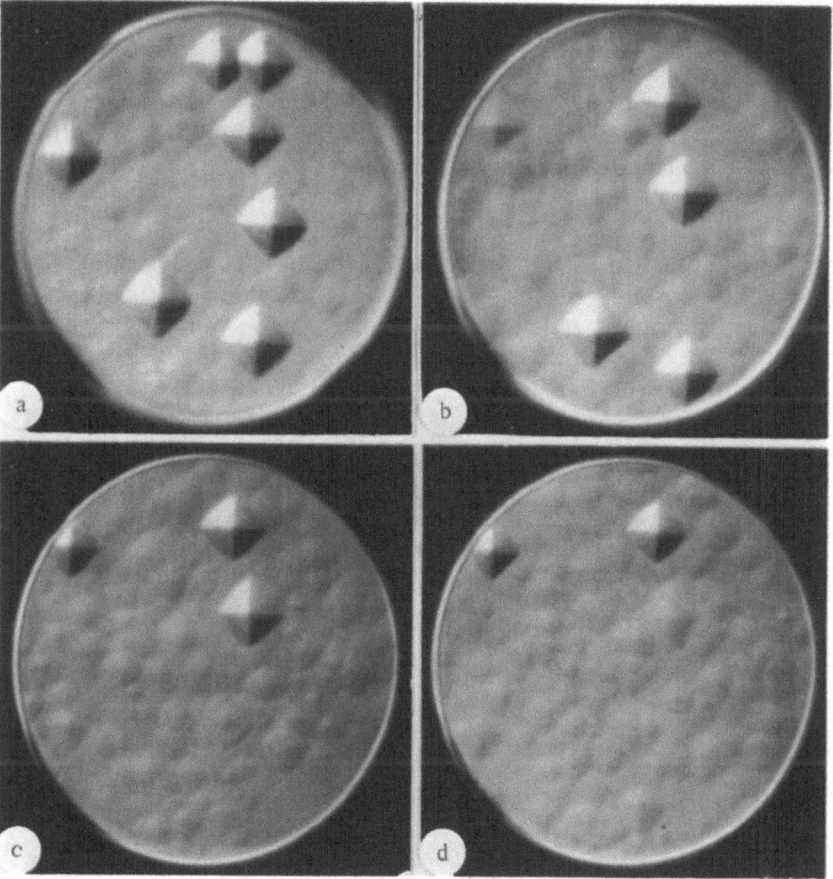

Fig. 3. Growth pyramids on a (100) face denoting emergence of dislocations. Parts a-d indicate the reduction in the number of screw
dislocations during growth.

Production of Single Faces Free from Dislocation [7-9]

If a silver single crystal is grown in a capillary with an appropriate orientation, one can provide conditions under which the frontal face of the crystal fills the entire capillary; these conditions involve the purity of the electrolyte, a high silver ion concentration, and super-position of alternating current on the dc.

Figure 1 shows the electrolysis cell, while Fig. 2 shows the frontal face of the crystal growing in the capillary during filling; if the amplitude of the ac is chosen appropriately during growth, the defects arising on this frontal face become smaller and may vanish entirely. Figure 3 illustrates the rise in uniformity of a cubic face, where the screw dislocations had first been developed by applying a short growth pulse to an initially smooth face.

Effects Observed in Growth of Dislocation-Free Faces at

a Given Current [2-10]

If a set current is used, growth of a dislocation-free face is accompanied by periodic variation in the overvoltage (Fig. 4). The variations are clearly related to two-dimensional nucleation.

When the current is switched on, the overvoltage rises rapidly because there are no growth sites; when a critical value is reached, nuclei are formed for a new layer, and at this instant the voltage begins to fall, while the length of the growth front increases, since incorporation is thereby considerably facilitated. The voltage begins to rise again when the growing layer has propagated over the entire face, and this continues until the critical value is reached and the process is repeated. The period of oscillation is inversely proportional to the current, and the product of the current and the period gives the amount of electricity equivalent to deposition of a monolayer, which provides evidence that the variations in overvoltage are related to the formation and propagation of monotonic layers.

Effects at Constant Overvoltage [2, 10, 11]

If the overvoltage applied to a dislocation-free face is kept constant and below the maximum on the $\eta-t$ curve for a given current (Fig. 4), i.e., below the critical value for nucleation, then no current should flow through the electrode; a dislocation-free face has no growth points,

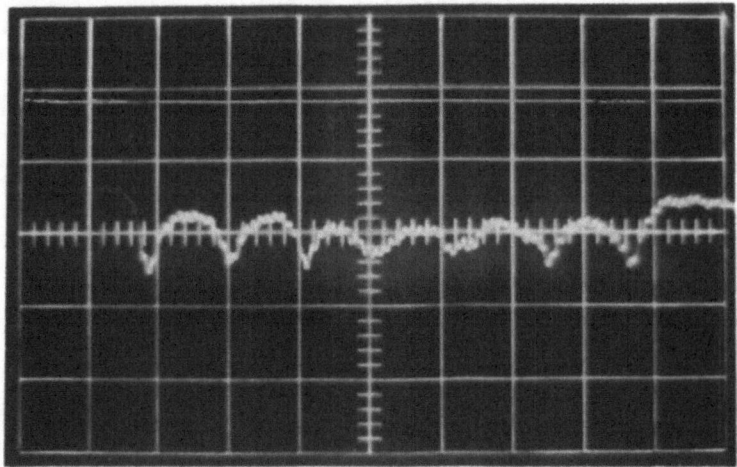

Fig. 4. Overvoltage variations on a dislocation-free (100) face with a steady growth current i = 5.8 · 10^{-4} A/cm^2. Y scale 5.2 mV/cm, X scale 0.2 sec/cm.

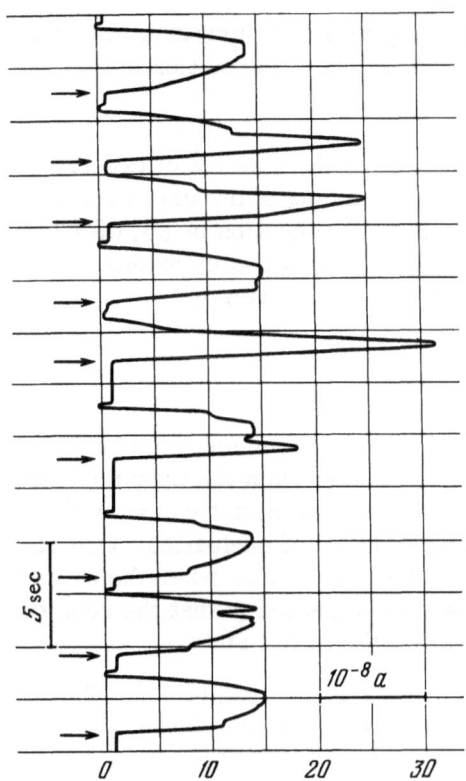

Fig. 5. Current as a function of time at η = 6 mV after a single pulse with η = 13 mV, τ = 1 msec, which produced a two-dimensional nucleus on a dislocation-free (100) face in a cylindrical capillary.

and the overvoltage is insufficient to nucleate a new layer. This has been confirmed by experiment and is one of the characteristic features of dislocation-free faces.

If a short rectangular voltage pulse is applied to such a polarized electrode, a single nucleus for a new layer may be formed if the amplitude and length are appropriate; from that time onwards there is a flow of current, which at first rises, but then falls again to zero. The phenomenon is repeated if a fresh pulse is applied (Fig. 5). This is undoubtedly related to the formation and propagation of a new layer over the undamaged surface of the face, and it may be explained as follows. After a two-dimensional nucleus has been formed, the applied potential is sufficient to cause the new layer to begin to grow, which is seen as a current flow. The current increases with the length of the growth step. When the propagating layer reaches the capillary wall, the growth front begins to decrease, and so does the current. The different shapes taken by the successive current curves are due to the different sites of nucleation

TABLE 1. Exchange Current in Relation to Surface Preparation

Preparation conditions		Step density given by (18) $N_S = L_S$, cm^{-1}	Measured exchange current, i_0, A/cm^2	Exchange current for atoms on steps, $i_0(L)$, A/cm^2	$\dfrac{i_0(L)}{\eta}$, $\dfrac{A}{mV \cdot cm^2}$
current density in mA/cm^2	overvoltage correction, mV				
1.1	0.4	$1.52 \cdot 10^4$	0.14	0.08	$0.20 \cdot 10^3$
4.4	0.8	3.04	0.19	0.13	0.16
8.8	1.0	3.8	0.28	0.22	0.22
17.6	1.6	6.08	0.37	0.31	0.19
32.2	2.3	8.74	0.53	0.47	0.20
52.9	2.8	10.67	0.64	0.58	0.21
Dislocation-free face		0	$0.06 = i_{0,ad}$	0	0

Note: Cube face of a silver single crystal in 6 N $AgNO_3$ at 45°C.

(Table 1). Although the curves differ in shape, the time integral of the current is the same for each curve and corresponds to the amount of electricity equivalent to deposition of a monolayer.

Two-Dimensional Nucleation Rate [2, 10, 11]

The method described above is very convenient in studying the kinetics of two–dimensional nucleation; the different shapes of the curves indicate that the face of the crystal is uniform and the site of nucleation is random, so one can use the stochastic equation of (5), which indicates that the mean time required by the fluctuation mechanism for two–dimensional nucleation is $\tau = 1/SI$ (S is the electrode area in square centimeters), which is given by

$$\log \tau = -\log SI = -\log Sk_1 + \frac{k_2}{2.3\eta}. \tag{6}$$

When single pulses are applied, the nucleation does not occur immediately after the pulse starts but with a certain delay; if we neglect the correction arising from the transient involved in nucleation [5], this delay may be taken as τ; in accordance with this, one can vary the pulse duration with a given amplitude to find the time for 50% probability of nucleation. One can tell whether a nucleus has been formed by whether a current continues to flow when the overvoltage is below the critical value. One of the major advantages of this method is that the resulting nucleus propagates over the entire face and thereby restores the initial state.

Figure 6 shows $\log \tau$ as a function of $1/\eta$, where τ in sec is the length of the pulse of amplitude η in V giving 50% probability of nucleation. The slope of the straight line gives the constant $k_2 = 2.3d(\log \tau)/d(1/\eta)$, and from this we get ε in accordance with (4) and (5):

$$\varepsilon = \left(\frac{kTe_0k_2}{4f}\right)^{1/2} \cdot 10^7 \text{ ergs/cm}. \tag{7}$$

In this way we have determined ε for monotonic layers on cubic and octahedral faces of silver single crystals in 6 N $AgNO_3$ at 45°C; both gave the same value of

$$\varepsilon = 2 \cdot 10^{-6} \text{ erg/cm}.$$

The nucleation parameters differ only as regards the frequency factor k_1 [12]:

$$k_{1\,(100)} = 10^{12} - 10^{14} \text{ sec}^{-1} \cdot \text{cm}^{-2},$$
$$k_{2\,(111)} = 10^{10} - 10^{11} \text{ sec}^{-1} \cdot \text{cm}^{-2}.$$

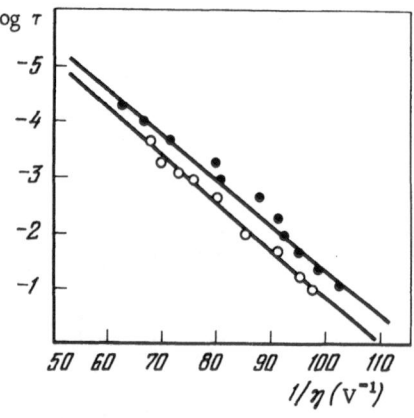

Fig. 6. Plot of time to formation of first nucleus against $1/\eta$ for a dislocation-free (100) face for two different electrodes.

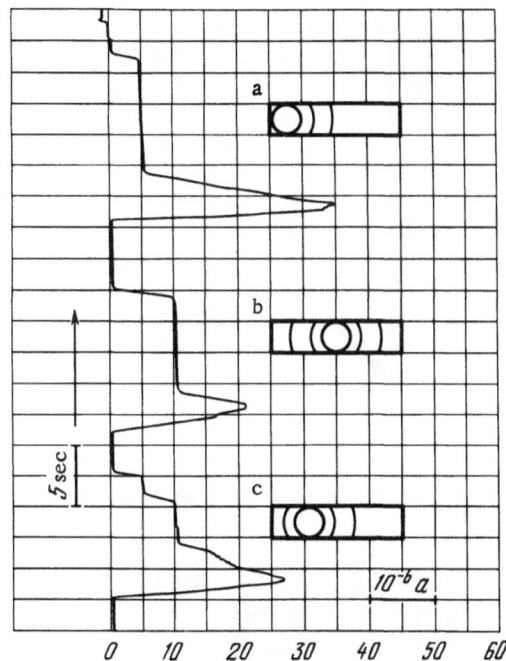

Fig. 7. Current as a function of time at a constant overvoltage after formation of a two dimensional nucleus on a (100) face bounded by the walls of a rectangular capillary 100 × 400 μm; a) nucleus formed at one end of face; b) nucleus formed at center of face; c) nucleus formed at center of one half of face. Traced from recorded curves.

Shape and Propagation of Monotonic Layers [13, 14]

The i—t curves for constant η indicate that i is approximately proportional to the length L (cm) of the growth front for the monotonic layer; it is also logical to assume that the current is proportional to the overvoltage, i.e.,

$$i = \varkappa L \eta \quad \text{[A]}. \tag{8}$$

The growth conditions in a circular capillary, which give a face of circular form, are insufficient to confirm this equation, and it is impossible to determine the value of $\varkappa$ exactly in ohm^{-1}-cm^{-1} [2, 10, 11].

One can examine the shape of the current curves after a nucleation pulse, if one uses a narrow capillary of rectangular cross section, which gives a rectangular face; here the current—time curve should have a flat range corresponding to motion of a step of constant length, and here there are several possible cases in relation to the site where the new layer arises: a) the nucleus arises at one end of a rectangular face (Fig. 7a), and the current curve should show a plateau of definite height; b) the nucleus is formed at the middle of the face (Fig. 7b), and the plateau on the curve should be of twice the previous height, which corresponds to two growth steps; c) the nucleus arises at the middle of one half of the rectangular face (Fig. 7c), and the current curve rises first of all to a plateau of double height, but then one of the fronts reaches the capillary wall, and the current is halved, persisting at this value until the new layer has spread over the entire face.

Figure 7 shows a unique recording of three curves taken one after another and corresponding to the above three cases.

The plateau heights on the curves of Fig. 7 are proportional to the overvoltage, as shown in Fig. 8; Eq. (8) may thus be considered as essentially proven. However, exact determination of the constant requires a knowledge of the growth front length, which cannot simply be taken as equal to the narrow side of the rectangular capillary, since the growing layer may be circular or polygonal, and with one of two possible orientations. To resolve this problem, the long

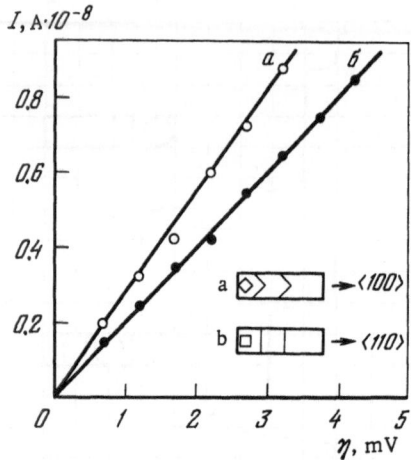

Fig. 8. Steady-state current (area height) as a function of overvoltage for two orientations of the longitudinal axis of the rectangular cross section of the capillary relative to the initial crystal: a) on $\langle 100 \rangle$; b) on $\langle 110 \rangle$.

axis of the rectangular cross section must be oriented on the $\langle 100 \rangle$ or $\langle 110 \rangle$ crystallographic axes of the initial crystal.

If the front is circular, the plateau height in both cases should be the same for a given overvoltage; if the front is square, with orientation in one of the $\langle 100 \rangle$ or $\langle 110 \rangle$ directions, the height should relate to the previous as $1 : \sqrt{2}$. Experiment shows that the front is polygonized, with orientation in the $\langle 110 \rangle$ direction (Fig. 8). This solves for the front length, and then $\varkappa$ in (8) can be calculated exactly. For a 6 N solution of $AgNO_3$ at 45°C

$$\varkappa = 2 \cdot 10^{-4} \text{ ohm}^{-1} \cdot \text{cm}^{-1}.$$

The layer is monotonic and the height h is known (h = $2.04 \cdot 10^{-8}$ cm), so $\varkappa$ can be calculated, together with the monolayer propagation speed v corresponding to a constant propagation rate $\varkappa = v/\eta$:

$$\varkappa_v = \varkappa \frac{v_M}{Fh} \quad [\text{cm} \cdot \text{sec}^{-1} \cdot \text{V}^{-1}], \quad \varkappa_v = 1.04 \text{ cm} \cdot \text{sec}^{-1} \cdot \text{V}^{-1}. \tag{9}$$

Growth of Dislocation-Free Faces at High Current Densities [15]

High current densities produce overvoltage fluctuations analogous to those at low densities, except that the oscillations are of transitional character and gradually die away towards a constant overvoltage (Fig. 9). Here again, the product of current and oscillation period is constant, and corresponds to the amount of electricity equivalent to a monolayer. The effect

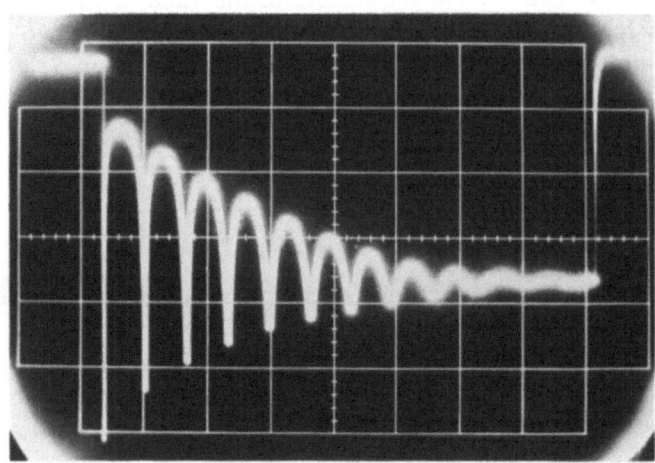

Fig. 9. Damped overvoltage oscillations for a dislocation-free (100) face at a high current density i = $5.5 \cdot 10^{-2}$ A/cm²; Y axis scale 5 mV/cm, X scale 5 msec/cm.

is clearly related to formation and propagation of a large number of nuclei, with several levels for their networks, which leads to gradual attainment of a steady-state profile on the surface. The two-dimensional growth of such a set of nuclei was first considered in [16]. More accurate equations for the steady state were given later in [17], and a mathematical simulation was presented in [18]. The equation for the electrochemical model was first derived by Armstrong and Harrison [19]; this gives the current density in A/cm² in the steady state as defined by

$$i_{\text{stat}} = q_{\text{mon}} \beta I^{1/3} v^{2/3} \ [\text{A/cm}^2].\tag{10}$$

where I is the nucleation rate in $\text{cm}^{-2} \cdot \text{sec}^{-1}$, v in cm/sec is the speed of the monotonic steps, and q_{mon} is the amount of electricity needed to deposit a layer in C/cm², while β is a constant that varies within narrow limits for the different theories.

It was pointed out in these studies that one gets slowly damped oscillations in the initial period, the number of which is related to the number of layers needed to produce a steady profile; this number has been determined roughly as 8-10, and experiment (Fig. 9) agrees closely with this.

If the nucleation rate is represented by (5), and if we use (8) for the speed, as determined by experiment, then (10) may be put as

$$\log i_{\text{stat}} \eta^{-2/3} = \log q_{\text{mon}} \beta k_1^{1/3} \varkappa_v^{2/3} - k_2/2.3 \cdot 3\eta.\tag{11}$$

A plot of $\log i_{\text{stat}} \eta^{-2/3}$ against $1/\eta$ serves to check the theory; Fig. 10 gives the observed relationship. The slope of the straight line gives k_2, while [7] gives the specific edge energy. The result is in complete agreement with the ε found from the two-dimensional nucleation behavior:

$$\varepsilon = 2.2 \cdot 10^{-6} \ \text{erg/cm}.$$

Figure 10 also gives k_1, and this is larger than that found from the nucleation rate by more than 4 orders of magnitude (1.5). The essential difference in the experiments in the two cases is that I refers to the steady state when many nuclei are formed, while we have the transient state for two-dimensional nucleation. Corrections from Zel'dovich's theory do not result in a change in the frequency factor by as many orders as that found by Stoyanov [6].

A second difference arises from aging of the surface; when there are only single nuclei, the newly formed surface exists for several seconds before fresh nucleation, while in the polynuclear method the time is only a few milliseconds. It may be that aging involving adsorption may here play a substantial part.

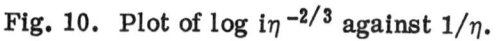 Fig. 10. Plot of $\log i \eta^{-2/3}$ against $1/\eta$.

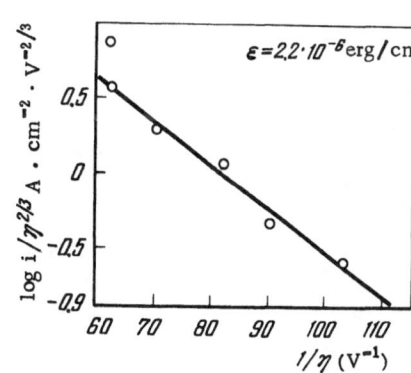

Growth in the Presence of Screw Dislocations

Theoretical Introduction

When there are screw dislocations, two-dimensional nuclei are not obligatory for growth, since each dislocation emerging on the face of the equilibrium form is a source of a step; it has been shown [20] that a growing step twists into a spiral and provides a sufficient number of growth sites. The spiral may be circular or polygonal, in accordance with the anisotropy in the propagation rate of the step.

In the optical microscope, these spiral formations are seen as cones or pyramids; the theory due to Barton, Cabrera, and Frank indicates that the distance between two successive steps is $\alpha = 4\pi\rho_c$, where ρ_c is the radius of a two-dimensional nucleus for a given supersaturation. When one has polygonized square spirals, one can take the factor $4\pi\rho_c$ as approximately 8, neglecting the length dependence of the front propagation speed [21].

The propagation speed is less at the center of a spiral on account of the short step length, so the factor may be considerably larger than 8; in any case, the distance between two successive turns of the spiral is proportional to the dimensions of a two-dimensional nucleus, as determined by the radius ρ_c of the circle inscribed in the nucleus:

$$d = b\rho_c = \frac{bv_M\varepsilon\cdot10^7}{zFh\eta} \quad \text{[cm]}, \tag{12}$$

where $b \geq 8$ for a polygonized square spiral and $b = 4\pi\rho_c$ for a circular spiral; v_M in cm^3/mole is the molar volume, while h is the step height in centimeters.

One can assume that (8) applies also for spiral growth, where one expects a deviation from this only in response to closely spaced steps in the path of the material transport. The total step length per unit surface L_s, in cm/cm^2, is inversely proportional to α, and it can be derived from the known quadratic relationships of the current (growth rate) to the overvoltage (supersaturation) as derived by Barton et al.:

$$i = B\eta^2, \tag{13}$$

where

$$B = \frac{\varkappa hzF}{bV_M\varepsilon\cdot10^7} \quad \text{[A}\cdot\text{V}^{-1}\cdot\text{cm}^{-2}\text{]}. \tag{14}$$

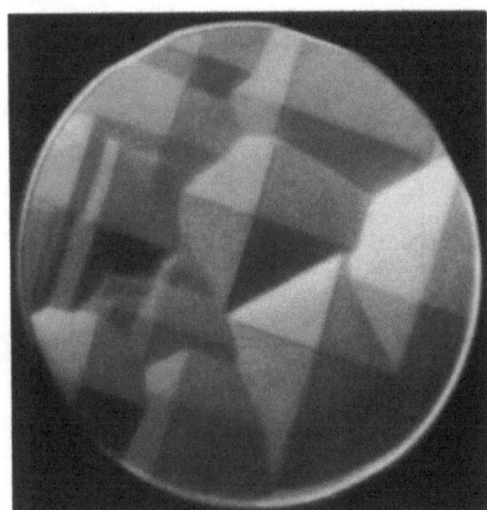

Fig. 11. Growth pyramids with square bases on a (100) face under steady conditions.

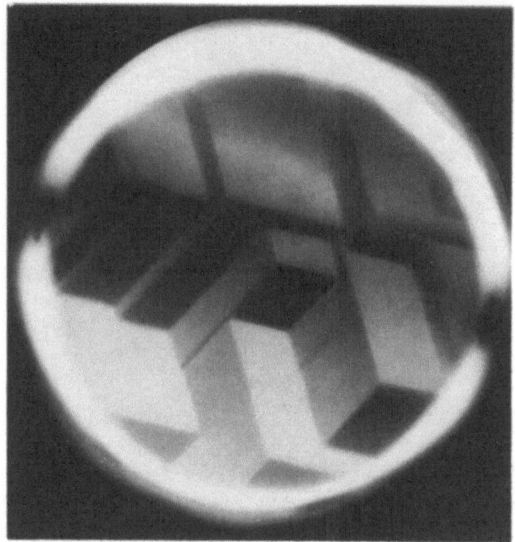

Fig. 12. Growth pyramids with triangular bases on a (111) face under steady conditions.

Growth and Dissolution Morphology of Faces with Screw Dislocations [8, 22-24]

Spiral growth theory implies that screw dislocations will produce pyramids on the surface; these pyramids are square on a cubic face (Fig. 11), while they are triangular on an octahedral (Fig. 12). The following experimental evidence indicates a relationship between the growth profile and the screw dislocations:

1. The steepness of the pyramid is dependent on the current, i.e., the overvoltage; if the current is low, the face becomes so smooth that the growth pyramids are scarcely visible.

2. When a cathode-polarity pulse is applied, one gets isolated growth pyramids at points where the vertices of pyramids were previously produced at a constant growth current (Fig. 13a).

3. If a face is smoothed by a low current and an anode-polarity pulse is applied, one gets etch pits at the same point (Fig. 13b). These pits expand and become deeper (Fig. 13c) if the pulse is repeated.

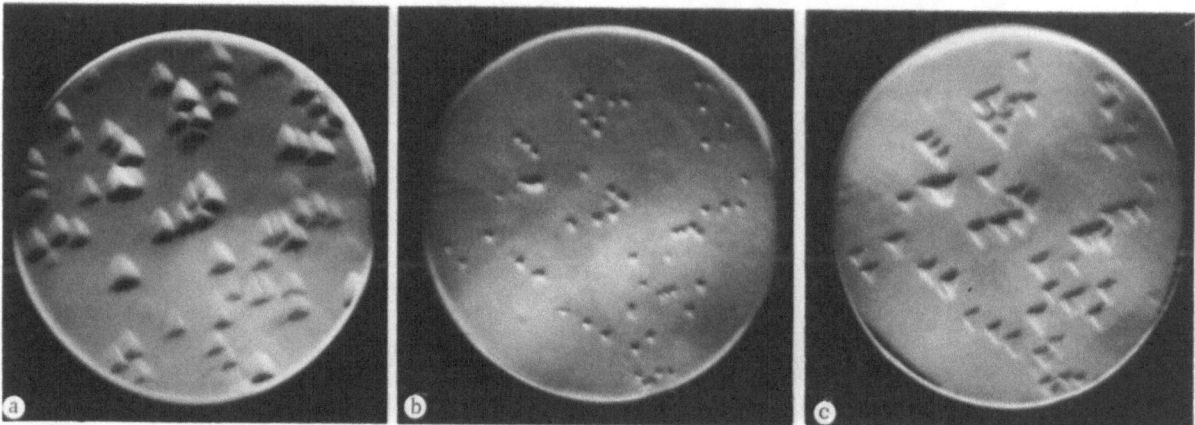

Fig. 13. Comparison of cathode and anode pulses: a) isolated growth pyramids on (111) after a cathode pulse; b, c) etch pits on the same face after an anode pulse. The face had been electropolished at a low current before use.

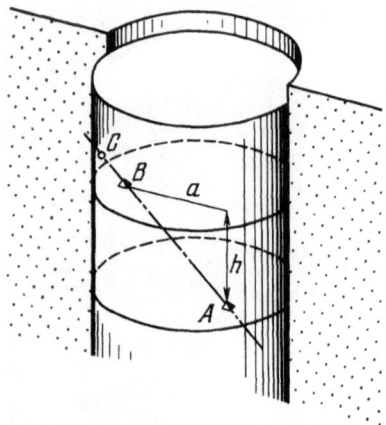

Fig. 14. Definition of the orientation of the Burgers vector of a screw dislocation during growth. The point of intersection between the screw dislocation and the surface of the face moves from A to B and vanishes at C.

The cathode pulse method can be used to locate the points of emergence of screw dislocations, and hence to determine the distribution of these on the crystal; the same method can be used to examine the motion of these over the surface during the growth. Figure 14 shows the essence of the method. If the point of emergence lies at A in the initial state, while it is displaced to position B after deposition of a layer of silver of thickness h, then the direction and magnitude of the displacement a at the surface of the face enable one to determine the direction of motion of the screw dislocation during the growth.

Figure 15 shows the shifts in the emergence points on a cubic face; each successive disposition of the points relates to deposition of 14 μm of silver. The theory indicates, and experiment shows, that the direction of propagation is close to that of the most suitable Burgers vector for silver, namely $\langle 110 \rangle$.

Relation of Pyramid Slope to Overvoltage [23, 24]

We have already shown that a previously smooth face containing a few screw dislocations gives rise to individual growth pyramids in response to a rectangular current pulse; the amount of deposited silver is easily determined from the length and duration of the pulse, and this

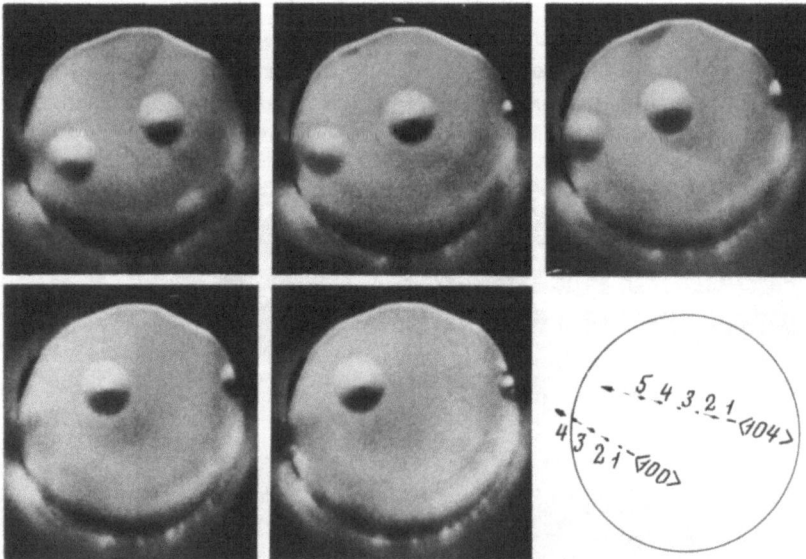

Fig. 15. Displacement of the points of emergence of screw dislocations on a (100) face during growth. Photos from left to right made at the end of sequential deposition of 14-μm silver films.

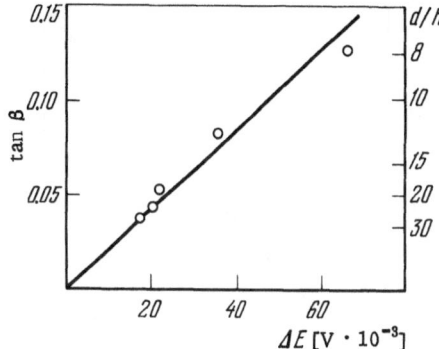

Fig. 16. Pyramid slope (tan β) as a function of η.

goes to form these pyramids. The volume of the pyramids is consequently known, while the base size can be measured directly under the microscope, which gives one the slope.

Figure 16 shows the observed relationship of the slope tan β to overvoltage; spiral growth theory gives

$$\tan \beta = \frac{h}{d} = \frac{Fh^2\eta}{bV_M\epsilon\cdot 10^{-7}}.$$ (15)

The slope of this straight line gives $b\epsilon = 17.6 \cdot 10^{-6}$ erg/cm, which with b = 8 gives ϵ = $2.2 \cdot 10^{-6}$ erg/cm, which at first appears to be good agreement, but which rather is to be considered as accidental, since various effects are involved in the experiment, and these cannot be evaluated quantitatively. Difficulties arise particularly with determination of the effective value of η, and also from some loss of current or electricity at the face parts not covered by the growth pyramids.

Spiral Growth Front Speed Determination [25]

If ordinary microscopy is employed, the individual fronts forming a spiral cannot be seen; the distance d between individual turns is dependent on the overvoltage, and this can be used to determine the shape and propagation speed for spiral fronts. If a short current pulse is applied under steady-state conditions (EEG, at a constant current), then a zone of more closely spaced steps is formed near the vertex of each pyramid, which is seen under the microscope as a zone differing in illumination. During growth this zone expands from the vertex to the base as a band of different slope, which is visible under the microscope as a band differing in intensity. The speed of this band clearly coincides with the propagation speed of the spiral fronts, and it can be determined directly. Measurements were made on cubic faces of silver in 6 N AgNO$_3$ solution at 45°C to give

$$\varkappa_V = \frac{V}{\eta} = 0.92 \ \text{cm} \cdot \text{sec}^{-1} \cdot \text{V}^{-1}.$$

whence we have as follows for $\varkappa$ from (9) on the assumption that the spirals are monotonic:

$$\varkappa = 1.77 \cdot 10^{-4} \ \text{ohm}^{-1} \cdot \text{cm}^{-1}$$

Steady-State Growth Rate with Screw Dislocations [23-25]

Spiral growth theory indicates that the current should be a quadratic function of the overvoltage; experiment is hindered by the fact that various polarization effects arise with standard methods of measuring the electrode potential, such as concentration polarization and voltage drop in the electrolyte. If the changes in the potential are small, we can assume that

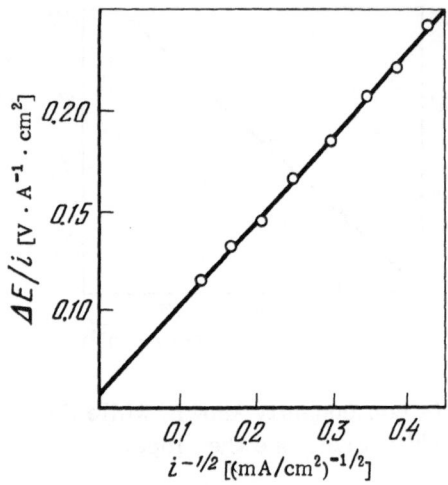

Fig. 17. Observed plot of $\Delta E/i$ against $i^{-1/2}$ for steady-state growth on a (100) face by the dislocation mechanism.

the effects sum linearly in the total potential change:

$$\Delta E = \eta + \eta_c + \eta_\Omega, \tag{16}$$

where η is overvoltage and η_c and η_Ω are the concentration polarization and voltage drop in the electrolyte, the latter two being linearly dependent on the current (in view of the low polarization potential), while η varies as $\sqrt{i}$, as given by (11), so (16) can be put as

$$\frac{\Delta E}{i} = \frac{1}{\sqrt{B}} \cdot i^{-1/2} + R_p, \tag{17}$$

from which one can determine the constant B by plotting $\Delta E/i$ as a function of $i^{-1/2}$ (Fig. 17). Here R_p is the constant resistance due to all the polarization effects linearly dependent on the current, including the resistance of the electrolyte. For a cubic face of silver in 6 N AgNO$_3$ at 45°C we get

$$B = 6.7 - 7.7 \cdot 10^3 \ \text{A} \cdot \text{V}^{-1} \cdot \text{cm}^{-2}.$$

The constant B of (14) involves the specific conductivity $\varkappa$ of the steps and the specific edge energy; the geometrical factor B can be derived from $\varkappa$ and ε derived previously by other methods. The value of b found for a cubic face is 24, which is in principle acceptable, but which is difficult to interpret at the present stage of development of the theory.

Determination of Step Density [26, 27]

One can use the quadratic dependence of the current density on the overvoltage to determine the step density in spiral growth; from (8) and (13), which are experimental relationships, it follows that the total step length per cm^2 of surface, i.e., the growth step density N_s, is determined by

$$N_S = L_S = \frac{B\eta}{\varkappa} \quad [\text{cm}^{-1}]. \tag{18}$$

For a cubic face with $B = 7.2 \cdot 10^3 \ \text{A} \cdot \text{V}^{-1} \cdot \text{cm}^{-2}$ and $\varkappa = 1.77 \cdot 10^{-4} \ \text{ohm}^{-1} \cdot \text{cm}^{-1}$ we get

$$N_S = L_S = 4.1 \cdot 10^7 \eta \quad [\text{cm}^{-1}].$$

It is important for a quantitative study of electrodeposition kinetics and mechanisms to be able to determine step densities for single crystallographic faces in relation to production conditions, or more precisely in relation to the overvoltage giving the surface relief.

Electrochemical Deposition Mechanism for Silver

Theoretical Introduction

A second topic in crystal growth kinetics involves the propagation and growth mechanism of monatomic layers formed by dislocation or two-dimensional means. Two essentially different incorporation mechanisms can be envisaged theoretically:

(a) Direction incorporation, in which the ions from the solution are discharged directly at the growth points, with simultaneous incorporation into the lattice, as first suggested by Volmer [28].

(b) The surface-diffusion mechanism, in which the ions are discharged initially at an arbitrary surface part of the crystal; then they take the form of adsorbed ions, which diffuse to the growth points and are incorporated into the lattice.

Brandes [29] was the first to suggest that surface diffusion is the rate-limiting step in electrochemical deposition of metals; the first theoretical study is found in [30], where the theory due to Barton et al. [20] was applied to electrical crystallization. Subsequent studies are to be found in [31-33], in which the theory was developed to its existing level.

This theory gives the current as a function of overvoltage as

$$i = i_{0,\,ad}\left[e^{\frac{\alpha z F \eta}{RT}} - e^{-\frac{(1-\alpha)z F \eta}{RT}}\right]\frac{\lambda_0}{x_0}\tan\frac{x_0}{\lambda_0}, \tag{19}$$

where

$$\lambda_0 = (z F D_S c_{,\,ad}^0/i_{0,\,ad})^{1/2}e^{-\frac{\alpha z F \eta}{2RT}} \quad [\text{cm}] \tag{20}$$

is the penetration depth for surface diffusion, D_S in cm^2/sec is the surface diffusion coefficient, c_{ad}^0 in $moles/cm^2$ is the equilibrium anion concentration, $i_{0,ad}$ in A/cm^2 is the ion exchange current involving the anions, and $2x_0$ in cm is the mean distance between two growth steps on the surface. This distance can also be expressed via the step density N_S in cm^{-1} and via the total step length L_S in cm/cm^2 on unit surface:

$$\frac{1}{2x_0} = N_S = L_S. \tag{21}$$

Here there are two limiting cases: (1) the penetration depth is greater than the mean distance between growth steps: $\lambda_0 \gg x_0$; then (9) coincides with the Butler−Volmer equation, i.e., the rate is determined by ion discharge; (2) the relationship is $\lambda_0 \ll x_0$, and (19) becomes

$$i = i_{0,\,ad}\left[e^{\frac{\alpha z F \eta}{RT}} - e^{\frac{-(1-\alpha)\,z F \eta}{RT}}\right]2\lambda_0 N_S. \tag{22}$$

Surface diffusion here plays an important part. The current density affects very greatly the state of the face, and especially N_S.

The direct incorporation mechanism was first suggested by Volmer [28] and was subsequently discussed in other papers by various workers [32, 34, 35], etc.

It is comparatively easy to derive a quantitative expression given in [36]; if we use the symbol $i_{0,\,st}$ for the density of the exchange current between the ions in the solution and the atoms at the growth steps, we get for the current density i that

$$i = i_{0,\,st}\delta L_S\left[e^{\frac{\alpha z F \eta}{RT}} - e^{\frac{-(1-\alpha)\,z F \eta}{RT}}\right], \tag{23}$$

where δL_S is the part of the face surface on which the transition occurs, while δ is the width in centimeters of one row of atoms incorporated in the growth step.

If the supersaturation is small ($\eta \ll RT/zF$), (22) and (23), which correspond to the two mechanisms, give a linear relationship of current to η and step density $N_S = L_S$, provided that in the first case $\lambda_0 \ll x_0$:

$$i = \varkappa L_s \eta,$$

for

$$\varkappa = i_{0,\,ad} \cdot 2\lambda_0 \frac{zF}{RT} \tag{24}$$

for surface diffusion and

$$\varkappa = i_{0,\,st} \cdot \delta \frac{zF}{RT} \tag{25}$$

for direct incorporation.

It is clearly impossible from steady-state polarization curves to determine which of the mechanisms is the dominant one; in both cases, the current for low overvoltages is linearly dependent on the latter and is proportional to the total length of the growing front. In the case of surface diffusion, this is true only while λ_0 remains less than x_0, namely half the distance between two adjacent steps.

Consequently, the experimental linear law of (8), and the quadratic relationship observed with screw dislocations, namely (13), cannot be used to solve this problem; but all the same we can say that if surface diffusion is the decisive mechanism, then λ_0 will be less than half the distance between turns throughout the range of overvoltage for which the quadratic relationship applies. The quadratic law has been demonstrated for a cubic face of silver and overvoltages in the range $0-3 \cdot 10^{-3}$ V, for which (18) gives $d = 8 \cdot 10^{-6}$ cm, which implies that λ_0 must be less than $4 \cdot 10^{-6}$ cm. If this is not so, the quadratic law should give way to a linear law $i = \text{constant} \cdot \eta$, where the current is not dependent on the surface profile.

A decision on the mechanism for silver can be obtained only by determining $i_{0,ad}$, $i_{0,st}$, and λ_0, and then comparing the result for $\varkappa_v$ with the values obtained by experiment.

Determination of the Anion Exchange Current in

an Electrolyte [37, 38]

To determine $i_{0,ad}$ one can measure the impedance given by dislocation-free faces at high frequencies, where the behavior of such a face may be represented by an equivalent circuit consisting of parallel frequency-independent resistance R_p and capacitor C, in series with which is the electrolyte resistance R_{el}. The parallel resistance R_p corresponds to the resistance R_F in the Faraday reaction, while C corresponds to the capacitance C_D of the double electrical layer.

In that case, one can calculate the exchange current from R_F:

$$i_{0,\,ad} = \frac{RT}{F} \cdot \frac{1}{R_F}. \tag{26}$$

The following values were obtained for a cubic face of silver in 6 N $AgNO_3$ at 45°C:

$$i_{0,ad} = 0.06 \ \text{A/cm}^2; \quad C_D = 30 \ \mu\text{F/cm}^2.$$

Determination of Ion Exchange Current with Atoms

in Growth Step Edges [26, 27]

We have seen above that the spiral step density on a face containing growth pyramids (arising from screw dislocations) is dependent on the overvoltage that produced the profile; the step density can be calculated from (18). Impedance measurements on such faces indicate that the resistance in the Faraday reaction is very much dependent on the step density (column 4 of Table 1). A surface element of new form such as a step edge will make a distinctive contribution to the total exchange current, which means that this current has two components: the exchange current with the anion $i_{0, ad}$ and the exchange current $i_0(L)$ with the atoms in the step edges:

$$i_0 = i_{0,ad} + i_0(L). \tag{27}$$

The exchange current amplified in this way should be proportional to the step density; column 5 of Table 1 gives $i_0(L)$ for various step densities, and (18) shows that L_S is proportional to η so $i_0(L)/\eta$ should be constant, and Table 1, column 5, shows that there is proportionality between $i_0(L)$ and L_S, while column 6 shows that $i_0(L)/\eta$ is constant throughout the overvoltage range used. The following is the mean value of $i_0(L)/\eta$ derived from several experiments:

$$\frac{i_0(L)}{\eta} = 0.21 \cdot 10^3 \ \text{A} \cdot \text{V}^{-1} \cdot \text{cm}^{-2}.$$

The exchange current for the step atoms may be calculated from

$$i_{0, st} = \frac{i_1(L)}{\delta L_S}. \tag{28}$$

The product δL_S is the part of the face involved in the exchange, where L_S is the total length of steps per unit surface of the face and δ is the width of one row of atoms in a step. For a cubic face of silver, $\delta = 2.88 \cdot 10^{-8}$ cm.

It follows from (18) and (28) that

$$i_{0, st} = \frac{i_0(L) \varkappa}{\eta B \delta}, \tag{29}$$

and then from the values for $i_0(L)/\eta$, B, and $\varkappa$ we have

$$i_{0,st} = 180 \ \text{A/cm}^2.$$

This value is larger by three orders of magnitude than $i_{0,ad}$; the exchange currents at different points on the face have been considered [39, 40], but the subject is hardly clear enough for a detailed discussion here.

Silver Deposition Mechanism under Steady-State

Conditions

This mechanism can be established by comparing the calculated values for $\varkappa_v$ and B in the quadratic law $i = B\eta^2$ with the observed values.

Column 6 of Table 1 gives the constant layer propagation rate derived from the propagation rates of single monotonic layers and from the propagation rates of spiral steps (column 4 of Table 2), and these can be compared with the calculated value from (24) or from (9) and (25):

$$\varkappa_v \lesssim \frac{2\lambda_0 V_M}{RT\eta} i_{0, ad} \quad \text{(surface diffusion)}, \tag{30a}$$

$$\varkappa_v = \frac{\delta V_M}{RTh} \cdot t_{0, st} \quad \text{(direct incorporation)}. \tag{30b}$$

Table 2 gives the values for $\varkappa_v$ derived from λ_0, $i_{0,ad}$, and $i_{0,st}$.

TABLE 2. Values of $\varkappa_V$ (Propagation Constant) and of B in the Quadratic Law

Mechanism	$\varkappa_V$, cm $\cdot$ sec^{-1} $\cdot$ V^{-1}	Coefficient B in quadratic relation, A $\cdot$ V^{-2} $\cdot$ cm^{-2}
	From (30)	From (33)
Surface diffusion	0.094	$0.74 \cdot 10^3$
Direct incorporation	1.01	$7.8 \cdot 10^3$
	Observed	
Monatomic layer propagation	1.04	-
Spiral growth	0.92	$6.7\text{-}7.7 \cdot 10^3$

Note: Cube face of a silver single crystal in 6 N AgNO$_3$ at 45°C.

The value of B in $i = B\eta^2$ has been found by experiment for cubic faces with small numbers of dislocations, and it can also be calculated for the two deposition mechanisms via $i_{0,ad}$ and $i_0(L)/\eta$ as given by the impedance method.

We get from (22) after linearization for small η and substitution for N_S from (18) that

$$i = i_{0,\,ad} \frac{F}{RT} \cdot 2\lambda_0 \frac{Fh \cdot \eta^2}{V_M b\varepsilon \cdot 10^{-7}} \cdot \tag{31}$$

The observed quadratic dependence of the current on the overvoltage indicates that the surface diffusion penetration depth, if relevant to the deposition, is less than the minimum value of half the distance between the spiral steps in the working overvoltage range:

$$\lambda_0 \leqq \frac{1}{2} d_{\min} = \frac{V_M b\varepsilon \cdot 10^{-7}}{2Fh} \cdot \frac{1}{\eta_{\max}}, \tag{32}$$

where $\eta_{\max}$ is the maximal overvoltage for which the observed $i = B\eta^2$ has been demonstrated by experiment. With this λ_0 we get from (31) that

$$i = B_{\mathrm{sd}} \cdot \eta^2 \leqq \frac{i_{0,\,ad}}{\eta_{\max}} \cdot \frac{F}{RT} \eta^2 \quad \text{(surface diffusion)}. \tag{33a}$$

One can calculate B for direct incorporation from (23) as linearized for small values by replacing L_s from (18) and $i_{0,st}$ from (29):

$$i = B_{\mathrm{di}} \cdot \eta^2 = \frac{l_0(L)}{\eta} \cdot \frac{F}{RT} \eta^2 \quad \text{(direct incorporation)}. \tag{33b}$$

Table 2 gives values calculated from (33a) and (33b) for B for two deposition mechanisms.

These two observed results have been obtained from the propagation speeds of single layers and the growth rates of spirals, and they require minimum material fluxes at least ten times those that at best ($\lambda_0 = 400$ Å) can be provided by surface diffusion; on the other hand, the flux of material obtained from the direct incorporation mechanism appears quite sufficient to provide the observed growth rate, so the question is decided in favor of direct incorporation for electrolytic growth of silver from concentrated solutions of silver nitrate.

Nanev and Kaishev have recently examined the kinetics of anode dissolution of silver under the same conditions, and the results provide a confirmation of the dominant effects of direct incorporation here also.

Conclusions

The results so far described for electrodeposition of silver may be formulated as follows:

1. The exchange current for metal ions in a 6 N solution of $AgNO_3$ at 45°C as against adsorbed ions and atoms on a cubic face of silver ($i_{0,ad} = 0.06$ A/cm^2) is less by at least three orders of magnitude than the exchange current for metal ions from the solution with atoms in step edges on the same face ($i_{0,st} = 180$ A/cm^2).

2. The penetration depth for surface diffusion, if it makes any contribution to the overall process, is in any case less than 400 Å.

3. The growth rate is adequately explained by the direct incorporation mechanism, and the contribution from surface diffusion (which has not been shown to be present) is less by at least an order of magnitude.

Literature Cited

1. R. Kaishev. Ezh. Sol. Univ., Fiz.-Mat. Fak., 42(2):109 (1946).
2. E. Budevskii, V. Bostanov, T. Vitanov, Z. Stoinov, A. Kotseva, and R. Kaishev. Elektrokhimiya, 3:856 (1967).
3. M. Volmer. Kinetik der Phasenbildung, Steinkopf-Verlag, Dresden (1939).
4. R. Becker and W. Döring. Ann. Phys., 24:719 (1935).
5. R. Kaishev and I. Stranski. Z. Phys. Chem., 26B:317 (1934).
6. St. Stoyanov. Izv. Otd. Khim. Nauk Bolg. AN, 3:491 (1970).
7. E. Budevskii and W. Bostanov. Electrochim. Acta, 9:477 (1964).
8. V. Bostanov, A. Kotseva, and E. Budevskii. Izv. Inst. Fiz. Khim. Bolg. AN, 6:33 (1967).
9. E. Budevskii, T. Vitanov, and W. Bostanov. Phys. Status Solidi, 8:369 (1965).
10. E. Budevskii, W. Bostanov, T. Vitanov, Z. Stoinov, A. Kotseva, and R. Kaishev. Electrochim. Acta, 11:1697 (1966).
11. E. Budevskii, W. Bostanov, T. Vitanov, Z. Stoinov, A. Kotseva, and R. Kaishev. Phys. Status Solidi, 13:577 (1966).
12. V. Bostanov and R. Rusinova. Private communication.
13. V. Bostanov, R. Rusinova, and E. Budevskii. Proceedings of the Fourth All-Union Conference on Crystal Growth [in Russian] (1972), p. 215.
14. V. Bostanov, R. Rusinova, and E. Budevskii. Chemie-Ingr.-Techn., Berlin, 93 (1971).
15. V. Bostanov, R. Rusinova, and E. Budevskii. J. Electrochem. Soc., 119:1346 (1972).
16. W. B. Hillig. Acta Metallurgica, 14:1868 (1966).
17. L. Borovinskii and A. Tsindergozen. Dokl. AN SSSR, 183:1308 (1968).
18. U. Bertocci. Surface Sci., 15:286 (1969).
19. R. Armstrong and J. Harrison. J. Electrochem. Soc., 116:328 (1969).
20. W. K. Burton, N. Cabrera, and F. C. Frank. Philos. Trans. Roy. Soc. London, A243:299 (1951).
21. R. Kaishev, E. Budevskii, and J. Malinovski. Z. Phys. Chem., 204:348 (1955).
22. V. Bostanov, Chr. Nanev, R. Kaishev, and E. Budevskii. Kristall und Technik, 2:319 (1967).
23. R. Kaishev and E. Budevskii. Contemporary Phys., 8:489 (1967).
24. E. Budevskii, V. Bostanov, T. Vitanov, Kh. Nanev, Z. Stoinov, and R. Kaishev. Izv. Otd. Khim. Nauk Bolg. AN, 2:479 (1969).
25. V. Bostanov, R. Rusinova, and E. Budevskii. Izv. Otd. Khim. Nauk Bolg. AN, 2:885 (1969).
26. T. Vitanov, A. Popov, and E. Budevskii. J. Electrochem. Soc. (1973).
27. E. Budevskii. J. Crystal Growth, 13/14:93 (1972).
28. M. Volmer. Das elektrolytische Kristallwachstum. Paris, Herman (1934).
29. H. Brandes. Z. phys. Chem., 142A:97 (1927).

30. W. Lorenz. Z. Naturforsch., 9a:716 (1954).

31. D. Vermilyea. J. Chem. Phys., 25:1254 (1956).

32. M. Fleischmann and H. R. Thirsk. Electrochim. Acta, 2:22 (1960).

33. A. Damjanovic and J. O'M. Bockris. J. Electrochim. Soc., 110:1035 (1963).

34. H. Gerischer. Protection against Corrosion by Metal Finishing, Forster Verlag, Zürich, (1967), p. 11.

35. N. Mott and R. Watts-Tobin. Electrochim. Acta, 4:79 (1961).

36. C. Feather: Electrochemical Kinetics [Russian translation], Khimiya, Moscow (1967).

37. T. Vitanov, E. Sevast'yana, V. Bostanov, and E. Budevskii. Elektrokhimiya, 5:451 (1969).

38. T. Vitanov, E. Sevast'yana, Z. Stoinov, and E. Budevskii. Elektrokhimiya, 5:238 (1969).

39. B. Conway and J. O'M. Bockris. Proc. Roy. Soc., A243:394 (1958).

40. A. Despic. Croat. Chim. Acta, 42:265 (1970).

THERMAL APPROACH OF ATOMS, PHASE TRANSITIONS, AND POLYTYPES*

A. A. Shternberg

The thermodynamic description of phase transitions leaves unanswered questions of the molecular kinetics, and this means that many aspects cannot be fully elucidated.

The most valuable information on this aspect of phase transitions comes from variations in interatomic distances, which reflect the forces between atoms.

The interatomic distances increase with temperature in the stability range of a given state of aggregation; it has been found [1-3] that transitions to higher-temperature states of aggregation always involve reduction in some of the interatomic distances (Table 1), apart from evaporation of the inert gases. This is the thermal approach between atoms in phase transitions, and we examine how it occurs. The energy of the thermal vibrations increases with temperature, and the interatomic distances increase, while the bond strengths are reduced, especially in relation to the increasing energy of the thermal vibrations. However, bond breakage is replaced by the substitution of shorter stronger bonds, which can exist at the given temperature. As a consequence, the coordination number is increased, i.e., the number of near bonds is reduced. Strengthened compact atomic groups arise, and the links between these groups are weakened or completely broken (melting and evaporation). Consequently, the thermal approch between atoms can be considered as a phenomenon directly preceding or accompanying melting.

Then phase transitions caused by temperature change are two-stage processes involving on the one hand approach between atoms in groups and on the other weakening of the links between groups. As a result, the overall volume change in a phase transition can have either sign or be zero. The high-temperature phase nucleates via a process of approach between the atoms to form pairs and triplets; this occurs readily. The converse processes involve moving apart these closer groupings, which is more complex and hence the nucleation probability is small for low-temperature phases.

Shortening of some of the interatomic distances is frequently observed also when the crystalline phase gives way to a high-temperature polymorphic modification (Table 2). In cer-

---

*The paper extends ideas presented by the author in Vol. 5A of this series (pp. 146-151, 1968) and in Vol. 9 (pp. 34-39, 1975). Although the points are debatable, the paper appears to be of considerable interest. The author's views derive largely from his conception of the important part played in growth processes not only by the Volmer adsorption layer (transition from the medium to the crystal) but also by a second transition layer in the crystal itself, which has a quasicrystalline or slightly liquid-type structure, in contrast to the deeper truly crystalline layers (Editor).

TABLE 1

| Substance | Interatomic distances (Å) and coordination numbers (CN) | | | | | |
| | in crystal | | in melt | | in vapor | |
	d	CN	d	CN	d	CN
LiCl	2.66	6	2.46	3.7	2.04	1
KCl	3.26	6	3.10	3.6	2.67	1
LiBr	2.85	6	2.68	5.2	2.17	1
CsI	3.94	8	3.85	3.5	2.32	1
MeH *	3.10	6(8)	2.95	4.5	2.55	1
Cu	2.556	12	-		2.22	1
Ag	2.884	12	-		2.53	1
Ar	2.495		-		2.44	1

*Mean for 48 halides; Me — metal; H — halogen.

tain instances, direct observations have been shown that the structure change arising from the approach between atoms is the direct cause of the phase transition [4-6]. It is of interest from this viewpoint to consider polymorphism and polytypism in the common structural transitions of sphalerite—wurtzite type, which have recently attracted much attention [7, 8].

Both of these structures involve anion close packing; we have either a three-layer cubic structure ...(ABC/ABC)... or else a two-layer hexagonal one ...(AB/AB)...; the cations fill half of the holes in each layer, namely those over which the anions of the next layer lie, these holes thus becoming tetrahedral gaps between spheres [9].

The cubic modification goes over to the hexagonal one by adjustment of four of each six layer pairs (sulfur and zinc) in the networks perpendicular to one of the three-fold symmetry axes, this occurring in the direction belonging to $\langle 112 \rangle$ and involving a change of 1/3 lattice parameter. The hexagonal close packing differs from the cubic one in that the octahedral cavities lie one above the other and form channels along the c axis. Parallel to these lies one of the four sets of falling tetrahedra making up the structure of wurtzite. The other three lie at angles of about 71° to the channels [9, page 60]. Table 3 shows that this physical difference in height of the tetrahedra results in some degree of distortion as the temperature is raised. Observations show that the initial stages of thermal deformation of the tetrahedra occur in the cubic form of ZnS before the phase transition [4]. It is clear that approach between the atoms is associated with deformation of the tetrahedra, since the high-temperature hexagonal modifications are denser than the initial cubic ones for the compounds given in Table 3. Then the transition of the sphalerite modification to the high-temperature wurtzite one may be considered as a structural conversion that relieves the elastic stresses set up by the atomic approach. The reverse elastic deformation of wurtzite is by compression of the crystal along the c axis, and the properties become those of sphalerite [10].

The above deformation of the tetrahedra causes c_0/a_0 to deviate from the calculated value of 1.633; Table 3 gives the values of the difference. In CdS, ZnSe, SiC, and ZnS the dif-

TABLE 2

| Substance | Low-temperature form | | | Higher-temperature form | | |
	Name, type, structure	Å	CN	Name, type, structure	Å	CN
C	Diamond	1.542	4	Graphite	1.42	3
NH_4Cl	CsCl	3.55	8	NaCl	3.27	6
NH_4Br	CsCl	3.51	8	NaCl	3.45	6
NH_4I	CsCl	3.78	8	NaCl	3.65	6

TABLE 3

Substance	CdS		ZnSe		SiC			ZnS		
Structure, polytype	3C	2H	3C	2H	3C	6H	2H	3C	6H	2H
a_0	5.818		5.667		4.3596			5.412		
a_0	4.116*	4.136	4.007*	3.996	3.0822*	3.08065	3.0763	3.827*	3.824	3.8826
c_0	-	6.713	-	5.818	-	5.039**	5.048	-	6.242**	6.2605
c_0/a_0		1.622		1.624		1.636	1.641		1.634	1.638
$c_0/a_0 - 1.633$		-0.011		-0.009		0.003	0.008		0.001	0.003
$a_0 3c - a_0 2H$	-0.02		-0.011		0.006				-0.004	
h_1	3.361	3.356	3.272	3.265	2.517	2.520	2.524	3.125	3.124	3.124
h_2		3.388	3.263	3.263		2.515	2.513		3.122	3.118
$h_1 - h_2$		0.032		0.002			0.011			0.006
Sp. gr.	4.87	4.89			3.210	3.211	3.214	4.090	4.099	4.101

h_1 = heights of tetrahedra parallel to c axis (in hexagonal structure)
h_2 = heights of tetrahedra inclined at 71° to the c axis
$h_1 - h_2$ = deformation of tetrahedron
* = a_0 in hexagonal setting
** = $c_0/3$

ferences are small, and at ordinary temperatures they occur in the cubic modification and in the metastable hexagonal one.

We can compare the differences in the a_0 for the cubic form (in the hexagonal setting) and the hexagonal form; we denote the difference by $a_{0c} - a_{0h}$. In CdS and ZnSe, the values are considerable, and therefore it is unlikely that one will get epitaxial overgrowth of the different modifications in one crystal as occurs in polytypism. The differences are small for SiC and ZnS, and so polytypism is possible, with the cubic and hexagonal modifications coexisting in one crystal. Then polytypism is possible when the deformation of the structures is not too large; the phase transition lies a considerable way off, while the structure parameters in the interface planes are similar.

Any structure found in a polytype is merely the best of a set of a few allowed forms, and it imposes its own restrictions on the dispositions of the atoms; as our measure of the permissible discrepancy between the structures in a polytype we can use the extent of the temperature range of existence, or the range of existence for ferroelectric domains [11, page 30], or else the varying structure parameters. As $a_{0c} > a_{0h}$ in the polytypes of SiC and ZnS, the layers with cubic packing (C) in the (111) plane are somewhat compressed, while the hexagonal (H) ones are somewhat extended in the (0001) planes, and hence it is possible to combine the C and H layers by bringing the structure parameters into correspondence with the thermal deformation of the tetrahedra. Polytypism arises in a transitional temperature region, where the cubic modification has become unstable but the deformation of the tetrahedra is as yet insufficient to produce the hexagonal 2H phase. The positive or negative sign of the deformation determines whether a C or H layer will arise after n layers on the growing surface of the crystal; deformation accumulates when the stretched C layers or compressed H layers accumulate, or even when there are very simple combinations of these. For this reason, complex polytypes with long repeat distances are produced on the basis of simpler polytypes. The additional layers that complicate the structure of the polytypes have little effect on the parameters of the growing crystal, and bring them into correspondence with the thermal deformation of the tetrahedra. This has been confirmed by numerous statements as to good ordering of the structure in complex polytypes [7, page 253], and this is characteristic of stable growth conditions.

It has also been shown that the growth mechanism itself under certain temperature conditions can produce substantial deviations from the structures stable at a given temperature.

A structure already formed is stable against certain temperature variations, since any modification involves displacing 2/3 of the atoms and should involve as a minimum 6 atomic layers; intermediate states (polytypes) are particularly stable, since the thermal deformations for them are more restricted than for the 3C $\leftrightarrow$ 2H transition.

External factors such as compression or tension can produce elastic deformations in the crystals comparable with those produced by the thermal approach between atoms in this group of compounds, especially for SiC and ZnS; structural modifications may perhaps be activated by combining external deformation with suitable temperatures.

This occurs widely in nature. Crystals of ZnS growing under hydrothermal conditions or from gases differ from all other substances in growing free from cracks; the growth stresses result in twinning, packing defects, and change in polytypes, and these effects occur widely in SiC crystals and have been called syntactic coalescence [8, pages 27, 78].

All structural varieties of zinc sulfide at 1024°C go over to the 2H hexagonal form, which melts at 1850°C and 150 atm; on cooling, the reverse transition to the cubic modification is retarded and does not go to completion. Mixed or layered crystals are formed. The content of the hexagonal phase in these varies from 10 to 40%. The pure cubic modification is formed in the range 300-600°C under hydrothermal conditions. Gas-transport processes at 600-900°C produce polytypes, of which 50 are known. Single crystals of the 2H hexagonal phase are formed in nature as a metastable type at low temperatures, and also in marly clays [12], as well as on cooling pure saturated solutions of zinc sulfide in phosphoric acid or alkali [13].

The ZnS polytypes are produced in the range 600-1024°C, but below 100°C we get the metastable 2H hexagonal phase.

If we neglect the effects of impurities, we get a generally analogous picture for silicon carbide, but the polytypes are formed in a range from 2000°C to the melting (decomposition) point at 2830°C, so the polytypes 6H and 15H are formed in equilibrium with the melt, instead of the high-temperature 2H phase, and these polytypes contain 33 and 40% of the hexagonal phase [14]. In the range 1430-2800°C one gets the low-temperature 3C sphalerite phase. Below 1430°C, the wurtzite modification 2H of SiC is produced metastably [15]. However, below 1000°C we again get the low-temperature 3C phase [8, pages 106-110].

To elucidate these conflicting phase formations, we need to consider the initial stage of crystal growth incorporating the atom approach.

The thermal vibration amplitudes in the surface layers of a crystal are much larger than those in deeper layers [16], so here the thermal approach between systems should occur at lower temperatures; consequently, the surface may have a layer of high-temperature or metastable phase at any temperature. Crystals of nucleation consist entirely of surface, and therefore they arise as high-temperature modifications, and remain in that form up to certain critical sizes. Platelets [17] and needles [18] of ZnS have the 2H structure, while simultaneously growing larger crystals are polytypes or even sphalerite as a result of conversion of the surface structures to the stable phase. Similarly, silicon carbide whiskers of thickness less than 1 μm, as produced by reduction of SiO_2 by carbon at 2400°C, have the wurtzite structure [19], while bulk crystals are represented by polytypes. If the growth rate exceeds the rate of conversion to the stable modification, one first of all gets the high-temperature metastable phase, which subsequently, and often only when seeds are inserted, goes over to the stable form (Ostwald's rule).

When a crystal grows at low temperature, the conversion of the surface high-temperature layer to the stable modification may be completely inhibited, so one gets bulk crystals of the high-temperature modification; for this reason, wurtzite is found in clays, while crystals of 2H silicon carbide, which should be formed only at very high temperatures, in fact grow below 1430°C. However, temperature does restrict the scope for producing high-temperature modifications at the surface of the crystal also in the case of silicon carbide. For instance, one again gets the cubic phase, not the hexagonal 2H one, below 1000°C. These apparent contradictions complicate elucidation of polymorphism and polytypism, and further complexities are introduced by the major part played by impurities in these phenomena.

Impurities affect lattice parameters and consequently alter phase transition temperatures [20-22]. The conditions for phase transitions in ZnS and SiC are determined by the small thermal deformations of the tetrahedra; the parameters of the polymorphic forms are similar, and therefore even very small impurity concentrations or small deviations from stoichiometry can greatly displace the temperature ranges for stability of the basic modifications, and especially those for the intermediate polytypes, and the effects are larger than those that can be produced by temperature alone. For instance, excess of silicon provides empty undeformable tetrahedra in the crystal, and this stabilizes the cubic modification at any synthesis temperature [8, page 72]. Similarly, excess sulfur stabilizes the cubic modification of sulfides, while excess zinc goes to some of the free tetrahedral cavities and deforms the corresponding tetrahedra, thereby making the crystals more hexagonal [23, 24].

Impurities act in much the same way as deviations from stoichiometry: anionic ones increase the sum of the anions and hence the number of empty tetrahedra, so they stabilize the cubic modification; conversely, ones that act as cations increase the deformation of the tetrahedra and expand the stability region of the hexagonal phases. These features must be borne in mind in considering heterovalent substitutions.

Conclusions

The variations in interatomic distance with temperature show that transitions to high-temperature states of aggregation are accompanied by shortening of some of the distances, namely thermal approach between atoms. Therefore, phase transitions are to be considered as processes consisting of two stages: approach between the atoms and separation of groups. The overall volume effect may have either sign. The approach effect enables one to explain the ease of nucleation for high-temperature phases and the difficulty in producing low-temperature ones.

Many polymorphic transitions are also due to preliminary approach between atoms.

Polymorphic transitions of sphalerite —wurtzite have been analyzed on this basis via lattice-parameter data to solve various problems connected with this transition. For instance, in this way it is possible to do the following:

1. To relate the thermal deformations in the structures of compounds to the character of the transitions and to determine the conditions under which polytypism is possible.

2. To give an explanation of the long-range (1500 Å) information transmissions that determine the combination of polytypes.

3. To relate polymorphism or polytypism to isomorphism.

4. To explain the relationship between crystal size and stability of polymorphic forms.

5. To define the growth conditions for particular polymorphs of single crystals.

6. To explain the reasons for syntactic coalescence.

It is clear from this that the thermal approach between atoms enables one to relate various phenomena.

Thermal approach between atoms is amongst the important factors in structural transitions, along with effects such as rotation of groups and increase in interatomic distance, the latter tending sometimes to increase the crystal symmetry, as in oxides and borates.

The thermal conditions for occurrence of such effects may be established, for example, by precision X-ray measurement of interatomic distances, and this should provide new evidence for elucidating this category of effect.

I am indebted to Academician N. V. Belov and Professor N. N. Sheftal' for advice and assistance in this work.

Literature Cited

1. K. Meyer. Physicochemical Crystallography [Russian translation], Metallurgiya, Moscow (1972), p. 196.
2. E. A. Moelwyn-Hughes. Physical Chemistry, Vol. 1 [Russian translation], Izd. Inost. Lit., Moscow (1962), pp. 492-501 [Second English edition, Pergamon Press, New York (1964)].
3. B. V. Nekrasov. Principles of General Chemistry [in Russian], Vol. 3. Izd. Khimiya, Moscow (1970), pp. 26, 28.
4. W. L. Buck and L. W. Strock. Am. Mineral., 40:192 (1955).
5. T. Baars and J. Brand. J. Phys. Chem. Solids, Vol. 34, No. 5 (1973).

6. S. K. Filatov. Uch. Zap. LGU, Ser. geol. nauk, No. 14 (1973).

7. A. Verma and P. Krishna. Polymorphism and Polytypism in Crystals [Russian translation], Izd. Mir, Moscow (1969).

8. Silicon Carbide (G. Heslin and R. Roy, eds.) [Russian translation], Izd. Mir, Moscow (1972).

9. N. V. Belov. Structures of Ionic Compounds and Metal Phases [in Russian], Izd. AN SSSR (1947).

10. G. L. Bir, G. E. Pikus, L. G. Suslina, D. L. Fedorov, and E. B. Shodrin. Fiz. Tverd. Tela, Vol. 13(12) (1971).

11. A. A. Shternberg. Crystals in Nature and Technology [in Russian], Uch. Ped. Izd., Moscow (1961).

12. Minerals Handbook [in Russian], Vol. 1, Izd. AN SSSR, Moscow (1966), pp. 206-213.

13. A. A. Shternberg. Abstracts for the Fourth All-Union Conference on Crystal Growth [in Russian], Izd. AN Arm. SSR, Erevan (1972).

14. W. Knippenberg. Phil. Res. Rept., 18:161 (1963).

15. W. Kohn and Z. J. Sham. Phys. Rev., 140A:1133 (1965).

16. A. Maradudin and A. Berander. J. Phys. radium, 24:89 (1963).

17. E. F. Gross, G. G. Sushina, and E. B. Shadrin. Fiz. Tverd. Tela, 10(4):1036 (1968).

18. A. G. Fitzgerald, M. Mannamim E. M. Pogson, and A. D. Joffe. J. Appl. Phys., 38:3303 (1969).

19. W. E. Knippenberg, H. B. Maanstra, and J. R. Dekkers. Phys. Techn. Rev., 24:181 (1962-1963).

20. M. L. Kite and O. F. Tuttle. In: Experimental Researches in Petrography and Ore Formation [Russian translation], Izd. Inost. Lit (1954).

21. V. G. Nozdrina. Trudy VNIISIMS, 3:2 (1960).

22. A. A. Shternberg. Kristallografiya, Vol. 7, No. 3 (1963).

23. E. T. Allen. Amer. J. Sci., Ser. 4, 34:341 (1912).

24. S. A. Stroitelev and A. D. Stroitelev. Izv. AN SSSR, Neorg. Mat 10(1):186 (1974).

SIMULATION OF CRYSTAL GROWTH

N. N. Sheftal' and A. N. Buzynin

Methods, principles, and theories have been developed for growing large almost perfect single crystals of various substances, as well as large-area epitaxial films and substantial ordered systems of whisker crystals; these applications are basic to scientific crystallography at the present time. Technical progress is very much dependent on solution of the numerous technical problems involved.

Three major purposes are envisaged by those researching single-crystal growth: the production of very homogeneous crystals, very large ones, and rapid growth. At present, there are no firmly established ways of attaining these three purposes simultaneously.

However, we do have a system of principles for growing perfect crystals mainly as a result of experience on growth processes, and the available evidence in many cases provides a correct qualitative solution to problems of this type·

We deal with certain principles that appear of particular importance for simulation purposes* :

1. The uniformity of a crystal is very sensitive to the growth conditions, especially to effects from harmful impurities and from the supercooling [2, 3].

2. The essence of damage to a crystal is entry of the disordered medium into it [2]. A general cause of this entry is elevated supercooling, or rather the occurrence of supercooling differences at the surface, the supersaturation differences increasing with the supercooling and the size of the crystal [7].

When there are supersaturation differences, the medium is incorporated where the supersaturation is inadequate; one can eliminate nonuniformity largely or completely by forced motion of the medium relative to the crystal, which eliminates supersaturation differences. Random motion of the medium is particularly favorable to uniformity, in accordance with Curie's principle [1]; and similarly, nonuniform incorporation is particularly harmful. The latter arises from variations in growth rate, and it particularly produces bulk stresses of purely growth origin.

3. We may note here a paradoxical principle: a high supercooling (supersaturation) facilitates the formation of a perfect crystal. This principle was enunciated in relation to studies on the crystallization pressure, which increases with the supercooling; this is a mechanism that hinders the entry of the medium into the crystal. A sharp temperature gradient, which means that the supersaturation is localized in a thin layer near the crystallization front,

---

* Many of these principles are to be found in [1], and also in [2-6].

in fact enables one to grow crystals of elevated perfection while simultaneously accelerating the growth [4].*

4. An important principle is involved in the relationship between the crystallography of the growing single crystal and the anisotropy as regards capacity for perfect growth. A crystal has directions optimal for perfect growth and optimal for face growth; these are represented by the major equilibrium faces, which have the highest capacity to repel impurities. A crystal during growth should be disposed very precisely in an optical orientation, and the growth rate should be kept constant and not exceeding those giving uniform growth on the faces with the optimal directions employed [1].

5. The actual growth of a crystal does not occur from single atoms but from groups, submicroscopic crystals, and even ones of visible size. The composition and structure of these complexes can be controlled to a certain extent via the composition of the medium, and this may enable one to improve the perfection and adjust the shape of the crystal in the desired direction [2-9].

The science of crystal growth has now reached the stage where these principles and certain others can be linked up into a consistent theory of perfect single-crystal growth. This theory should provide a means of quantitative description of crystal growth.

The present study is the first step in this direction.

As our basic principle for the purpose we use the one formulated by Kossel [10, 11], which has not been used to any great extent: the energy advantage of regular defect-free growth. Regular growth with sequential attachment of particles is most favored by energy and most probable in the absence of interfering factors; deposition of particles out of sequence is improbable, while defects accidentally arising are almost always corrected. Kossel considered that a physicist setting out to construct a perfect crystal would construct it in the same way as the crystal actually grows. The purpose here is regular construction, and in nature it is a consequence of the tendency to minimum free energy in all the individual links of the process.

Kossel's discussion contains no analysis of the relationship of the crystal to the medium that leads to contradictions between the tendency of a crystal to perfect structure and the external conditions. These contradictions amount to the production of concentration flows in response to the growing crystal itself, with supersaturation differences at the surface, which lead to incorporation of the external medium.

Experience with crystal growth shows that one can control crystal development only via the appropriate laws; the basic laws of crystal formation are ones of self-control, and they are most clearly seen in the trends for perfect crystals. When one comes to consider real processes, it is important to know not only the principles for perfect growth but also the limits to the action of these principles, i.e., it is important to know what are the critical conditions for damage to a crystal, and how to bring about the transition from one growth mechanism to another, as well as why one finds structure and composition defects in crystals, such as dislocations, inclusions, and other impurities.

One can answer these questions by using the reliably established trends in crystal growth together with the cybernetic method, which corresponds to the nature of a growing crystal, which provides a self-regulating mechanism to a certain extent.

---

*A particular case of use of this principle is the means of eliminating damage to the crystal arising from concentration supercooling by increasing the temperature gradient [8]; this method involves increasing the supercooling of the melt near the crystal and superheating the latter at depth.

Crystallization is a discrete process, which consists of steps of various sizes: nucleation, formation on new layers of faces, and production of new rows on layers. The attachment of single particles also represents steps, namely unit growth steps.

If we have a simple set of rules for the behavior of particles in migration and so on in relation to the binding energies between adjacent atoms, then we can represent crystal growth as a sequence of operations with integers; this cybernetic model can be used with a computer to follow step by step the production of a crystal. One can also examine the courses and development of diseases, such as the onset and formation mechanism of imperfections, with their subsequent effects on the growth; in that way one can hope to find possible ways of restoring a defective crystal to the almost perfect state. One result of this should be to establish reliably the principles of perfect growth in the form of a reasonably strict mathematical theory.

Crystallization

Energy Characteristics of the Process

Crystallization is subject to minimum free energy; preferential passage of particles from the medium into the crystal (crystallization) is provided by the difference in the free energies ΔW in the two states, which is the driving energy of the process:

$$\Delta W = W_1 - W_2,$$

(1)

or in terms of the chemical potential, the free energy for 1 mole of substance μ:

$$\Delta\mu = RT \ln \frac{c}{c_0}.$$

(2)

The crystallization mechanism and sequence are subject to the principle of minimum free energy for the individual steps; energy considerations favor attachment of a particle at a site on the crystal where it will have the minimum energy. The energy advantage means that attachment is more probable at such a site in comparison with others; the relative magnitude of the energy due to the unsatisfied bonds at a given site determines the probability of particle attachment, i.e., the probability of a unit step.*

Then the probability of particle attachment at a given site may be considered quantitatively; it is dependent on the number of unsatisfied bonds at a given site. We use the following models and symbols to reduce the physical problem of crystal growth to a mathematical one.

Symbols. We consider the growth of a (100) face of a Kossel crystal via individual atomic cubes as a sequence of events on elementary growth areas (growth sites):

l is the order of a site, which equals the number of unused bonds (the number of faces of the cube atom in contact with the growing surface, and l = 1, 2, 3, 4, or 5 (Fig. 1).

φ is the energy of the phase transition per bond.

The growth sites of various orders on the surface of a real crystal are accompanied by active sites, which have more unsatisfied bonds, whose energy is higher, than does the basic lattice.

---

* In the calculations we will use the relative frequencies in place of the probabilities of events; these are simpler to calculate, while qualitative arguments for the probabilities apply also to the relative frequencies. The more probable an attachment, the greater the frequency with which it occurs.

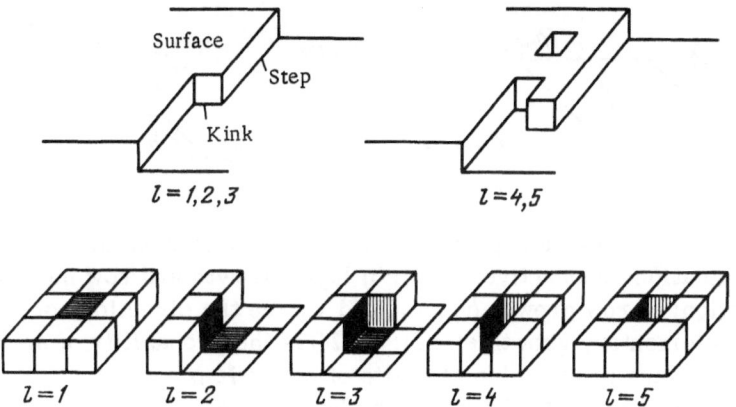

Fig. 1. Fig. 1. Sites of different orders on the surface of a
Kossel crystal.

The behavior of a particle at each possible growth site is characterized by the relative frequency:

ν is the frequency of the events (ν_l at a site of order l, or ν_a at an active site);

p is the relative frequency of the events.

(a) The following characterize the exchange of particles between the medium and the adsorption layer:

1. p ↓ entry of particles from the medium;

2. p ↑ loss of particles to the medium.

(b) The following characterize the motion of particles in the adsorption layer:

3. p ← migration entry;

4. p → migration loss.

(c) The following characterizes the result of event sequence 1-4:

5. P incorporation or attachment of a particle:

$$P = p \downarrow - p \uparrow + p_{\leftarrow} - p_{\rightarrow} \tag{3}$$

The relative frequency of the events of each type, for instance p →, is

$$p_{\rightarrow} = \nu_{\rightarrow}/\nu \downarrow \ .$$

The frequency of the events will be expressed in terms of the heat of evaporation $A = 3\varphi/kT_0$ and the supercooling $\Delta T = (T - T_0)/T_0$, where T_0 is the saturation or equilibrium temperature, for which purpose we use the distribution for the nonequilibrium particle transfer derived from kinetic considerations, which in form coincides with the Boltzman distribution for equilibrium:

$$\frac{\nu_1}{\nu_2} = \exp\left(-\frac{\varepsilon_1 - \varepsilon_2}{KT}\right). \tag{4}$$

In the simplest cases, we can best use a two-dimensional Kossel crystal as our model [12]; this enables one to establish the sequence of growth processes (layer formation and

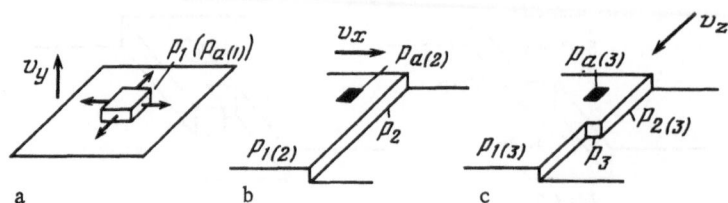

Fig. 2. Crystal face growth scheme: a) layer nucleation (by fluctuation or at active sites); b) layer growth and row nucleation; c) row growth (v_x, v_y, and v_z are the directions of the elementary growth steps); $p_l(l')$ is the probability of an elementary growth step at a site of order l, $p_{a(l')}$ at an active site; l' is the site of highest order.

growth), using simple energy considerations, which enables one to define in general terms the trends in the phenomena.

The model grows via unit square atoms; l is the order of a site in this model, being the number of sides of the square in contact with the growing surface, so $l = 1, 2,$ or 3.

If one is to approximate to real crystallization, we have to replace growth via single atoms by growth from complexes and small crystals, as it actually occurs under real conditions.

The same approach can be used to simulate crystal growth for material of any structure.

Crystallization Stages. Any crystal runs through the following basic stages of development, which follow one after the other (Fig. 2).

The onset of crystallization can be controlled by introducing a seed, for which purpose one can use the following: a crystal of the same substance, a foreign crystal similar in structure in the growth plane (epitaxial growth), or an amorphous surface, which is structureless in a microscopic respect [13].

Slowly growing faces decelerate their growth in the presence of rapidly growing ones on account of particle flow via the adsorption layer into incorporation sites more favorable in energy. In other words, the probability of particle attachment at sites less favored by energy is reduced in the presence of more favorable sites. Consequently, the presence of an incomplete layer on a surface should hinder the formation of a new layer, while an incomplete row should hinder the formation of a new row. The general conclusion is that the probabilities of attachment at sites of different orders, and also at active sites,* are determined by the presence of sites of higher order, i.e., the values will differ according to the growth stage.

Then the events at sites of a given order will occur with different probabilities at the various growth stages (Fig. 2).

Growth of an Ideal Crystal

The growing surface of an ideal crystal (here a Kossel crystal) may have sites of various orders (Fig. 1).

Within certain limits, the crystal itself controls its growth conditions, via the energy advantage or particle attachment at certain sites, which automatically leads to sequential

---

*The role of the sites that determine the particle incorporation sequence may pass from the sites of the highest order to the active sites or back again in the different growth stages.

layerwise defect-free growth. There is an energy advantage (i.e., higher probability) of particle attachment at sites of the highest order relative to ones of lower order, namely:

(1) Kink on a step, which is a repeating site, where the phase-transition energy is completely released;

(2) A step (in the absence of a kink);

(3) There is no advantage in particle attachment on the smooth surface of an ideal crystal.

Omissions are also possible, i.e., errors or growth defects ($l = 4$, $l = 5$), but the energy of advantage of attachment at the sites of highest order produces a high probability of correction, while the probability of their remaining uncorrected is negligible. The essence of the atomic growth mechanism (minimum energy in the individual growth steps) forms the basis of the tendency for a perfect defect-free crystal to grow [10, 11].

Crystallization is a discrete process consisting of various stages: layer nucleation, layer growth, row nucleation, and row growth. The stages follow one after the other, so the crystal growth rate is determined by the rate of the slowest stage. However, one has to bear in mind that the presence of an incomplete layer retards the formation of a new one, i.e., the rate of formation of new layers is itself dependent on the presence or absence of incompleted ones.

Consequently, if the entire process is to occur at an appreciable rate (not too low), one needs a supersaturation that will provide a finite rate for the slowest stage. Calculations show that the relevant supersaturation should be about 50% [14], whereas real crystals in nearly all cases grow at supersaturations of 1% or even 0.8%.

Growth of a Real Crystal

A real crystal contains imperfections: structure defects, inclusions of the medium, and foreign particles, which make themselves felt and can be taken into account at the active sites and in the bulk of the crystal, i.e., these are sites with excess energy in the incomplete bonds over the normal lattice value.*

The energy of an active site determines the role it plays in the crystallization and how it makes itself felt in the various growth stages.

In some cases, active sites merely accelerate the growth without disturbing the mechanism; for instance, they may eliminate the need for a fluctuation-type nucleation origin for new layers without interfering with layerwise growth. In other cases, at a given energy level they may not only interfere with the layerwise growth mechanism but also result in defects, and the various types of defects are closely interrelated. For instance, dislocations facilitate the formation of point defects. Inclusions arising during growth are usually accompanied by dislocations. Local point defects often lead to cracks and dislocations [15]. However, this does not mean that the crystallization is governed by the defective structure of the crystal.

The significance of active sites can be overestimated, as was done at one time for dislocations.

Crystallization is controlled by the basic lattice of the crystal; in other words, the basic lattice may act by the active sites when these are present, i.e., the defective structure relates to real conditions of operation for the program built into the basic lattice.

---

*We use the term active sites in the broad sense, and not merely to denote point defects, which, strictly speaking, may not be active, or may be less active at least than sites in the main lattice (compensated defects, isomorphous impurities, and so on).

The minimum free energy principle for the unit steps applies to a real crystal; the energy advantage persists for attachment at sites of minimum energy.

In a real crystal, the role of these sites may be played at the various stages either by sites of the highest order or by active sites.

Various relationships are possible between the energies of sites of various orders and active sites, so there can be various relationships between the attachment probabilities at these sites (high energy means high attachment probability), and these probabilities correspond to the particle attachment sequence and a definite growth mechanism.

In the limiting ideal case considered above there are no active sites, and the minimum energy principle for the individual steps indicates energy advantage of sequential layerwise defect-free growth.

The following situations can occur in relation to the energies of the active sites when present:

1. Attachment probability for an active site greater than that for a defect-free part on a smooth surface but less than that at a step:

$$P_1 < P_a < P_2.$$

In that case we get the growth sequence of Fig. 2: (a) layers arise at active sites, while layer growth is as in the ideal case, (b) nucleation of rows, and (c) row growth.

The layerwise growth mechanism therefore persists; the crystal grows via layers, and the layers arise at active sites. These active sites on the surface reduce the energy barrier to layer generation, and hence new layers will arise at an appreciable rate at a lower supersaturation than that required in the ideal case, i.e., less than 50%. We have $P_a \gg P_1$, and here we have the reason for the growth of real crystals at appreciable rates for supersaturations of 1% or less; and here also we have the reason why an imperfect crystal tends to erode a perfect one (higher rates of formation of new layers and hence of growth) [1].

Active sites of this type can occur where the crystal is attached to the holder, at intergrain boundaries, at foreign inclusions [16], and at dihedral reentrant angles on the intergrowth planes of twins [17]; they provide elevated growth rates and are frequently seen as vicinals, especially when P_a is close to P_2.*

The layerwise growth rate is exponentially dependent on the supercooling [18]:

$$v = K_1 \exp\left(-\frac{K_2}{T\Delta T}\right) \tag{5}$$

and quadratically dependent on the supersaturation [19]:

$$v = \beta \Pi^2, \tag{6}$$

where $\Pi = (c - c_0)/c_0$ and β is the kinetic crystallization coefficient.

2. Attachment probability at an active site greater than that for a step but less than that for a kink:

$$P_2 < P_a < P_3.$$

This can arise when a screw dislocation emerges on the surface.

---

* In this case the vicinal is the steeper the closer P_a to P_2.

The growth sequence is as follows (Fig. 2): nucleation at active sites; spiral growth or with formation of various growth figures particle attachment is more probable when a step is fairly close to an active site, while a remote step, or one that has vanished, means that particles are attached not to the step but to the active site itself, in which case the growth occurs by a dislocation mechanism or in accordance with the features arising from the specific growth figure.

The rate of such growth is higher than the layerwise value; the value for spiral growth is quadratically dependent on the supercooling [20]:

$$v = K_3 \, (\Delta T)^2, \qquad (7)$$

where K_3 is a constant, which is dependent on the number of active sites and the energies of these as well as on the nature of the material:

$$\text{Number of active sites} \; = \; \left| \frac{\text{Excess free energy}}{\text{Energy of active site}} \right.$$

3. Probability of particle attachment at an active site greater than that at a step:

$$P_3 < P_a < P_4 \, (P_5).$$

The unified principle for all growth processes persists: minimum free energy in the individual steps, and therefore we have the following face growth sequence: any face begins to grow by attachment at its active sites of particles that develop into nuclei; these nuclei expand and link up, which produce sites of fourth and fifth orders,* which are more favorable in energy for particle attachment than are the active sites. The later particles will become attached at these sites and enable the layer to grow over the entire face.

Then here again the surface of the face will remain planar or nearly so; the rate of such growth is linearly dependent on the supercooling [16]:

$$v = K_4 \Delta T, \qquad (8)$$

where K_4 is a constant dependent on the energy and number of the active sites.

Then in all the above cases the relation between the particle attachment probabilities at the various growth sites is determined not by the supercooling or supersaturation, but by the perfection of the crystal face, i.e., by the presence and energy of the active sites; this applies also to the growth sequence and mechanism. The active sites go with the supersaturation to set a definite growth rate. The presence, concentration, and energy of the active sites determine the growth sequence and mechanism, and stability exists over wide ranges in these; these quantities are related, and it is possible for one growth mechanism to give way to another even on different parts of one face.

For instance, supersaturation differences at the surface of a crystal reduce the advantage of sequential defect-free growth; attachment occurs out of turn at sites of elevated supersaturation, while the medium is incorporated where the supersaturation is inadequate, as well as particles of foreign impurities. The uniformity is disrupted, as is the layerwise growth mechanism; the reverse transition is also possible, with return to layerwise defect-free growth, or recovery of the crystal [16].

---

* In this case, the sites of fourth and fifth orders are not accidental improbable errors in sequential growth, but are due to the natural growth sequence, in accordance with the principle of minimum free energy.

Growth Front Complexes and Small Crystals

A crystal grows under real conditions as a rule, but not always [21], not only from single atoms or ions but also from complexes and microscopic crystals [1, 22-24]. This must be borne in mind in simulating real crystal growth.

Frenkel' has shown that a saturated vapor contains a considerable number of complexes, each containing a thousand or more molecules [25], i.e., fairly large groups are present in the medium before crystallization starts. The medium is not structureless, and the crystals grow to a considerable extent from structured elements or complexes [22].

The complexes have a shape that makes them most stable under the given conditions; Khodakov [26] has given a simple method for approximate calculation of the binding energy of a complex ion.

These fairly large complexes involved in crystal growth do not replace the layerwise growth mechanism; layer growth begins with three-dimensional nuclei, which expand, join up, and form a continuous smooth layer [27].

The principle of minimum energy for the individual steps applies here also (it is merely that the steps are larger); the sequence of attachment for the complexes is determined by the number of bonds formed [22], and the probability of incorporation increases with the number of bonds used. Long-range bonds play a particular part in growth via complexes.

The supersaturation is not of itself the cause of deviation from layerwise growth; the medium contains complexes, and if we consider the supersaturation in relation to the complexes, we find that the size of the equilibrium complexes varies with the supersaturation. This size may be such that the complexes cannot combine with the existing continuous layer, and layerwise growth then ceases. There are two means of avoiding this: 1) one maintains a strictly constant supersaturation during growth (then one retains a relatively constant size for the complexes); 2) to provide a rapid crystallization one produces high supersaturation only in the thin layer adjoining the crystallization front [4] (then the large complexes do not have time to accumulate).

Simulation

Role of Particle Migration in the Adsorption Layer

It has been found [28, 29] that a growing or dissolving crystal is surrounded by an adsorption layer, which provides a link between the different sites on the surface. Particle flow occurs in this layer to sites of most favorable incorporation, i.e., in fact the layer transmits information to the various parts of the surface and is responsible for the distribution of supplies in accordance with the bond energies at the sites.

The adsorption layer thus acts as an independent second medium or microscopic medium, i.e., a thin shell surrounding the growing or dissolving crystal; it has its own force distribution, and the particle distribution occurs in accordance with the relevant laws, namely those of a potential field.

These arguments are now applied with the superposition principle to the particle distribution in crystal growth: (a) between the medium and the adsorption layer; and (b) between the adsorption layer and the various sites on the surface, for which purpose we use in both cases the Boltzmann distribution of (4), which applies to any potential force field.

Studies on crystal growth simulation by computer have neglected the scope for particle migration, and the incorporation probability was found as dependent only on the order of a site, no matter whether sites of other orders are present [30, 31]. However, this is true only if all sites on a surface are of the same order, i.e., are equivalent in energy; then migration has no

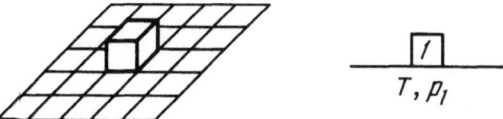

Fig. 3. Attachment of a particle to a smooth surface.

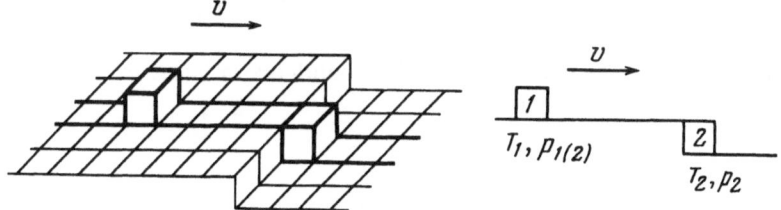

Fig. 4. Attachment of a particle to a growing surface in the presence of a step.

preferred direction, and so it makes no contribution to the particle attachment frequency, and $\nu_{\leftarrow} = \nu_{\rightarrow}$ at any site, or $\nu_{\leftarrow} - \nu_{\rightarrow} = 0$:

$$v_0 = v\downarrow - v\uparrow + v\leftarrow - v\rightarrow = v\downarrow - v\uparrow. \tag{9}$$

Migration does have a preferential direction if there are sites of different orders: from sites of lowest order to those of highest order. Also, migration does not produce new particles on the surface, merely displacing those present from some sites to others, so this reduces the probability of attachment at low-order sites and increases that at high-order ones.

The extent of this probability reduction or increase is dependent on the presence of sites of the highest order (or active sites), and so the attachment probability for sites of a particular order varies with the growth stage.

The following cases are possible:

1. Probabilities in the absence of second-order sites (Fig. 3):

$$P_1 = p\downarrow - p\uparrow + p\leftarrow - p\rightarrow = p\downarrow - p\uparrow = p\downarrow \left[1 - \exp\left(A - \tfrac{\varphi}{kT}\right)\right]^*. \tag{10}$$

If the surface has active sites, the calculation is as in the following case.

2. Probabilities when the surface has sites of first and second orders in the growing layer (Fig. 4). At a site of first order ($l = 1$, $T = T_1$). Deposition from the medium $p\downarrow$; loss to the medium $p_1\uparrow = p\downarrow \exp\left(A - \tfrac{\varphi}{kT_1}\right)$, as given by (4); remainder from exchange with the medium $p\downarrow\left[1 - \exp\left(A - \tfrac{\varphi}{kT_1}\right)\right]$, of which the following depart for second-order sites:

$$p_1\rightarrow = p\downarrow\left[1 - \exp\left(A - \tfrac{\varphi}{kT_1}\right)\right]\left[1 - \exp\left(\tfrac{\varphi}{kT_1} - \tfrac{2\varphi}{kT_2}\right)\right].$$

---

*The expression for P_1 gives a somewhat excessive attachment probability relative to the real value, i.e., the value taking into account the need for two-dimensional nucleation. The two values come close together as the supersaturation or supercooling increases.

There remain at first-order sites

$$p \downarrow \left[1 - \exp\left(A - \frac{\varphi}{kT_1}\right)\right] \exp\left(\frac{\varphi}{kT_1} - \frac{2\varphi}{k\Gamma_2}\right).$$

Entry from sites of second order:

$$p_1 \leftarrow = \frac{p_2 \rightarrow}{n} = \frac{p \downarrow}{n}\left[1 - \exp\left(A - \frac{2\varphi}{kT_2}\right)\right]\exp\left(\frac{\varphi}{kT_1} - \frac{2\varphi}{kT_2}\right)$$

(where n is the number of first-order sites per second-order site, or the length of a row expressed in terms of the number of particles).

Incorporated

$$P_1 = p \downarrow - p_1 \uparrow + p_1 \leftarrow - p_1 \rightarrow$$

$$P_1 = p \downarrow \left[1 - \exp\left(A - \frac{\varphi}{kT_1}\right)\right]\exp\left(\frac{\varphi}{kT_1} - \frac{2\varphi}{kT_2}\right) + \frac{p\downarrow}{n}\left[1 - \exp\left(A - \frac{2\varphi}{kT_2}\right)\right]\exp\left(\frac{\varphi}{kT_1} - \frac{2\varphi}{kT_2}\right). \qquad (11)$$

At second-order sites ($l = 2$, $T = T_2$). Entry from the medium $p \downarrow$, loss to the medium $p \uparrow = p \downarrow \exp\left(A - \frac{2\varphi}{kT_2}\right)$. Residue from exchange with the medium $p \downarrow\left[1 - \exp\left(A - \frac{2\varphi}{kT_2}\right)\right]$. Of these, loss to sites of first order

$$p_2 \rightarrow = p \downarrow \left[1 - \exp\left(A - \frac{2\varphi}{k\Gamma_2}\right)\right]\exp\left(\frac{\varphi}{kT_1} - \frac{2\varphi}{kT_2}\right).$$

Remainder at sites of second order

$$p \downarrow \left[1 - \exp\left(A - \frac{2\varphi}{kT_2}\right)\right]\left[1 - \exp\left(\frac{\varphi}{kT_1} - \frac{2\varphi}{k\Gamma_2}\right)\right].$$

Entry from first-order sites*:

$$p_2 \leftarrow = m p_1 \rightarrow \left[1 - \exp\left(A - \frac{\varphi}{kT_1}\right)\right]\left[1 - \exp\left(\frac{\varphi}{kT_1} - \frac{2\varphi}{kT_2}\right)\right],$$

(where m is the number of particles in twice the mean-free path):

$$\lambda_s \lambda_s = a \exp\left(\frac{3\varphi}{2kT}\right) \quad [14].$$

Incorporated

$$P_2 = p \downarrow - p_2 \uparrow + p_2 \leftarrow - p_2 \rightarrow$$

$$P_2 = p \downarrow \left[1 - \exp\left(\frac{\varphi}{kT_1} - \frac{2\varphi}{kT_2}\right)\right]\left\{1 - \exp\left(A - \frac{\varphi}{kT_1} + 2\exp\left(\frac{3\varphi}{2kT_1}\right)\left[1 - \exp\left(A - \frac{\varphi}{kT_1}\right)\right]\right\}. \qquad (12)$$

The expression for the attachment probability at an active site takes the same from, except that 2φ is everywhere replaced by the excess energy of an active site.

---

* Here we calculated P_2 for the particular case where the length of the row serving one step is greater than or equal to twice the mean free path λ_3; if the row length is less, one should put

$$m = \frac{l \text{ (row length)}}{a \text{ (particle edge length)}}$$

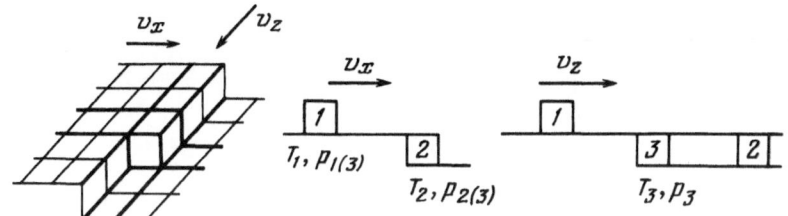

Fig. 5. Attachment of a particle to a growing surface in the presence of a kink.

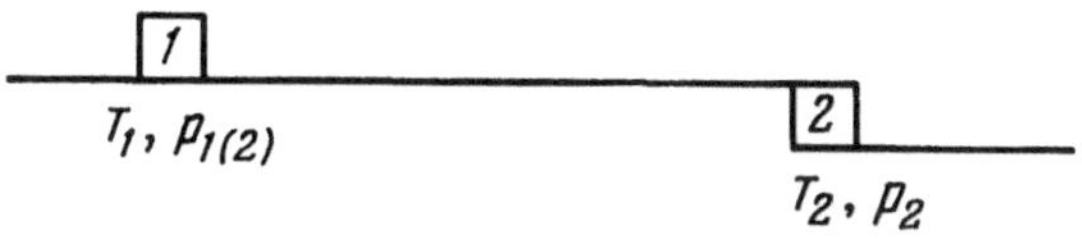

Fig. 6. Diagram for problems 1 and 3.

3. When there are third-order sites at steps (Fig. 5).

The principle is the same; one takes into account additionally the result of exchange between sites of orders 1 and 2 on the one hand and those of order 3 on the other.

Solutions to Limiting Problems

We consider the following problems:

Problem 1. Determine the relative probability of particle attachment at sites of first order in relation to those of second order, this to be found as a function of the supercooling (Fig. 6); $T_1 - T_2 = T$ is the surface temperature, i.e., the supercooling at the surface is constant; A is taken as 6 and $P_{1(2)} = ?$:

$$P_2 = \frac{P_{1(2)}}{P_2} = \frac{P \downarrow \left[1 - \exp\left(A - \frac{\varphi}{kT}\right)\right]\exp\left(-\frac{\varphi}{kT}\right)}{P \downarrow \left[1 - \exp\left(\frac{\varphi}{kT}\right)\right]\left\{1 - \exp\left(A - \frac{2\varphi}{kT}\right) + 2\exp\left(\frac{3\varphi}{2kT}\right)\left[1 - \exp\left(A - \frac{\varphi}{kT}\right)\right]\right\}}.$$

We calculated $P_{1(2)}/P_2$ as a function of $\Delta T = (T - T_0)/T_0$, where T_0 is the equilibrium temperature, by computer with ΔT taking a discrete series of values from 0.1 to 50% with a step of 0.1%.*

We get the following relationship as shown in Fig. 7, which indicates that high supercooling is not the cause of deviation from layerwise growth sequence,† as the probability of particle attachment at the step may actually increase relative to that for attachment at the surface of the growing layer as the supersaturation increases: for instance, we have as follows:

$$\Delta T = 7.5\% \quad \frac{P_2}{P_{1(2)}} \simeq 200,$$

$$\Delta T = 28\% \quad \frac{P_2}{P_{1(2)}} \simeq 1000,$$

$$\Delta T = 50\% \quad \frac{P_2}{P_{1(2)}} \simeq 20000.$$

---

*The value of $P_{1(2)}/P_2$ for small supercoolings or supersaturations gives somewhat too high a result, since $P_{1(2)} > P_{nuc(2)}$, which is acceptable for the subsequent calculations; the higher the supercooling, the closer this ratio to the true value, since $P_{1(2)} > P_{nuc(2)}$ $P_{1(2)} \longrightarrow P_{nuc(2)}$.

†As the supercooling or supersaturation increases, the area served by a single two-dimensional nucleus decreases, but this does not interfere with layerwise growth.

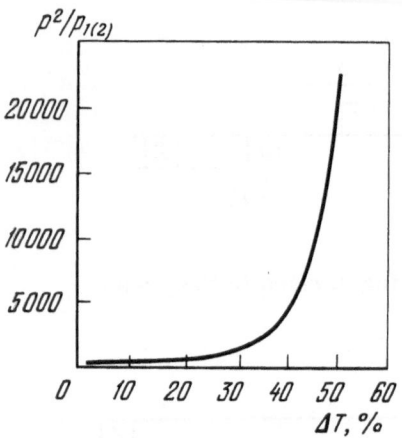

Fig. 7. The relation of $P_2/P_{1(2)}$ to super-cooling.

Then the energy advantage of sequential defect-free growth actually increases with the supersaturation, as qualitative considerations indicate [4].

Problem 2 . Determine the increase in the attachment probability for first-order sites in a completely constructed layer relative to an incomplete layer, i.e., determine the retarding action of an incomplete layer as a function of supercooling (Fig. 8). Here

$$\Delta T = \text{const}, \quad T_1 = T_{1(2)} = T, \quad \frac{P_1}{P_{1(2)}}(T) = ?,$$

$$\frac{P_1}{P_{1(2)}} = \frac{p \downarrow \left[1 - \exp\left(A - \frac{\varphi}{kT}\right)\right]}{p \downarrow \left[1 - \exp\left(A - \frac{\varphi}{kT}\right)\right]\exp\left(-\frac{\varphi}{kT}\right)} = \exp\left(\frac{\varphi}{kT}\right). \tag{13}$$

This shows that the inhibition increases with the supercooling, and also with the bond energy. Here again, then, high supercooling actually facilitates layerwise sequential growth.

Problem 3 . Determine the critical supercooling different Δ at the surface of a growing crystal for which new deposition on an incomplete layer becomes inevitable. Determine also the dependence of this on the minimal supercooling at the surface (Fig. 6). Here $T_1 < T_2 < T_0$, $P_{1(2)} = P_2$ or $P_{1(2)}/P_2 = 1$,

$$\Delta(\Delta T_2) = \Delta T_1 - \Delta T_2 = ?$$

The problem was solved by computer as for problem 1; the solution gives an answer to two practical problems:

1. The permissible supercooling difference Δ' per unit length at the surface of a growing face for a given specified face size r:

$$\Delta' = \frac{\Delta}{r}. \tag{14}$$

2. The critical or maximum permissible face size for a given specific supercooling difference attainable under given conditions:

$$r_{\text{cr}} = \frac{\Delta}{\Delta'} \tag{15}$$

Fig. 8. Diagram for problem 2.

Conclusions

1. A method has been proposed for deriving quantitatively the attachment probabilities for various sites on a growing surface with allowance for the particle migration in the adsorption layer.

2. The attachment probability is dependent on the presence of higher-order sites and active sites; the value varies in the different growth stages.

3. The active sites are points of action of the main lattice, and their concentration and energy can influence in various ways the particle incorporation sequence, the growth mechanism, and the growth rate.

4. The growth sequence and mechanism are determined by the crystal perfections; they are stable over wide ranges in supercooling and supersaturation, although the latter influence the growth rate.

5. Computer calculations on the probabilities for a Kossel crystal have shown that:

a) A high supercooling or supersaturation, in the absence of supercooling differences at the surface, is not the cause of deviations from correct sequential growth and may even favor the latter;

b) Deviations occur from regular particle deposition on account of supercooling differences; in the particular case $m = 2e^{3\varphi/2kT}$, the critical supercooling difference has been found as a function of the supercooling at the surface, together with the related critical specific difference in the supercooling, for a face of a given size as well as the critical size of the face for a specific supercooling difference.

We are indebted to V. V. Voronkov for a discussion of this work.

Literature Cited

1. N. N. Sheftal'. This volume, p. 185.
2. N. N. Sheftal'. Dokl. AN SSSR, 31(1):33 (1941).
3. N. N. Sheftal'. In: Growth of Crystals, Vol. 1, Consultants Bureau, New York (1959), p. 5.
4. N. N. Sheftal'. In: Growth of Crystals, Vol. 3, Consultants Bureau, New York (1962), p. 3.
5. N. N. Sheftal' and I. V. Gavrilova. In: Growth of Crystals, Vol. 4, Consultants Bureau, New York (1966), p. 24.
6. N. N. Sheftal'. In: Growth of Crystals, Vol. 5A, Consultants Bureau, New York (1968), p. 25.
7. P. S. Vadilo. Zh. Éksp. Teor. Fiz., 8:1218 (1938).
8. B. Chalmers. Principles of Solidification, Wiley, New York (1964).
9. B. M. Bulakh. J. Crystal Growth, 7:196 (1970).
10. W. Kossel. Nachr. Ges. Wiss. Göttingen. Math.-phys. Kl., 135 (1927).
11. W. Kossel. Naturwissenschaften, 18:901 (1930).
12. N. N. Sheftal'. Priroda, No. 4, p. 42 (1964).
13. N. N. Sheftal' and A. N. Buzynin. Vestnik MGU, ser. geol., No. 3, p. 102 (1972).
14. W. Barton, N. Cabrera, and F. Frank. In: Elementary Crystal Growth Processes [Russian translation], IL, Moscow (1959), p. 11.
15. T. G. Petrov, E. B. Treivus, and A. N. Kasatkin. Crystal Growth from Solution [in Russian], Nedra, Leningrad (1967).
16. A. A. Chernov. Usp. Fiz. Nauk, 73(2):277 (1961).
17. N. N. Sheftal'. Kristallografiya, 16(2):394 (1971).

18. Ya. I. Frenkel'. Zh. Éksp. Teor. Fiz., 9:123 (1939).
19. A. A. Chernov. Kristallografiya, 16(4):842 (1971).
20. W. Hilling and J. Turnbull. In: Elementary Crystal Growth Processes [Russian translation], IL, Moscow (1959), p. 293.
21. E. Budevskii, V. Bostanov, and T. Vitanov. This volume, p. 231.
22. B. M. Bulakh. This volume, p. 88.
23. Chang Yuan-lung and Chang Kuei-fen. In: Growth of Crystals, Vol. 6A, Consultants Bureau, New York (1968), p. 37.
24. N. P. Yushkin. Theory of the Microblock Growth of Crystals in Natural Heterogeneous Solutions [in Russian], Syktyvkar (1971).
25. Ya. I. Frenkel'. Collection of Selected Papers [in Russian], Vol. 3, Izd. AN SSSR, Moscow (1959).
26. Yu. V. Khodakov. Elements of Electrostatic Chemistry [in Russian], ONTI, Moscow (1934).
27. E. A. Krivorotov, Yu. G. Sidorov, and L. N. Aleksandrov. Zh. Neorg. Mat., 7(11):1947 (1971).
28. M. Volmer and G. Adhikari. Z. Phys., 35:170 (1926).
29. Z. Kyulai and S. Bieleck. Acta Phys. Hung., 1:199 (1952).
30. A. A. Chernov. Usp. Fiz. Nauk, 100(2):1 (1970).
31. V. V. Solov'ev and V. T. Borisov. Dokl. AN SSSR, 202:2 (1972).

A CRYSTAL AS A MEDIUM THAT ORDERS PHENOMENA*
N. N. Sheftal'

Curie's symmetry superposition principle is used to consider the superposition of the symmetry of order and disorder by reference to the interaction of a crystal with x rays and light.

Order and disorder are considered as characteristics of the state of inorganic matter and radiation; we do not discuss the transformation of these concepts when they are applied to the organic world and information.

It is important to establish how the interaction with a crystal will occur when a radiation flux passes through it; the various artificial and natural single crystals go with single-crystal films to provide particularly a variety of changes in microscopic and macroscopic energy fluxes under many different circumstances.

We naturally examine this topic from the most general viewpoint, i.e., from a crystallographic or especially symmetry viewpoint, with emphasis on the changes that occur in the energy flux on passage through a crystal. In other words, we seek to establish how the symmetry of the energy flux is affected by the interaction with a crystal, for which purpose we use Curie's principle [3].

Curie's Principle

The more obvious principle for superposition of the symmetries of geometrical figures is a prototype of Curie's superposition principle for the symmetries of physical phenomena.

Shubnikov gives the following definition [4]: If one combines two or more nonidentical symmetrical figures into one compound figure, the latter retains only the symmetry elements common to all the constituent figures in a given mode of arrangement. For instance, the

---

* The paper completes a study begun in two previous papers [1, 2]. The first paper was concerned with the definition of symmetry, while the second dealt with the reasons for crystal symmetry. The cause of crystal symmetry has been shown to be the tendency for chemical bond energy to be transformed to thermal energy, which is then dissipated, i.e., crystal symmetry arises as an unexpected and additional consequence of the tendency to disorder. The second half of this contradiction is examined here, namely the character of the effects occurring in crystals. The problem discussed is whether this is inherent in the medium in the way that a gas tends to disorder, or whether there is here also a tendency to ordering in the phenomena. Originally presented on February 4, 1969 at a meeting in the Department of Crystallography and Crystallochemistry at Moscow University, this paper was presented at the Fedorov meeting at Leningrad Technological Institute on May 20, 1974, under the title "Curie's principle and superposition of the symmetries of order and disorder."

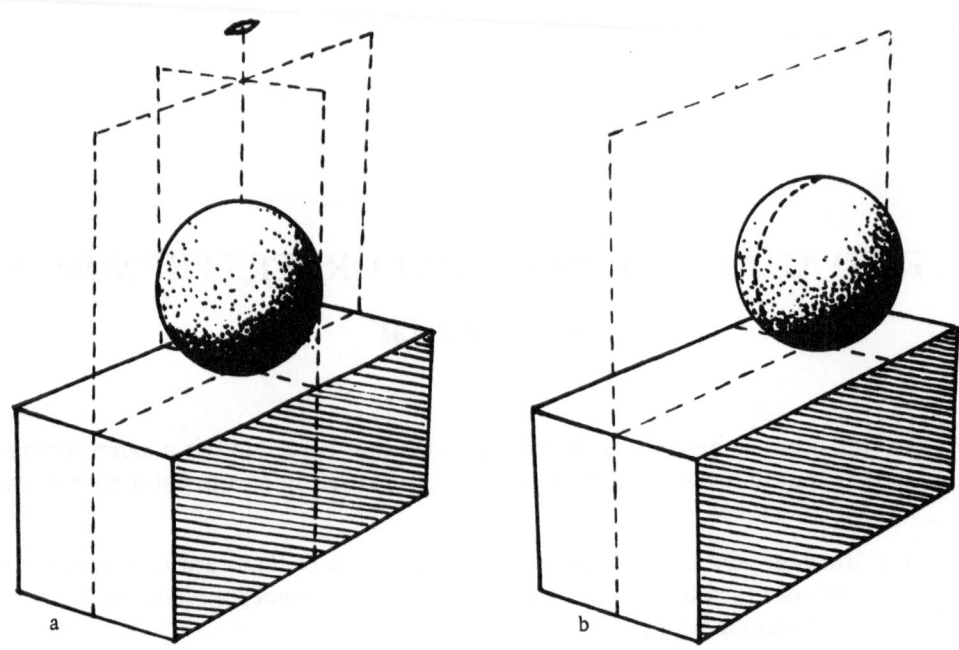

Fig. 1. Superposition of the symmetries of geometrical figures.

symmetry of a sphere is $\infty/\infty/$mmm, while that of a rectangular parallelepiped is 2/mmm; the symmetry of a figure composed of a sphere touching the center of a face of such a parallelepiped is 2mm, while that of a figure in which the sphere is displaced to the center of a face but is not in the symmetry plane is only m (Fig. 1).

The physical properties of an isotropic medium are described by the symmetry of a sphere $\infty/\infty/$mmm; such symmetry occurs for example in a compact mass of randomly oriented particles of iron. Application of a magnetic field, which has the symmetry $\infty/$m of an axial vector, eliminates all the symmetry elements except those in common with the field, and it magnetizes the iron and causes it to acquire the symmetry $\infty/$m.

When a physical field acts on a crystal having the property described by a definite tensor surface (indicatrix), it produces a new symmetry in this property, which is dependent on the direction of the applied field; for instance, the symmetry $\infty/$mmm applies to the positive or negative mechanical stress set up by two parallel planes. Stress applied perpendicular to the principal axis of a crystal in one of the middle systems, whose optical properties are described by an ellipsoid of rotation (symmetry $\infty/$mmm), transforms the latter to a triaxial ellipsoid of mmm symmetry.*

Curie's principle is applicable also to crystal growth phenomena; for instance, the concept of random motion in the medium, which has been shown to be optimal for growth of homogeneous single crystals [5], may be considered as a consequence of Curie's principle, for under the conditions of such motion (approximate symmetry $\infty/\infty/$mmm for randomly moving medium), the crystal should be produced with all unaffected symmetry elements of its form, and hence most uniformly and perfectly.

The shape of a crystal growing in a gravitational field (field symmetry ∞mm) loses all symmetry elements disposed obliquely relative to the vertical direction [6].

---

*The same result is obtained on deforming the crystal in any direction oblique to the principal axis of the ellipsoid, with the stress vector resolved into two: along the principal axis and perpendicular to it.

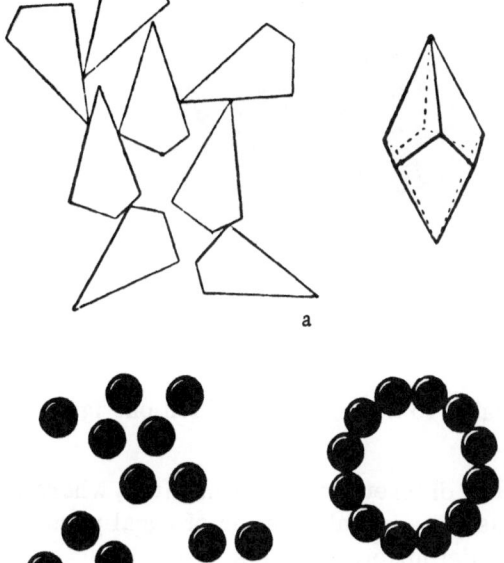

Fig. 2. Unequal and equal disposition of finite figures.

The shapes of ice crystals evolved in a layer of snow on account of differences in the gradients in the vapor pressure, temperature, and vapor movement, so the medium may be considered as asymmetric. Under these conditions, the gradual entry of the medium causes the number of defects in the external and internal symmetry to increase. The process terminates with break-up of the crystal into fragments that are completely asymmetric, i.e., the asymmetry of the source medium ultimately completely eliminates the symmetry of the crystal [7].

As Curie's principle is applicable to many different phenomena, we can use it to consider the symmetry change in the microstructure of energy fields interacting with crystals.

We first of all consider x rays, whose interaction with crystals has been studied in the most detail. First of all we deal with the essence of symmetry not only with the crystal itself but also of the radiation field.

Symmetries of Order and Disorder: The

Ordering Phenomenon

The symmetry of a crystal may be defined via the equal location of equal parts or particles [1]. This definition is convenient in the present case because it does not involve the following formal concepts*: symmetry elements, symmetry operations, general and special positions, and so on. Curie in his time defined symmetry by a single word: order [8]. Our definition characterizes the essence of this order for a discrete medium consisting of equal parts. Then the concept of order is identical with that of symmetry for a discrete static mate-

---

*In such an array, the dispositions of all the particles around each are superimposable or mirror equivalent. An observer transferred from one equal particle to another either sees no changes in the environment (near and far) or finds them in a relationship of mirror images. This is the geometrical condition for complete satisfaction of all free short-range and long-range bonds for all particles in the crystal, i.e., the condition for minimum in the volume free energy.

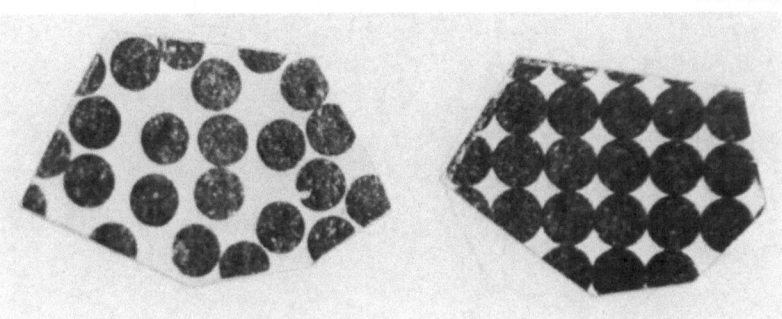

Fig. 3. Unequal and equal disposition of infinite figures.

rial medium; the two concepts are completely applicable to a crystal for which one ignores the thermal motion.

The opposing concept is that of disorder in static discrete material medium, where one again ignores the thermal motion; it can be defined as the unequal location of equal parts or particles. We can illustrate the correctness of these definitions via Figs. 2 and 3.

An instance of discrete disorder is an idealized amorphous medium that has no short-range order in the disposition of the particles, which differs from an ordinary amorphous medium.

When one includes the discussion of a disordered medium, the concepts of symmetry and order combined by Curie have to be separated; if the above idealized amorphous medium is isotropic, this means that the geometrical and physical properties as defined from any view-point along different directions are equal, since one has equal mean static distances between nearest neighbors in all directions. Therefore, the symmetry of such a medium corresponds to that of a sphere. However, we can define the character of the symmetry more closely by saying that this is the symmetry of static disorder.

The structure of an ideal gas in thermal equilibrium is an example of dynamic disorder; here again we have equality of the properties in different directions, as in the previous case, but with the difference that the gas atoms are constantly and randomly moving in all directions, and the velocities at any point in these directions are equally probable. This can be seen more clearly by imagining the velocities of the atoms at some definite moment as a bundle of vectors emerging from one point and forming a sphere around it. The density of the vectors at the surface of a sphere will be constant, so the symmetry of the gas in equilibrium will also be described by the symmetry of a sphere. However, this is a higher symmetry, the symmetry of dynamic disorder. This representation of the gas is known as the hypothesis of elementary disorder [9].

It is possible also to have lower degrees of macrosymmetry in dynamic disorder; for instance, a flow of gas has symmetry ∞mm, while a medium in which there predominate vari- ♦ ous unbalance directions of macroscopic motion has symmetry I.

However, the general essence of dynamic disorder is that the microscopic motions of the individual particles are independent and uncorrelated; dynamic disorder is characteristic also of electromagnetic radiation. Rayleigh [10] was the first to represent white natural light as a completely irregular and nonperiodic process. Vavilov [11] made a detailed study of disorder in light over a period of many years, and he stated that one can distinguish between macrooptics and microoptics the first being the optics of high light fluxes, long times, and large sources, while the second differs from the first much as the molecular theory differs from thermodynamics. A light function varies randomly not only in magnitude but also in direction on account of the statistical character of the molecular perturbations in the source. The chaos in light is similar to that in x rays and gamma rays, as measured by ionization chambers and Geiger counters. Statistical chaos is present already in the elementary emission of an individual atom.

When one characterizes a disorder in light, one compares it with noise in electrical circuits, but it is far more complex, in that it is not of zero dimensions but is complicated by involving several parameters of the three-dimensional form of random noise [12].

The radiation function varies randomly in phase, frequency, and polarization direction, on account of the statistical character of the molecular emissions in the source; the symmetry of a parallel beam or flow of x rays in general is as for a rectilinear flow of gas, i.e., the symmetry is that of dynamic disorder and is represented by a group ∞ mm.

When one superimposes the symmetry of a crystal and that of an x-ray beam emerging after interaction with the crystal, Curie's principle indicates that the latter should retain the symmetry elements common to the beam and the direction in the crystal along which it passes.

As a result, the symmetry of the radiation, which is characterized by axes of infinite order, should be reduced to the symmetry of the crystal, which has no such axes.

We then naturally have to examine whether this reduced symmetry of the radiation retains the symmetry of disorder, or whether there is any transformation toward the symmetry of order.

To consider this we consider the phenomenon of ordering generally, first with reference to a disordered material medium and then for radiation.

The unequal location of equal particles is indicated by the above definition as being disorder, but unequal location of several sets of different particles represents this to an even larger degree, especially if the particles from the different sets are unequal.

Therefore, extraction of one such set from a medium composed of two or more such disordered sets, represents only reducing the degree of disorder, but it is not yet ordering.

For instance, if one removes the black spheres from a disordered pile of black and white spheres, the remaining set of white spheres remains disordered, although the degree of disorder in the pile has been reduced.

We now have to consider what is the ordering of dynamic disorder.

One can assume [13] for phenomena at the atomic and wave levels that the order in motion is greatest when the speeds of the particles are identical in magnitude and direction, i.e., when the motion of all the particles forms translational motion of a body. This motion of all the particles as a single whole can be called macroscopic motion. In contrast, we can give the name microscopic motion to the motion of particles differing in velocity and direction, which is the basis of the independence and lack of correlation between the particles.

The second law of thermodynamics deals with the scope for macroscopic motion going over to microscopic motion with a uniform distribution of the latter; the basis is the high probability of disorder.

The reverse effect is ordering of dynamic disorder, namely transformation of microscopic motion into macroscopic, either partial or complete.

The essence is the conversion of independent, uncorrelated, and unequal motion of the particles into correlated, uniform, and equal motion in velocity and direction.

This ordering is to be distinguished also from reduction in the degree of disorder arising from the separation of unequal disordered sets mixed into one whole, in which each set as isolated from the mixture remains disordered. An example is the separation of a mechanical mixture of several gases into component parts. The degree of disorder is reduced but order does not arise. The state of each gas corresponds to dynamic disorder, namely uncorrelated, unequal in direction and velocity.

Interaction of X Rays and Light with a Crystal

We now consider the interaction of x rays with a crystal via diffraction in order to establish whether partial or complete ordering occurs in this interaction in such a way as to satisfy the above definitions.

We first of all consider interference, which is the combination of radiations equal in wavelength arising as a rule from two sources that must be coherent. The coherent sources are usually formed by splitting one point source into two parts by reflection or refraction. The difference in path length traversed by the radiation from each source results in a phase shift between the waves; if the shift equals an even number of half-waves, one gets summation, whereas if the difference is an odd number one gets mutual cancellation. A very simple example of optical interference is the sequence of light and dark bands when monochromatic light is reflected from the upper and lower surfaces of a thin film of oil on water.

Here the interference appears as the result of a single addition of waves provided that the wavelength corresponds with the thickness of the reflecting film in the direction of the beam.

On the other hand, x-ray diffraction in a crystal is a result not of the addition of one or a few waves but a vast number of additions of waves scattered in millions of directions not randomly but regularly and with regular repetition by the periodically disposed atoms of the crystal.

There is strict equality in the distances between the various equivalent atoms in the crystal, and periodicity in the disposition, which provides very high sharpness in the interference maxima and minima relative to that found in simple interference.

The diffraction conditions are defined by Laue's equations, which relate the monochromatic wavelength λ to the angles α_0, β_0, and γ_0 between the incident x-ray beam and the crystallographic axes X, Y, and Z together with the angles α_p, β_q, and γ_r, which are the angles between the same beam and the directions in which one gets diffraction. Let a, b, and c be the parameters of the crystal along the crystallographic axes, while p, q, and r are small integers; then Laue's equations take the form

$$a\,(\cos\alpha_p - \cos\alpha_o) = p\lambda, \qquad b\,(\cos\beta_q - \cos\beta_o) = q\lambda, \qquad c\,(\cos\gamma_r - \cos\gamma_o) = r\lambda.$$

The diffraction results in a large accentuation of the energy of the scattered x rays in the diffraction directions by depletion of all other directions. This involves summation of the waves in each burst, which is provided by the coherent scattering of the periodically disposed atoms in respect of the diffraction directions, the phase shift being an integral number of wavelengths along these directions, while along all other directions the value is not an integer. This is the reason for the amplification in the diffraction direction and suppression of the electromagnetic field in all the others.

This is illustrated by the scheme of Fig. 4 for the reflection of x rays from single atomic series, and also by Fig. 5 for the vector scheme for addition of the scattered waves from an idealized simple model for a defect-free crystal consisting of identical equal atoms [14].

The atomic network that scatters the radiation produces diffraction peaks only for those directions of the series for which Laue's second condition, $b\,(\cos\beta_q - \cos\beta_0) = q\lambda$, is met,

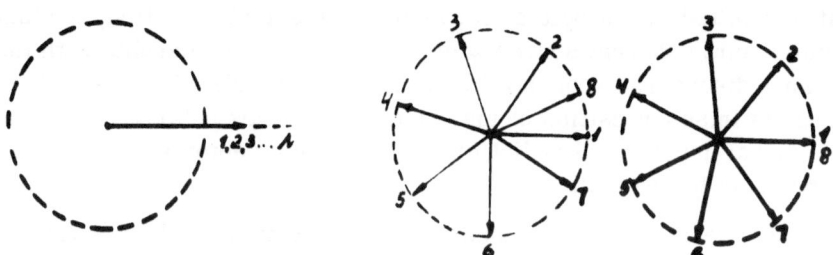

Fig. 4. Scattering of x rays by a single row of atoms.

while a three-dimensional lattice in turn additionally accentuates the radiation of a single
atomic net only when the additional paths between the first net and the sets of parallel ones
are integer values of the wavelength, i.e., when Laue's third condition, $c(\cos\gamma_r - \cos\gamma_0) = r\lambda$,
is met. Reflections not meeting this condition are also suppressed.

A net gives an accentuation of a second-order relative to a row, while a lattice gives
third-order accentuation.

This means that a monochromatic x-ray beam interacting with a crystal in the reflecting
position results in ordering of the radiation microstructure. The independent uncorrelated
wave motion is transformed by the lattice into correlated and equal motion.

An x-ray pattern is a record not only of the structure of the crystal but also of the macroscopic motion of the radiation. The symmetry of the radiation as ordered after passage
through a crystal may be defined via Curie's principle.

When electrons or neutrons interact with a crystal, one gets ordering in the microstructure in the same way, as is clear from electron-diffraction and neutron-diffraction patterns;
the ordering mechanisms are analogous if one uses wave concepts for the particle fluxes.

Curie's principle indicates that the symmetry of the disordered x rays is transformed
into the symmetry of order corresponding to the crystal, i.e., the radiation is ordered.

So far we have discussed an interaction between the radiation and the crystal in which
each packet of radiation is ordered in phase; for this purpose one needs a correspondence
between the wavelength and the interatomic distance in the appropriate direction. We turn now
to polarization, which is a phenomenon not involving this correspondence.

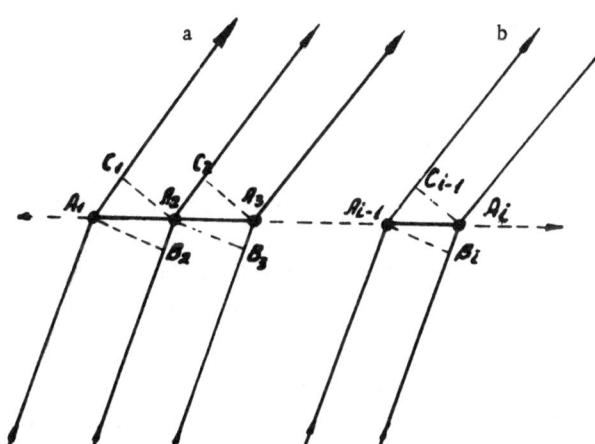

Fig. 5. Vector schemes for: a) radiation
amplification in the diffraction direction; b)
suppression in all other directions.

The radiation emitted by a single atom, including thermal radiation, is plane-polarized; the emission from a point coherent source whose diameter is comparable with the wavelength consists of numerous disordered sequential or simultaneous acts of emission by the atoms present. This results in superposition of many bursts of waves polarized in different planes, which results in natural unpolarized radiation. The electric vector vibrates in all possible directions perpendicular to the beam.*

If the radiation enters a crystal of medium or low symmetry along a direction not coincident with an optic axis, the polarization vectors are resolved into two mutually perpendicular vectors lying only in two planes corresponding to the structural features of the crystal. One gets summation of the vectors along one direction together with suppression of all the initial ones. Therefore, there is also a change from microscopic motion to macroscopic, i.e., ordering in the radiation, though this is weaker than ordering in phase.

The ordering increases as one passes from triclinic crystals to orthorhombic ones; in a triclinic crystal, all three symmetry planes of the indicatrix involved in vibration of the polarized waves for different frequencies are somewhat differently disposed, while in monoclinic crystals one plane is identical for the indicatrices for all frequencies, and in orthorhombic crystals all three planes coincide.

No polarization is observed when two or more atomic series of least optical density lie perpendicular to the beam propagation direction (the directions of the optic axes in crystals, or all directions in cubic crystals), as though the radiation sensed (an expression due to Born [16]) the molecular granularity of the material and detected the screw axes (rotation of the plane of polarization in enantiomorphic crystals).

In one form or another, this radiation ordering occurs via polarization in most crystals; this ordering occurs, although to only a small extent, for thermal radiation on the same footing as other forms.

However, polarization occurs also in amorphous media, such as glasses under stress, which leads (in accordance with Curie's principle) to increase in the symmetry of the radiation, but on the above basis it should not lead to replacement of the symmetry of disorder by the symmetry of order. We consider that there is an essential difference at the atomic level between polarization of radiation produced by crystalline and amorphous media, and that scattering by randomly disposed atoms in an amorphous body differs in microdisorder from scattering by the atoms of a crystal.

Bearing this in mind, we may assume that ordering is more probable than disordering for phenomena occurring in crystals.

The microscopic motion is ordered in accordance with the symmetry of the direction in which the energy propagates [17], and if summed over considerable time intervals can lead to appreciable effects.

We have already noted a contradiction existing in the essence of crystallization [18]. Crystallization is a spontaneous process directed to completion of all long-range and short-range bonds between particles. The geometrical condition for this bond satisfaction is equal location of equal particles, i.e., formation of a symmetrical construction [2]. If we move the particles from unequal positions to equal ones, we have a conversion of the binding energy of the particles into thermal microscopic motion of the surroundings, and this requires removal and dissipation of this heat, i.e., increase in the entropy.

_____

* More precisely, the natural unpolarized light is the result of superposition of identical very thin ellipses with randomly disposed axes [15].

Consequently, there is an essential connection between the tendency to chaos and order (or between the probabilities of these). At one pole lies chaos, while at the other we get an increase not only in the volume of ordered medium, but also in the probability of ordering for phenomena.

One gets a steady increase in the ordering of electron-diffraction patterns from a low-density monatomic gas, an associated gas, a liquid or amorphous body, a liquid crystal,* and finally a solid crystal; this shows that the increase in relationship between the particles goes in hand with the symmetry change for media composed from very simple symmetrical molecules on the one hand to the symmetry of associated closely spaced complexes of high order, i.e., media with a short-range order, as in liquids and amorphous solids, and finally to the symmetry of infinitely extended media known as crystals, which have long-range order.

This sequence should correspond to gradual increase in the ordering of phenomena in such media, i.e., ordering in accordance with the above definitions.

Crystals, Plants, and Animals

In the first textbook of crystallography (1783), De Romé de l'Isle [19] referred to the famous Baron Haller, who expressed the view that there exist three degrees of life: animal life, plant life, and crystallization.

To this we may add that plant and animal forms of life represent activity directed to ordering, i.e., purposeful organization of the medium for existence. Plants extract from the environment elements needed for their growth and functioning, while animals, and particularly man, adopt the environment to their requirements, and man in recent decades has had enormous effects on the environment in this way.

Solar energy is the source for the vital activities of these various forms; plants use solar energy in photosynthesis, while animals and man extract it from food.

Crystals differ from plants and animals in being in a state to order energy directly, without the need for growth. This is their main distinctive feature amongst the self-regulating systems listed by Galler. Crystals here constitute nonliving models for living organisms, and they can serve as material for creating artificial intelligence, which possesses a number of features and some particular advantages over plants and animals.

There is a marked difference between plant or animal life on the one hand and crystal growth on the other, since the living things require an influx of heat, whereas crystal growth needs removal of heat.

Crystals have approximate symmetry rather than ideal symmetry [4]. Therefore, the ordering of phenomena performed by a crystal is not ideal, in accordance with the structure; it is the more perfect the nearer the symmetry of the structure to idea. This is the reason for researches conducted throughout the world designed to improve the perfection of artificial single crystals, while giving them an ordered substructure to modify the basic properties.

Conclusions

The evidence may be summarized as follows. Curie's principle may be used with the following definitions: order (static) and disorder (static and dynamic), and also the symmetries of order and disorder, as well as the definition of ordering. This gives a new interpretation for the familiar phenomena of x-ray diffraction and electromagnetic radiation polarization, in

---

* Liquid crystals can be considered as condensed systems with crystallization initiated but uncompleted.

particular for thermal radiation; as regards crystals, the effects are ones of partial ordering in the microstructure of the electromagnetic field. It is shown that an ordered character is more likely for the processes in crystals than a disordered one, and also that the crystallographic essence of processes in crystals consists in changing the microstructure of energy and transforming the symmetry of disorder into the symmetry of order; the transformation is in accordance with the symmetry of th direction in the crystal in which the energy propagates. These phenomena provide the basis for the technical use of crystals at present and in the future.

A view is presented of a crystal as a purposive individual with its own interaction with the environment, and also of the world of crystals as a specific section of inorganic nature adjoining biological objects. It has been shown that this view arose in the very early days of crystallography.

Literature Cited

1. N. N. Sheftal'. Growth of Crystals, Vol. 4, Consultants Bureau, New York (1966), p. 183.
2. N. N. Sheftal'. Growth of Crystals, Vol. 4, Consultants Bureau, New York (1966), p. 190.
3. P. Curie. J. Phys., 3(3):393 (1894).
4. A. V. Shubnikov. Usp. Fiz. Nauk, 59(4):591 (1956).
5. N. N. Sheftal'. Growth of Crystals, Vol. 1, Consultants Bureau, New York (1959), p. 5.
6. N. N. Shabranovskii. Lectures on Crystal Morphology [in Russian], Izd. Vyssh. Shkola (1968).
7. N. N. Sheftal' and E. G. Kolomyts. Acta Physica Academiae Scient. Hungar., 33(3-4):335 (1973).
8. P. Curie. Oeuvres, pp. XII, 56-57 (1908).
9. L. V. Radushkevich. Textbook of Statistical Physics [in Russian], Prosveshchenie (1966).
10. J. W. Strutt (Lord Rayleigh). The Wave Theory of Light (1880).
11. S. I. Vavilov. The Microstructure of Light [in Russian], Izd. AN SSSR, Moscow (1950).
12. R. Clark Jones. J. Opt. Soc. Amer., 43:138 (1953).
13. A. A. Akopyan. Chemical Thermodynamics [in Russian] (1963).
14. G. B. Bokii and M. A. Porai-Koshits. X-Ray Structure Analysis [in Russian], Izd. Mosk. Univ. (1962).
15. W. Shercliffe. Polarized Light [Russian translation], Izd. Mir (1965).
16. M. Born. Optics [Russian translation], GIZ, Moscow (1937).
17. I. S. Zheludev. Kristallografiya, 16(2):273 (1971).
18. N. N. Sheftal'. The Structure of Crystals [in Russian], ONTI, Moscow (1933).
19. De Romé de l'Isle. Cristallographie, Paris (1783).

IN MEMORY OF A. S. SHEIN
N. N. Sheftal'

Arkadii Sergeevich Shein died at the age of 60 on 18 November 1972; he was a doctor of physicomathematical sciences and one of the most outstanding workers in technical crystallography, particularly synthetic single crystals, and was closely associated with the establishment of this discipline in our country.

I was acquainted with him and worked with him from his early years, in the period when he rose rapidly from being a research student to a principal engineer in the first plant in our country for making synthetic single crystals. However, in subsequent years I was in fairly frequent contact with him, via his contacts with the Institute of Crystallography, and especially his visits to A. V. Shubnikov, which continued for many years.

In 1939, the Laboratory of Crystallography of the Academy of Sciences completed the development of the first commercial method of making homogeneous single crystals of Rochelle salt, and practical uses were investigated; the Minstry of the Electrical Industry organized a Rochelle salt group, and in September 1939 Shein joined this as a research student.

283

The first conversation with him produced a good impression; he appeared to be the type of man who could bring crystals to bear on life. His formulations were clear, and he immediately understood and extended ideas put to him on the great prospects of this new technology, as well as on the abundance of ideas and the need for energetic realization. This impression was found to be completely justified. A basic feature of his work was his energy, and he was literally burning to check and realize his ideas at all times. He had many interesting and realistic ideas, and he worked on them with enthusiasm.

This is perhaps best expressed by something he wrote after his first six months in the laboratory (23 March 1940).

"I was exceptionally interested in work on ferroelectrics and considered that realistic ways of protecting them from water could not be found without serious research, since Rochelle salt piezoelectric elements were urgently needed in industry and technology; I therefore devoted my efforts to designing devices in which such crystals could play an important part.

In the laboratory I encountered the following researches:

1. The dielectric constants of complex dielectric and mechanical mixtures.

2. An oscilloscope with a storage screen.

3. Original methods of determining the porosity of glass, porcelain, and other complex dielectrics.

4. Methods of electrical integration for pulses, etc.

During my work in the laboratory I developed the following:

1. Methods of obtaining large voltages from piezoelectric elements.

2. A frequency meter with direct readout on a scale.

3. A piezoelectric instrument for examining nonuniform movements in the parts of gear transmissions.

4. An instrument for determining the damping of vibrational systems such as quartz or Rochelle salt with direct readout on a scale.

5. A meter for indicating condenser capacitance.

6. Detailed ways of using wavy and twinned quartz.

The basic task of providing the necessary electrical properties in piezoelectric elements was mine, as were all studies related to this topic."

In a paper published in Izvestiya on 30 April 1941 under the heading "An Outstanding Discovery," A. V. Shubnikov wrote a description of American researches on the preparation and commercial use of piezoelectric crystals for making various piezoacoustic devices:

"The creation of a piezoelectric crystal from which one could make Soviet piezoelectric elements is a task that has been placed before the Laboratory of Crystallography of the Academy of Sciences of the USSR." Having stated that the task had been successfully completed, Shubnikov wrote further:

"In the laboratory of one of the sections of the Committee for the Electrical Industry of the USSR, which has worked on this topic in collaboration with the Laboratory of Crystallography, it has proved possible to create a Soviet piezoelectric element from these crystals. The principal responsibility for the advance in this region is due to A. S. Shein; he suggested a new method of making Soviet piezoelectric elements from large Rochelle salt crystals. A number of prototype devices have been produced that have passed their tests. The Soviet piezoelectric telephone made by Shein is unusually sensitive and gives very clear sound. Very good specimens have also been made of piezoelectric adapters, microphones, and so on. It is now possible to make general use of piezoelectric elements in Soviet electronics, the electrical industry, radio communication, defence, electroacoustics, and monitoring and measuring technology.

The number of different styles of piezoelectric elements for various purposes prepared by Shein greatly exceeds the number of styles stated in the catalogues of foreign firms."

While Shubnikov was writing these lines, work was going rapidly ahead to organize the first plant for growing single crystals to produce the Shein piezoelectric elements.

In November 1941 the plant came fully into production, and at the start of 1942 another plant began to operate and rapidly grew into a large undertaking nearby [1].

Many millions of various sensitive ferroelectric acoustic devices were made during the Second World War from Shein elements.

While he was in the hospital and knew that his condition was hopeless, he completed the text of a patent submission before his death.

Shein was a favorite student of Shubnikov's, who saw in him an outstanding scientific engineer, who would open up the road for single crystals in technology.

He will always be remembered as an outstanding pioneer in the technical use of synthetic single crystals in the USSR, especially for the role he played during the vital period of the Second World War.

Literature Cited

1.   N. N. Sheftal'. Trudy Inst. Krist., No. 4, p. 231 (1948).

POSTSCRIPT
N. N. Sheftal'

Eighteen years have passed since the publication of the first volume of "Growth of Crystals," and this has been a period of rapid advance in researches in that area.

Synthetic single crystals may be considered as one of the sections of a large, new, and as yet not very clearly defined but rapidly growing area which is known abroad as materials science. This area relates to the production of new metals and alloys, ceramics, synthetic plastics and fibers, materials for electronics and optics (especially single crystals), and various other solid synthetic materials of technological significance.

The surveys of [1] represent the first attempt to analyze the development of this area, the scale of the researches at the present time, the relationship to older areas, and the likely rates of progress in this area in future.

Some of the results given in [1] relate directly to crystal growth proper.

The present time would appear to justify an outline survey of the past history of the area and an estimate of the prospects for development in crystal growth.

Single crystal growth as a scientific discipline is in the same position as materials science, i.e., it constitutes a complex problem including physics, chemistry, physical chemistry, mineralogy, and engineering. However, the core of the subject is crystallography as modernized by introduction of a genetic growth element. Researches on crystal growth are now performed in geological, physical, chemical, physicochemical, and technological institutes and organizations, and also recently in biological institutes; for this purpose, an essentially new aspect of crystallography has been introduced into these sciences, into technology, and into production.

In geology, studies on crystal growth represent an essentially new method of providing industry with inexhaustible resources of scarce or entirely lacking large specimens of natural single crystals, and also of similar or more remotely related analogs, which may not yet have found industrial use but which represent considerable practical interest.

Again, researches on growth of mineral crystals and analogs have provided a new and reliable experimental basis for understanding the origins of various types of deposits, which have provided details and basis for fundamental geological concepts.

The situation is somewhat analogous to geological methods of working: first of all one seeks to define crystals with the necessary properties, then establishes the physical properties of all specimens, defines closely the new properties, and finally develops a method for commercial production.

Researches on crystal growth and solid-state physics introduce genetic elements, since a profound understanding of the physical properties of single crystals can be attained only by taking into account the history of the crystals, which substantially influences the physical properties.

In chemistry, as is shown by the articles in this volume, controlled crystal growth via reaction requires a crystallographic analysis of the complexes in compounds that arise, i.e., a study of the reaction on a structural basis, which determines the morphology, growth mechanisms and homogeneity of the growing crystals.

In physical chemistry, the concepts of stable and metastable equilibrium find material expression in the equilibrium and nonequilibrium shapes of structures and crystals.

In technology, the manufacture of instruments and machines such as computers involves crystals particularly as one transfers from instruments controlled by human hands to ones controlled by molecular processes, of which the main anisotropic process is crystal growth itself.

However, crystallography always has been a complex science, so penetration of its methods and ideas into other natural sciences and into technology represents not isolation of these ideas but mutual combination of them on a more general structural and morphological basis.

We now turn very briefly to researches in this period on the commercial production of single crystals.

In the 1950s, there were rapid advances in the semiconductor industry using single crystals of vanadium and silicon; solutions were obtained in 1955-60 to the production of artificial single crystals of quartz, diamond, mica, and fluorite.

In 1955, epitaxy was first used successfully in the technical production of single crystals [2].* The p−n junctions made by epitaxy have been used in integrated-circuit technology, which represented a larger step forward in electronics than the invention of the transistor itself [4, 8]. Films enabled one to combine organically in a single material either sharp or smooth junctions with new properties, which provided scope for microminiaturization of electronic devices and improved computer design. In the early 1960s, rubies were made for the first lasers.

Recent years have seen advances in the production and use of oriented systems of whisker crystals [9, 10].

At the same time there have been advances in the production of single crystals of metals and alloys, semiconductor compounds, ferrites, ferromagnetics, borates, garnets, tungstates, and many other crystals. There have also been exceptionally vigorous searches in all parts of the world for new crystals with valuable physical properties.

The scope of studies on crystal growth has continuously and rapidly expanded throughout the world; there have been considerable advances in growth technology: there have been considerable improvements in the accuracy of temperature control, the smoothness and accuracy of devices for moving crystals, and extension of the temperature ranges and pressure ranges for growth. Crucible-free zone melting and solvent-zone methods have in certain cases eliminated the crucible problem. Much work has been done on automating the processes. There have been discussions at a practical level on the scope for performing experiments in space, where the weightlessness, purity, and very high vacuum go with possibilities of optical heating to provide facilities for making very pure single crystals [11].

----

*The priority of [2] has been confirmed in [3, 4]; the forerunners of crystallographic researches on technical epitaxy were studies in chemical technology, in which a method was devised for obtaining polycrystalline films for semiconductors using the reduction of silicon tetrachloride or germanium tetrachloride in a current of hydrogen. The relevant reference for work abroad is [5], while for the USSR it is [6]. See also [2, 7] on this topic.

Epitaxial films have extended to all uses of semiconductor single crystals and also to other areas.

The greater the range of single-crystal materials available to engineers and others, the greater the scope for combining them; this is one of the reasons for the rapid advances in the production of new crystals and the vigorous searches for valuable properties.

The regular lattice structure of a single crystal goes with the symmetry in the anisotropy and identity in the properties in parallel directions to make such crystals unique material for various active devices, i.e., devices that in a controlled fashion and with minimal loss transform a variety of forms of energy in accordance with the symmetry of the propagation direction. The greater the structural perfection of a crystal, the more suitable it is for use in such devices.

A perfect single crystal also represents the optimal initial material for doping, namely adjustment of the properties of a crystal in a desired direction; in that case, one needs to obtain a regular distribution of the dope.

As regards the economic significance of researches on making single crystals, any such material is of practical value only when industrial devices are put into production; this applies not only to materials previously available such as ruby, diamond, emerald, optical fluorite, and quartz, but also all new crystals, and sometimes the latter may involve a delay of decades before large crystals can be produced. The road is sometimes a long one, and far from all crystals have yet found application. It is stated [12] that the mean time lapse between the laboratory researches on an instrument and commercial production may range from 10 to 20 years, and of 10 completed developments, only 5 will survive commercial and market tests, and of these only 2 will have complete success. If one bears in mind that some researches are never brought through even to laboratory completion, the number of research studies that never reach general use may increase by as much as an order of magnitude. This, of course, does not mean that they are valueless; some of them later become important initial stages for successful researches, which involve improvements of detail, while also contributing to the experience and insight of researchers.

As regards the profile of a researcher concerned with single crystal growth, one can say that he combines to various extents the characteristics of scientist, engineer, and inventor; those with purely practical experience have made considerable contributions in this area.*

The last decade has seen a marked strengthening of the relationship between science and technology; this has led to great modifications in instruction in higher teaching institutions. The engineer entering training does not need to know the details of technology, but rather the principles of the fundamental sciences and the ways of approaching technological decisions [13]. The details of patenting have also been rationalized; researches have been done on methods, techniques have been classified, and ways of reaching decisions in technological problems have been elucidated, for which purpose particular use has been made of computers [14].

In view of the rapid technical progress in crystal growth, great importance attaches to the ways in which specialists can, if desired, continue their education and improve their qualifications throughout life. Various solutions to this problem have been devised [15].

Rapid advances in the growth and use of crystals requires the type of specialist familiar in detail with methods of growing crystals and techniques for structural and morphological

---

* Among those working on crystals there are many without training in this area, although their research tends to be laborious, lengthy, and not always successful; in evaluating results, the first place, and sometimes the only place, is taken by working devices.

examination, as well as with practical uses. In other words, one needs specialists familiar with crystallography, crystallochemistry, and mineralogy, thereby combining knowledge of natural crystals, optical and x-ray methods of examination, and the principles of electronics and solid-state physics.

Finally, considerable interest attaches to the trends in technical progress in crystal growth at the present time and in the near future.

Electronics constitutes a very important area where single crystals are essential; the very existence and advancement of this area must demand single crystals to act as the active components in electronic devices and microminiature integrated circuits.

The scientific and technological revolution has been based particularly on electronics; the industrial revolution began about 200 years ago with the use of steam, which increased man's physical power, while facilitating and replacing physical labor by mechanical labor; the new revolution now in progress will result in replacing human labor in many spheres by electronic machines.

Automated production control has advanced considerably in some branches of industry, and is designed in essence on the capacity of computers, which in various working respects exceed the capacity of man; there is now no especial difficulty [16] in teaching them to recognize errors and to take the appropriate measures to deal with them.

Single crystals, especially in combination with miniature integrated circuits, are the most suitable way of designing electronic devices and computers, since they represent the most highly organized and stable form of nonliving matter, while in perfection of structure and sensitivity to external conditions they are only a little inferior to the simplest organism.

The design of better computers and integrated robots using these is a very important and very promising task for the immediate future [17, 18].

Literature Cited

1. Materials Science and Engineering in the United States (E. Roy, ed.), Pennsylvania State Univ. (1970).
2. N. N. Sheftal', P. P. Kokovish, and A. V. Krasilov. Izv. AN SSSR, ser. fiz., 21(1):147 (1957). (Proceedings of the Eighth All-Union Conference on Semiconductors, Leningrad, 15-21 November 1955).
3. E. F. Cave and B. R. Czorny. RCA Review, 24(4):523 (1963).
4. W. M. Feist, S. R. Steel, and D. W. Ready. Physics of Thin Films (G. Hass and R. Thun, eds.), Vol. 5, (1969), p. 237.
5. G. K. Teal, J. R. Fisher, and A. W. Treptow J. Appl. Phys., 17(11):879 (1946).
6. A. I. Mel'nikov. Zh. Neorg. Khim., 2(2):233 (1957).
7. L. S. Palatnik and I. I. Papirov. Epitaxial Films [in Russian], Nauka, Moscow (1971), pp. 9-10.
8. Motorola Semiconductor Products. Integrated Circuits. Design Principles and Fabrications. McGraw-Hill, New York (1965).
9. R. S. Wagner and W. C. Ellis. Appl. Phys. Letters, 4:89 (1964).
10. E. I. Givargyev and Yu. G. Kostyuk, Growth of Crystals, Vol. 9, Consultants Bureau, New York.
11. K. M. Polyakov and L. N. Rashkevich, Proceedings of the Fourth All-Union Conference on Crystal Growth. Growth and Structure of Crystals [in Russian], Part 1, Izd. AN ArmSSR, Erevan (1972), p. 180.
12. E. Mansfield, Economics of Scientific Progress [Russian translation], Progress, Moscow (1970).

13. J. Dixon, Systems Design: Invention, Engineering Analysis, and Decision [Russian trans-
     lation], Mir, Moscow (1969).
14. G. Al'tshuller, An Invention Algorithm [in Russian], Mosk. Rabochii, Moscow (1969).
15. N. N. Sheftal' [Seftalj], Pedagógiai Közleményck, 1:9 (1973).
16. S. Yanov, Japanese Economics on the Threshold of the 21st Century [Russian translation],
     Progress, Moscow (1972).
17. N. Nilson, Artificial Intelligence [Russian translation], Mir, Moscow (1973).
18. M. V. Keldysh, "Space travel and scientific advance," Pravda, 9 October 1973.